A PRACTICAL GUIDE FOR THE PREPARATION OF SPECIMENS FOR X-RAY FLUORESCENCE AND X-RAY DIFFRACTION ANALYSIS

A PRACTICAL GUIDE FOR THE PREPARATION OF SPECIMENS FOR X-RAY FLUORESCENCE AND X-RAY DIFFRACTION ANALYSIS

EDITED BY

VICTOR E. BUHRKE
The Buhrke Company

RON JENKINS
International Centre for Diffraction Data

DEANE K. SMITH
Pennsylvania State University

WILEY-VCH

New York / Chichester / Weinheim / Brisbane / Singapore / Toronto

Copyright © 1998 by John Wiley & Sons, Inc.

Library of Congress Cataloging in Publication Data:

A practical guide for the preparation of specimens for x-ray
   fluorescence and x-ray diffraction analysis / edited by Victor E.
   Buhrke, Ron Jenkins, and Deane K. Smith.
         p.   cm.
      Includes bibliographical references (p.   —   ) and index.
      ISBN 0-471-19458-1 (alk. paper)
      1. X-ray spectroscopy.   2. X-ray crystallography.   3. Sampling–
   Technique.   I. Buhrke, Victor E.   II. Smith, Deane K. (Deane
   Kingsley)   III. Jenkins, Ron, 1932–
   QD96.X2P73   1998
   543$'$.08586—dc21   97-16687

Printed in the United States of America

10 9 8 7 6 5 4 3 2

This book is dedicated to the people who gave so much of their valuable time to write their section, to the people who encouraged us to produce such a book, to the users of XRF and XRD equipment all over the world who constantly face the problem of proper sampling and preparation of a specimen from a sample, and to the Denver X-ray Conference for making it all possible.

CONTRIBUTORS

RANDOLPH BARTON, JR., DuPont Company, Central Research and Development, P.O. Box 80302, Wilmington, Delaware 19880, USA.

WILLIAM A. BASSETT, Cornell University, Department of Geological Sciences, Snee Hall, Ithaca, New York 14853, USA.

THOMAS N. BLANTON, Eastman Kodak Company, Research Laboratories B82A, Rochester, New York 14650, USA.

VICTOR E. BUHRKE, The Buhrke Company, 10 Sandstone, Portola Valley, California 94028, USA.

LARRY E. CREASY, Titanium Health Technologies, Hemlock Road, Morgantown Business Park, Morgantown, Pennsylvania 19543, USA.

J. F. CROKE, Philips Electronic Instruments, Mahwah, New Jersey 07430, USA.

BRYANT L. DAVIS, Davis Consulting Group, 4022 Helen Court, Rapid City, South Dakota 57702, USA.

THOMAS A. ERICKSSON, Department of Structural Chemistry, Arrhenius Laboratory, University of Stockholm, S-10691, Stockholm, Sweden.

VICENTE ESTEVE-CANO, Universitat Jaume I, Departamento de Química Inorgánica y Orgánica, Apdo. 224, 12080 Castellon, Spain.

FRANK FERET, Alcan International Limited, P.O. Box 1250, Jonquière, Québec, Canada G7S 2P7.

SONA HAGOPIAN-BABIKIAN, Lafarge Canada, Inc., Corporate Technical Services, 6150 Avenue Royalmount, Montréal, Québec, Canada H4P 2R3.

RICHARD F. HAMILTON, Air Products and Chemicals, Inc., 7201 Hamilton Boulevard, Allentown, Pennsylvania 18195, USA.

SAMPATH S. IYENGAR, Technology of Materials, 2922 De La Vina Street, Suite C, Santa Barbara, California 93105, USA.

RON JENKINS, International Centre for Diffraction Data, 12 Campus Road, Newton Square Corporate Campus, Newton Square, Pennsylvania 19073, USA.

HOWARD M. KANARE, Construction Technologies Laboratories, 5420 Old Orchard Road, Skokie, Illinois 60077, USA.

VLAD KOCMAN, Domtar Inc. Innovation Center, Senneville (Montréal), Québec, Canada H9X 3L7.

CHARLOTTE K. LOWE-MA, CLM XRD Mentoring, 3617 Oak Drive, Ypsilanti, Michigan 48197, USA.

GREGORY J. MCCARTHY, North Dakota State University, Department of Chemistry, North Dakota Water Resources Research Institute, 104B Ladd Hall, Fargo, North Dakota 58105, USA.

HOWARD F. MCMURDIE, National Institute of Standards and Technology, A215, MATLS, Gaithersburg, Maryland 20899, USA.

DIANE E. NELSON, Air Products and Chemicals, Inc., 7201 Hamilton Boulevard, Allentown, Pennsylvania 18195, USA.

PAUL K. PREDECKI, University of Denver, Engineering Department, Denver, Colorado 80208, USA.

SVEND E. RASMUSSEN, Aarhus University, Department of Inorganic Chemistry, DK 8000 Aarhus C, Denmark.

JACQUES RENAULT, New Mexico Bureau of Mines and Mineral Resources, 801 Leroy Place, Socorro, New Mexico 87801, USA.

DAVID F. RENDLE, Forensic Science Service, Metropolitan Laboratory, 109 Lambeth Road, London SE1 7LP, United Kingdom.

GERALD I. D. ROACH, Alcoa of Australia Limited, Research and Development, P.O. Box 161, Kwinana 6167, Australia.

EARLE R. RYBA, Pennsylvania State University, 304 Steidle Building, University Park, Pennsylvania 16802, USA.

PETER SALT, formerly of English Clays, Inc., United Kingdom, deceased.

DEANE K. SMITH, Pennsylvania State University, Department of Geosciences, University Park, Pennsylvania 16802, USA.

JOSEPH E. TAGGART, JR., U.S. Geological Survey, Mineral Resources Team, Box 25046, M.S. 973, Denver Federal Center, Denver, Colorado 80225, USA.

YANBIN WANG, University of Chicago, Consortium for Advanced Radiation Sources, Sector 13, Building 434, 9700 South Cass Avenue, Argonne, Illinois 60439, USA.

WINNIE WONG-NG, National Institute of Science and Technology, A215 MATLS, Gaithersburg, Maryland 20899, USA.

CONTENTS

PREFACE

X-ray fluorescence spectrometry and X-ray diffraction methods are used in thousands of laboratories all over the world. However, few colleges teach the subjects and almost none teach how to avoid introducing large systematic errors caused by improper selection of a sample from a bulk or improper preparation of a specimen from a sample. Because only a few teaching institutions exist for learning the theory, users have to learn by studying the X-ray texts available and by relying on help from friends and associates. Fortunately, many textbooks on X-ray spectrometry and X-ray diffraction are available for learning the theory of these two methods; however, detailed practical information about proper specimen preparation is not included in these texts. This book was written to fill this void.

VICTOR E. BUHRKE
RON JENKINS
DEANE K. SMITH

INTRODUCTION

About This Book

This is a different kind of textbook. First, it is not a theoretical text. It was not meant to replace the classic texts written by Bertin, Klug and Alexander, Cullity, and Jenkins and Snyder. It is a book you will want to read because it provides information not available in conventional texts. When you want to know how to prepare a specimen for XRF or XRD analysis and avoid systematic errors, you will find help in this book.

Second, unlike many other texts, this one is task oriented. Most people buy a textbook to learn the theoretical facts about a technique. This book assumes you already know the theory but you want to know the right way to do the analysis.

What Is This Book?

This book is the result of the work of many experienced and dedicated people. It is the response to the voices of hundreds of analysts who attended the Workshop on Specimen Preparation at the annual Denver X-ray Conference over the past 18 years. It is a quick reference for both the beginner and the experienced analyst in need of help in the proper selection of a sample from a bulk material and the preparation of a specimen from a sample. It is the first book that deals exclusively with specimen preparation. Perhaps it is also the catalyst for the creation of more books of this kind. We hope it will encourage authors of articles on XRD and XRF and authors of future texts to include details (not just a few words) about the proper methods to extract a sample from a bulk material and prepare a homogeneous specimen out of the sample.

Who Should Use This Book?

Any beginner, intermediate, or advanced X-ray analyst who wants to avoid making a serious and costly error in the analysis of a material.

How Is This Book Organized?

Each chapter of this book is organized around a major task likely to occur in an X-ray analysis. Each chapter contains the necessary theoretical material to support the specimen preparation procedure described. Step-by-step methods are given along with suggestions regarding where to find the equipment needed and what literature references are recommended.

How to Use This Book

Start with the table of contents. It contains a description of every type of material discussed in the book. If the exact material you are working on is not listed, you might find something closely related.

Good luck and much success in your analysis.

VICTOR E. BUHRKE
RON JENKINS
DEANE K. SMITH

A PRACTICAL GUIDE FOR THE PREPARATION OF SPECIMENS FOR X-RAY FLUORESCENCE AND X-RAY DIFFRACTION ANALYSIS

CHAPTER 1

GENERAL INTRODUCTION

CONTRIBUTING AUTHORS: HAGOPIAN-BABIKIAN, HAMILTON, IYENGAR, JENKINS, RENAULT

1.1 SAMPLING

1.1.1 What Is Sampling?

In most analytical laboratories the analyst is presented with a *sample* for analysis, the sample, in turn, having been taken from the *bulk*. The bulk may be an ore deposit, a test probe taken from an industrial process, a finished product, a product in process, or, indeed, an almost limitless range of starting material. An *aliquot* is typically taken from the submitted sample, then the aliquot is prepared to give an analysis *specimen*. We depart from rigorous sampling terminology which takes *specimen* to be nonrepresentative. Four assumptions are typically made:

1. The *primary sample* is representative of the *bulk*.
2. The *aliquot* is a subsample representative of the *primary sample*.
3. The *specimen* is representative of the *aliquot*.
4. The *analyzed volume* of the prepared specimen is representative of the whole *specimen*.

The basic purpose of this section is to investigate these four assumptions. While we will concentrate mainly on the third and fourth of the listed assumptions, the analyst should be aware of the potential pitfalls in the areas of sampling, both from the sample as well as the bulk. Some sections of this text will be devoted specifically to the areas of sampling from the bulk (section 1.3) and taking an aliquot from the sample (section 1.4). Much confusion arises because of lax use in terminology. Most of the processes described in this text have to do with *specimen preparation*. Specimen preparation typically involves the taking of an aliquot from a submitted sample, then preparing a specimen for analysis from this aliquot, in such a way as to ensure that the relatively small volume of sample actually *analyzed* by the X-ray beam is representative of the total specimen (see Figure 1.1). If, and only if, the analyzed volume of the prepared specimen and the sample are of the same composition, then the analyst has really provided an analysis of the submitted sample. We continually

A Practical Guide for the Preparation of Specimens for X-Ray Fluorescence and X-Ray Diffraction Analysis, Edited by Victor E. Buhrke, Ron Jenkins, and Deane K. Smith. ISBN 0-471-19458-1 © 1998 John Wiley & Sons, Inc.

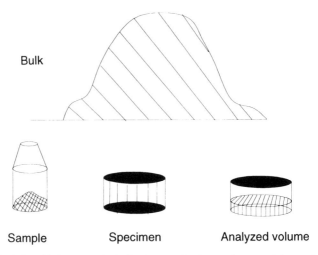

Bulk

Sample   Specimen   Analyzed volume

Figure 1.1. Relationship between the bulk, the sample, the specimen, and the analyzed volume.

sample our environment: the land in which we dwell, the buildings and machines we use, the humans and other creatures we encounter. Almost all of that sampling is done automatically by our sensory apparatus and our psychology. Some of it is done well and some not so well, but if we fail to correctly sample our environment, we can perish for it. Sample poorly the traffic on the street you wish to cross and you may never sample traffic again.

Sampling in the workaday world is similar to sampling in the analytical laboratory. By sampling, we observe a small part of a large system and take that observation as standing for the pertinent properties of the whole system. The sample is called representative when its pertinent properties are sufficiently like the whole system. *Sufficiently* means that the similarity of the sample to the whole serves our purposes. The nonrepresentative sample does, indeed, sample something, but it does not sample what we think it does, so it does not serve our purposes.

The results of poor analytical sampling in the laboratory are rarely so dramatic as a pedestrian missampling traffic, but they are frequently costly and usually troublesome to someone. Unlike the workaday world where we can count on our senses to automatically sample our environment, sampling in the analytical laboratory is conscious, deliberate, systematic, and rational. It is conscious because we are aware of what we are doing; deliberate because our procedures are under our control; systematic, because we follow a plan; and rational, because all our actions have a single objective, viz., to extract from the whole a representative sample.

One important aspect of sampling in the analytical laboratory that is different from our sensory sampling of the workaday world is subsampling. A subsample is a sample of a sample. It is drawn in order to have enough material to test the whole, yet retain enough sample for future and possible unforeseen use. Certified standards are of this sort. Each numbered container holds a subsample of a larger batch sample known to be of a certain composition representative of the whole. Many samples received by the analytical laboratory will be subsamples already, for the laboratory's clients frequently send only a portion of the original sample to the laboratory for analysis.

1.1.2   Kinds of Materials Sampled

Liquids and solids can be analyzed by both X-ray fluorescence (XRF) and X-ray diffraction (XRD). By far the most common materials analyzed by both methods are solid, but the fundamental limitations on sample size and state are those based on instrumental configuration and instrumental sensitivity to sample materials. Although liquids and glasses can be analyzed by XRD, crystalline solids are its usual fare.

Liquids and solids of almost any sort that will not damage the spectrometer can be analyzed by XRF. As with XRD, samples are usually brought to the instrument, but field portable XRF spectrometers are commonly used in many survey studies, for example, minerals exploration and environmental evaluation.

1.1.3   Record Keeping

Careful records have to be kept of samples received for analysis. These records require sample logging, sample labeling, and chain of custody procedures. It is crucial to good analytical procedure that accurate and retrievable records be kept of the history of samples that arrive in the laboratory. The sample that arrives will have a client number or description. This information must be encoded with a laboratory number defining its arrival sequence in a sample log book. Samples may pass through several hands in their analytical travels. This route along with accompanying procedures and results is called the chain of custody. These records are legal documents in courts of law and have to be maintained and treated with respect.

Of course, the analyst has no control over the client's sample numbering and labeling schemes. He/she must, however, have a systematic and understandable numbering and labeling system for the laboratory. A simple sequential numbering system is used by many laboratories. There are sophisticated systems available which produce a numbered adhesive label that corresponds to the logged number in a database.

1.1.4   Meaningful Sampling

There are two related questions that must be asked in order for sampling to be meaningful:

1. What is the purpose of the analysis?
2. How representative should the sample be?

The first question has two aspects for the analyst. If he or she is also a project scientist, then the purpose of the analysis is to illuminate the project, and primary sampling is undertaken so that the properties of the object of investigation can be analytically estimated. On the other hand, if the analyst is to provide only an analytical service to a third party, then he/she need only be concerned about secondary sampling of the samples received by the laboratory.

The answer to the second question has to begin with the analyst asking the submitter (or himself if directly involved in an investigative project) whether a qualitative or quantitative analysis is required. Each laboratory will have its own conventions, but a qualitative analysis will usually characterize the concentrations of analytes as present, trace, minor, or major. A quantitative analysis always attaches numerical concentrations to each analyte. If the analysis is to be quantitative, then the level of required accuracy and precision has to be known.

1.1.5 Sampling for Scientific Investigation

Primary sampling for an investigation is beyond the scope of this chapter, but it is useful to point out some of the considerations that the analyst should be aware of. The analyst is referred to standard statistical texts on the subject. For geological samples, *An Introduction to Statistical Models in Geology* by Krumbein and Graybill [1] covers the very brief material described later in much more detail. The sampling terminology used is theirs. The purpose of the *primary sampling* may be to search out a feature in the object of investigation; for example, one may search for an ore deposit by locating a geochemical anomaly in stream beds crossing a target area. This method is called *search sampling*. When an object of interest has been found, *purposeful sampling* may be undertaken. Specific examples of ore minerals could be collected for identification. Technically, this approach is not really sampling but rather purposeful specimen collecting. *Statistical sampling* may be undertaken at some point in the project to introduce a randomization element into the characterization of the object. For example, samples may be drawn at random intervals of time from a conveyor belt in order to determine the probability of the feed having certain properties.

There are a variety of sampling plans that can be implemented depending on the nature of the object of interest. In geochemical sampling, some of these procedures are simple random sampling, in which a device such as a random number table (see Table 1.1) is used to select samples along a traverse; systematic sampling in which samples are uniformly spaced; and stratified random sampling in which uniformly spaced cells have samples collected randomly within them. It is very important in any sampling plan that a sufficient number and size of samples be taken to characterize the object of interest. To be done correctly, an initial sampling program is conducted to give an estimated standard deviation and mean. Using those provisional values, the likely deviation from the mean at a particular probability can be calculated using Student's *t*-tables and then compared with the maximum allowable deviation. This subject is discussed more fully below in the section on statistics.

1.1.6 Sampling for Analytical Truth

Secondary sampling, that is, subsampling of material sent to the analyst for a blind test, can be called sampling for analytical truth. Almost all sampling in commercial and service laboratories is of this type. It is the task of the analysis to extract an

TABLE 1.1. A Sample from a Random Number Table

				Random Numbers					
92	47	32	92	55	53	52	44	0	57
53	34	74	21	23	55	39	67	83	3
36	80	60	52	73	41	91	50	39	86
46	37	47	89	86	90	90	71	33	99
39	28	76	53	51	94	96	68	11	21
15	30	94	90	77	63	57	38	48	28
50	22	42	41	55	96	70	19	53	27
24	50	13	17	69	88	17	1	78	75
25	73	78	84	96	96	30	91	16	37
27	69	97	70	24	11	94	34	89	71
74	37	77	39	1	91	35	32	53	69
24	17	60	87	78	56	68	14	72	76
28	2	30	63	10	46	47	85	44	6
45	36	80	17	46	5	81	15	75	74
34	3	87	34	93	31	16	37	39	60

aliquot, or subsample of the submitted sample that is representative of it and to determine its properties as requested. Depending on the degree of analytical truth that is required by the client, the analyst will plan his or her sampling approach. In most cases the submitted sample will have to be reduced to manageable size. The greater the analytical truth required by the client, the greater the care that the analyst must take in extracting this secondary sample.

1.1.7 Matching Sampling Methodology with Analytical Goals—Heterogeneity

The requirements of the client must be accommodated by the analyst or the job refused. Thus the requirements of both the analyst and client must be understood by each. Once the requirements are understood, the analyst can only then proceed with extracting a subsample for analysis. Very often, the client has little background in analytical procedures and possibilities. The experienced analyst may be able to advise the client in his choice between inexpensive and expensive determinations such as qualitative and quantitative analyses and thereby help the client to get the greatest value from the analyses. The client may not realize that sample heterogeneity impacts the meaningfulness of analytical results, and that sampling methodology may have to be elaborate if low analytical error is required.

To realize how the sampling methodology must match the analytical goal, consider the analysis of TiO_2 in a sample of basalt. The analysis of this oxide by XRF is commonly reported with a 5% relative error at the 1.00 wt% level. Reported concentrations and errors will be meaningful at those levels only if subsamples can be taken which are representative of the basalt sample with less than 5% error, because the analyses will range from 0.05 wt% less than the true concentration in the aliquot with the lowest TiO_2 to 0.05 wt% greater than the true concentration in the aliquot with the highest TiO_2.

The following illustration may be clearer to some after reading section 1.3. Suppose, for example, the heterogeneity of the sample caused the actual TiO_2 in all possible aliquots to range from 0.95 to 1.05 wt% with a standard deviation of 0.0167 wt%. If triplicate analyses are made, the standard error of the sampling at 95% confidence will be:

$$4.303 \times \frac{0.0167\%}{\sqrt{3}} = 0.0388\% \qquad (1.1)$$

where 4.303 is the small-sample multiplier (or t-value) as in Table 14.3 of Arkin and Colton [2] (see Table 1.2). The reported mean concentrations of triplicates, then, could range from 0.91 to 1.09 or ±9% relative. For duplicate analyses, the multiplier is 12.71. To halve the triplicate multiplier (4.303) requires 13 replicates.

Depending on the nature of the sample, it may be more cost effective either to replicate or homogenize. Brittle materials such as ceramics are more easily homogenized than flexible materials such as rubber or human hair. In the case of the basalt, homogenization would be the economical choice.

Consider now a mineralogical analysis. Perlite manufacturers are presently required to take certain actions if their product contains more than 0.1 wt% crystalline silica. It is possible to analyze quartz in perlite to the 0.05 wt% level [3, 4], but in many cases a qualitative scan may show that quartz is not detectable and thus below the regulatory limit.

However, suppose a 1 kg perlite sample containing 0.1 wt% quartz is received by the laboratory, and the grain size of the quartz is 1 mm. As quartz has a specific gravity of 2.65 g/cm^3, 1.0 g in the sample has a fictive volume of 0.377 cm^3 which is equivalent to 377 1 mm "cubic" grains per kg. A 10 g aliquot of the sample, then, contains less than 4 of these ideal grains if they are randomly distributed. If the grains were 2 mm cubes, there would be less than half a grain per 10 g aliquot! This effect is called the *nugget effect* [5], which should not be confused with the geostatistical term applied to variograms. When the nugget effect operates, there is a high probability that a small sample drawn from the bulk will contain a nonrepresentative amount

TABLE 1.2. Segregation of Particles

Property	Condition	Examples
Density	Vibration	Magnetite in diorite rate of flow
Particle size	Pouring, sieving	Hard and soft minerals ground or sieved together
Angle of repose	Pouring	Mixtures of mica and quartz
Magnetic fields	Ferrous containers	Magnetite, chromite
Moisture	Differential adsorption	Mixtures of clays and silts
Charge separation	Grinding, pouring	Powders in contact with plastics

of the analyte. The effect can be minimized by collecting large samples in the field and grinding and homogenizing them prior to splitting out subsamples for analysis. We leave it to the reader to determine how finely the 1 kg sample would have to be ground to achieve a level of 1000 quartz grains in a 10 g aliquot.

Sampling for preferred orientation in an alloy or rock requires a different methodology than for a cominuted sample. In many ways, it is like geologic mapping in an area of poor exposure where the structural relationships of different units cannot be understood until the map is completed. In both cases, the geometric distribution of inhomogeneities is the subject of investigation. Wenk [6] points out that pole figure data are usually presented as contoured intensities on a stereonet of some kind, and the statistical significance of the contours is questionable. He notes, however, that a contoured pole figure presents an acceptable nonparametric estimate of preferred orientation. Insofar as a pole figure can be characterized by a mean orientation, the means at several points on the sample can be statistically described.

The area sampled by the X-ray beam in a pole figure determination is small compared to the area of the mounted sample. The sampling can be thought of as a miniature field sampling problem and systematic, random, or stratified sampling might be used if it is necessary to characterize accurately the preferred orientation of the whole sample.

1.2   BASIC STATISTICAL DEFINITIONS

1.2.1   Introduction

Fortunately, few statistical principles are needed for the vast majority of sampling problems in the X-ray laboratory. The concepts of *representative sample, randomization, mean, variance*, and *t-distribution*, however, are indispensable to responsible sampling. A statistical reference book should always be available in the laboratory, but for a practical no-nonsense introduction, *Experimentation and Measurement* by W. J. Youden [7] is highly recommended. This wonderful little book was written for students and teachers of beginning science and very clearly and thoughtfully introduces the topics mentioned above. Another useful general reference is *Experimental Statistics* by M. G. Natrella [8].

1.2.2   The Representative Sample

The notion of a representative sample was introduced in the first paragraph of this chapter. But how can the analyst be confident that the subsample or aliquot is representative of the whole? The only way one can be absolutely certain is by knowing the composition of the whole—but that's what he's trying to estimate by analyzing only a subsample! The analyst may know from experience that certain procedures estimate adequately certain properties of the samples he usually receives. However, sample types differ, and the ease with which they can be homogenized and split varies. Furthermore, laboratory procedures evolve, and their validity has to be verified.

Randomization is the process of choosing or arranging things without structured bias. A randomly selected or drawn sample is one whose selection was not influenced by any of the ways used to draw it. Random sampling is often called probabilistic sampling because the probability of a suite of samples being representative can be determined. Paradoxically, random samples can only be drawn by highly systematic means, but this procedure means that the quality of sampling can be controlled. One of the simplest tools for randomization is the random number table found in laboratory handbooks and statistics books. Such a table is simply a list of unstructured, or random, numbers (see Table 1.1). A more sophisticated method of randomization might use the random number generator of a computer language such as BASIC to select a sequence of numbers. Simple random number generators have to be "seeded" in order to have a beginning point. If the seed is always the same, the random number sequence will always be the same, so a method such as reading the system clock can be used to select a different seed each time.

1.2.3 Mean

The *arithmetic mean* is the statistic that is usually reported in order to summarize replicate determinations. It is a technical term and not to be confused with the more loosely used term "average." Huff [9] in *How to Lie with Statistics* shows graphically the different meanings of "average" when used as mean, median, and mode. Equation (1.2) shows the expression for arithmetic mean:

$$\mu = \frac{1}{n} \sum_{i}^{n} x_i \tag{1.2}$$

where μ is the mean, n is the number of individual measurements, and x_i is the ith measurement. It is simply the sum of the measurements of the same entity divided by the number of measurements. The arithmetic mean assumes that if enough replicate measurements are made, they would be distributed in a symmetrical bell-shaped curve as in Figure 1.2 which is called the *normal distribution*. The equation for the

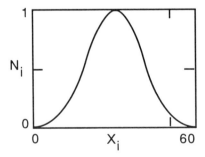

Figure 1.2. The normal distribution.

curve is

$$N_i = \exp\left(-\frac{(\mu - x_i)^2}{2\sigma^2}\right) \tag{1.3}$$

The curve might have a somewhat more peaked shape if the data were point occurrences like counter data, or pointed as for angular data. If enough measurements are taken, the mean will be in the most frequently measured interval of grouped data. In Figure 1.2, N_i is the number of observations normalized to 1.0. That is, the number of observations at any particular value of x_i is divided by the number of observations at μ; the mean σ is the standard deviation and is a measure of the breadth of the distribution as discussed below under *dispersion*. The figure shows there is a far less likelihood of a value of x being far from the mean than near to it. The range of values is about 60. In a population of this sort, the standard deviation is about $1/6$ of the range. Some distributions of measurements are not symmetrical. Trace element concentrations are frequently log-normal with long tails extending to the higher concentrations. The mean of such data sets has to be determined by taking the mean of the data transformed to their logs, then taking the antilog.

1.2.4 Dispersion

The next statistics to be considered are the *variance* and its child, the *standard deviation*. They are the most common measures of the dispersion of the mean. Variance is calculated by subtracting each measurement from the mean, squaring the differences, summing the squares, and dividing by the number of differences, n. The expression for calculating the variance is shown in equation (1.4):

$$V = \frac{1}{n}\sum_{i}^{n} d_i^2 \tag{1.4}$$

where d_i is the difference of mean and the ith measurement.

The variance has the very useful property of additivity. For example, if the analytical and sampling variances are both 0.0025 wt%, their sum is 0.005 wt% and their combined standard deviation is $\sqrt{0.05}$ or 0.071 wt%. A pioneering study by Shaw [10] thoroughly treats the combination of sampling, preparation, analytical, and true geochemical variances in a study of a granite analyzed for SiO_2 and TiO_2 by two different analytical methods. More recent work by Ramsey et al. [11] uses analysis of variance (ANOVA) to separate natural geochemical variance, sampling variance, and analytical variance from total variance in a geochemical exploration program. They suggest that for such a program, sampling variance plus analytical variance should not exceed 20% of total variance and point out that if analytical variance is less than about 1%, then perhaps unrealistic effort has been expended on analytical precision.

The standard deviation, σ, is the square root of the variance and is measured on either side of the mean. In normally distributed data, $\mu \pm 1\sigma$ will include about 68%

of the possible measurements, while 95% of the values are included in a breadth of $\pm 2\sigma$ and almost 100% fall within a breadth of $\pm 3\sigma$. Three standard deviations around the mean include effectively all possible measurements and approximate the range.

By convention, σ stands for the population standard deviation. This σ is almost always estimated by sampling, and then replaced by s and called "sample standard deviation." Sometimes "e.s.d." is used to designate estimated standard deviation. When s is computed from a small number of measurements, say, less than 30, n is replaced by $n - 1$. Samples from heterogeneous systems, for example contaminated soils, often show elemental distributions that poorly approximate normal distributions. For these, the absolute mean deviation (ADev) is usually a more robust measure of dispersion and should be used as a summary statistic. It is the mean absolute deviation from the mean as shown in equation (1.5):

$$\text{ADev} = \frac{1}{n} \sum_{i}^{n} |d_i| \tag{1.5}$$

where d_i is the deviation from the population mean as in the equation for variance, (1.5).

Just as individual measurements have a distribution, the sample means that can be measured also have a distribution. That is, a set of estimated means of successive samplings will have a spread around the true mean. The dispersion of the estimated means is called the *standard error of the mean*, s_μ, and is calculated from the variance or the standard deviation by either:

$$s_\mu = \frac{V}{n} \quad \text{or} \quad s_\mu = \frac{s}{\sqrt{n}} \tag{1.6}$$

In sampling, one should make the distinction between standard deviation and standard error. Standard deviation is the dispersion of individual measurements about an estimated mean, whereas standard error is the dispersion of estimated means about the true (and generally unknown) mean.

Multiples of standard error are used to place confidence intervals around a mean value. For small numbers of samples, the multiples are values of Student's t found in t-tables which tabulate them under column heads of percent confidence and by row heads of $n - 1$ as in Table 1.3. More extensive t-tables are found in standard statistics manuals and texts.

A word of caution: In some t-tables, the column heads are given in probability $\alpha/2$. The 95% confidence interval would then be found under 0.975, or $1 - (0.05)/2$. All t-tables are constructed from the t-distribution, which was first described by W. S. Gosset who wrote under the pseudonym "Student" so that the competitors of his employer would not realize its commercial applications. The quantity t is given

TABLE 1.3. A Section from a t-table

$n - 1$	50%	95%	99%
1	1.000	12.706	63.657
2	0.816	4.303	9.925
3	0.765	3.182	5.841
4	0.741	2.776	4.604
5	0.727	2.571	4.032
6	0.718	2.447	3.707
7	0.711	2.365	3.499
8	0.706	2.306	3.355
9	0.703	2.262	3.250

in equation (1.7):

$$t = \frac{\delta}{s/\sqrt{n}} \qquad (1.7)$$

where δ is the difference between the true mean and the estimated mean. The difference δ can be either plus or minus, so the t distribution is symmetrical about zero. The t distribution closely resembles the Gaussian distribution, but is more peaked and has higher tails. The equation for the t distribution can be easily solved for n after squaring both sides. It can then be used to estimate the number of samples required for a predetermined δ as described in the next section. The $n - 1$ heading the first column in t-tables is the degrees of freedom of a sample set. The degrees of freedom is a simple measure of the set's variability; $n - 1$ members of the set of x_i can vary independently, but the nth member is fixed by the sum of all the members minus the sum of the remaining:

$$x_n = \sum_i^n x_i - \sum_i^{n-1} x_i \qquad (1.8)$$

The sum, of course is required to determine the mean by dividing by n.

1.2.5 Sampling Errors

Sampling errors can usually be decomposed into *random* and *systematic errors*. In general, random and systematic errors have complimentary averages and variances [12]. The random error is characterized by the variance only and the systematic error by the mean only. When error is only random, repetitive sampling and analysis gives increasingly accurate estimates of the population mean. When error is systematic, repetitive sampling and analysis gives increasingly precise estimates of a value which differs from the population mean in some systematic way. These errors can

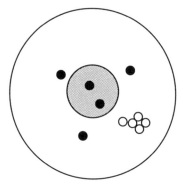

Figure 1.3. Closed circles illustrate random error alone and show accuracy with low precision. Open circles illustrate systematic error and some random error with high precision but low accuracy.

be illustrated in the bull's-eye diagram shown in Figure 1.3 where the center represents accuracy. The closed circles demonstrate good accuracy but poor precision, whereas the open circles represent good precision but poor accuracy. In other words, accuracy is a measure of analytical truth, and precision is a measure of analytical repeatability.

The analyst wants only random sampling errors to occur. Even if data are structured as, for example, Fick's law diffusion data would be, nonrandom errors cannot be treated without time-consuming and often inconvenient effort. Random errors, for the most part, the analyst can minimize and accept. For example, if the variance is too large, he may suspect the nugget effect, and can minimize it by grinding and blending. The nugget effect causes a random sampling error because the nugget containing a high concentration of the analyte causes random high concentrations when it is included and low when it is not. Note, however, that if the study is directed toward characterizing heterogeneity or systematic variation, nugget effects and structure should be preserved. Always be aware that sampling is usually a three-dimensional problem. In sampling to analyze compositional diffusion away from a weldment, it may be appropriate to sample randomly parallel to the weldment, but uniformly away from it as in Figure 1.4. In this example, random sampling parallel to the weldment accounts for random welding variations; whereas uniform sampling away from the weldment would be able to characterize the shape of the diffusion curve. In this example, if all the samples were collected near the edge of the plate, a systematic error due to differential cooling would be introduced. Note also that for the same reason, diffusion phenomena at the top of the plate will be different than at the bottom because of the "v" shape of the weldment cross section.

Systematic errors in sampling quasi-homogeneous materials are often difficult to identify because they only involve the mean, and one must know what the mean is beforehand to identify the error. Sometimes analysis by an alternative method or

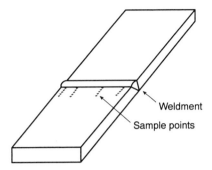

Figure 1.4. Sampling for diffusion from a weldment.

another laboratory will show that there is a systematic error. It is always prudent for the analyst to periodically obtain comparative analyses from other analysts. The client should always include blind duplicates to reassure himself of the analyst's reproducibility.

Systematic sampling errors occur when an underlying assumption about the sample is false and causes the analyst to draw a nonrepresentative sample. Usually the error occurs because the analyst has assumed random distribution of the analyte when the sample is segregated or the analyte defines a gradient as in the weldment example above. These errors are often due to improper extraction of an aliquot. For example, drawing from the top of a sample bottle without prior homogenization may bias the sample toward less dense particles.

Note that sieving, although it easily introduces a mineralogical bias, may be used to good effect in the preparation of standards. Ridley et al. [13] suggest that in the preparation of geochemical standards from heterogeneous materials, grinding and sizing to a mineralogically determined sieve interval greatly improves homogeneity and analytical precision.

Segregation of the physical components of a sample can be generated in many ways. Table 1.2 lists some of the more obvious ones. Systematic errors may occur due to sieving that biases against large or small particle sizes that have nonrepresentative compositions. High speed grinding can have the same result if significant quantities of dust are lost. Dusty samples should always be poured slowly to minimize loss (and inhalation) of dust.

1.2.6 Replication

It should be apparent from much of the foregoing discussion that it is important to replicate analyses. XRF and XRD analyses don't as a rule consume the sample, so analyses can be repeated. Repeatability serves as a check on instrumental reproducibility, but it doesn't have anything to do with sampling. Replication here means drawing several samples from the bulk and preparing them separately. It is only by replication that the analyst can know that the drawn sample is representative.

The number of replications necessary is always a question whose answer depends on the importance of the analytical result and the required accuracy of the analysis. Experience will usually help the analyst to know instinctively how much replication is necessary, but the analyst is always walking a tightrope between economy and accuracy. When new procedures are being tested, much more replication is necessary than with well established procedures.

Fortunately there is a statistical method for determining the number of samples required for a specific degree of accuracy. We will illustrate it with the TiO_2 example used previously. The equation is as follows:

$$n = \left(\frac{s \times t}{\delta} \right)^2 \qquad (1.9)$$

where the symbols have the usual meaning. The t is taken from the t-table, using the appropriate degrees of freedom and confidence interval, and δ is the acceptable error.

To begin, one needs an estimated standard deviation. The e.s.d. for the TiO_2 example is 0.0167 wt%. Three samples were drawn, so at 99% confidence, $t = 9.925$ is found in Table 1.3. Say we desire a sampling error of ± 0.05 wt%. Then $n = 11$. Three samples were already drawn, so eight more need to be taken. The reader may want to calculate the number of replications for 95% confidence so that the mean will be within ± 0.05 wt%.

1.2.7 Practicalities and Compromises

It is obvious the analyst cannot run statistical tests for every sample set. However, routine statistical testing should be a part of all laboratory procedures and comes under the category of quality assurance or Q/A. The basic statistical tool of Q/A is the e.s.d. To implement it, replicate determinations using all sampling and analytical steps are periodically made and the e.s.d.'s are plotted against date and time. It will be found that the e.s.d.'s oscillate. If they begin to exceed predetermined levels, then the cause must be determined and corrected. Usually it will be found that a new variable has crept into the procedures.

Outliers, or *runaways*, are values that deviate by large amounts from what is expected, and one always wonders if the value is a fluke that can be discarded, or if it should be included in a mean. Outliers are particularly troublesome when calibration curves are being constructed. As mentioned at the beginning of this section, 3σ on either side of the mean includes almost 100% of the normal range of a well-sampled target, so any value deviating $> 3\sigma$ is an outlier candidate. When small sample numbers are involved, three standard errors should be used, for it is a much more conservative measure than, say, 2σ.

A more rigorous criterion for outliers uses the Snedicor F test. The test compares the ratio of e.s.d.'s with and without the suspect value to the F value found in collections of statistical tables. The criterion was first proposed by Plesch [14] who illustrated its application to calibration curves.

The amount of sample received by the laboratory may strain the ability to draw standard size subsamples. Naturally, if a modification in the procedure is possible but results in a decrease in analysis quality, the client should be made aware of the circumstances. It is good to have alternative procedures available for such contingencies. For example, if standard practice in XRF analysis is to prepare infinitely thick pressed powder briquettes, then it may be possible to use several small weighed samples instead of a single infinitely thick one. Although the computations may require a different routine, recall that for many of the lighter elements only a thin surface layer is analyzed. Abandoning the practice of infinitely thick samples may allow the analyst to archive part of a valuable specimen and still be able to prepare duplicates.

Normally, a sample is drawn by some volumetric means, but sometimes the required amount of sample will be weighed by pouring directly from a bottle onto weighing paper in the balance. This method is one of the worst possible sampling procedures; it is as bad as scooping directly out of a SRM (standard reference material) bottle. It assures that a segregated sample will be weighed and that it will not be representative.

1.3   OBTAINING A REPRESENTATIVE SAMPLE FROM THE BULK

While the analyst is rarely involved in the actual process of taking the representative sample from the bulk, it might be useful to briefly review the major stages. There are two kinds of bulk: the whole object to be sampled and a large sample or specimen taken from it. The whole object might be a volcano, a coal seam, the contents of the hold of a ship or a rail car, a storage pile, a run of I-beams from a mill, the material in motion on a conveyor belt, a production run of bricks, and so forth. A representative sample of the whole which can be delivered to the analytical laboratory has to be extracted. This sample is the primary sample. Outcrop samples would be collected from the volcano, channel samples would be collected from the coal seam, a cylindrical probe sample from the ship's hold, a front-loader scoop from the storage pile, a number of steel I-beams from the mill run, periodic cuttings from the conveyor belt, a random selection of bricks from the kiln. Sampling for environmental studies has to be particularly stringent, for errors can easily become political and legal issues.

All these samples have to be collected so that operator bias plays no part and inhomogeneities are accounted for. Once a large object is in place, it is extremely difficult to sample its interior. Probes inserted into the contents of holds, rail cars, and storage piles rarely sample the bottom where water collects and fines settle. These kinds of bulk are best sampled during transfer from one site to another. Random truck loads of coal, for example, can be spread in a narrow linear pile from which a front loader can randomly cut a complete segment in one scoop. That scoopful then has to be resampled, say by the alternate shoveling method described below.

Objects with discrete dimensions such as I-beams and bricks are best drawn randomly during production, because it makes no economic sense to tear apart a large lot in order to gain random access to the interior. Special tools and techniques are required for specific materials. The reader should consult Pitard [12] for many theoretical and practical details in this regard. An I-beam will be sawn in two at a random point and a slice extracted for analysis. The slice may be drilled randomly and the cuttings chemically analyzed or the fabric of the slice analyzed geometrically.

Geological sampling in the field often requires accurate surveying. Core may be extracted from a rock outcrop or the interior of a mine by drilling. Channel sampling of rock strata has to be done normal to strata and from channels with parallel sides. Sampling of granular material on moving conveyor belts is particularly tricky. Sample cutters have to be carefully designed with the velocity and the physical properties of the feed in mind. The delimitation and extraction errors of sampling as defined later are easy to make when drawing a sample from moving materials [12]. Finally, our task is to reduce the bulky primary sample drawn as above into a secondary sample that can be prepared in the laboratory for analysis. The analyst will draw the ultimate random samples for analysis from material properly prepared.

Pitard [12] describes three fundamental kinds of errors in obtaining a sample from the bulk: the *delimitation error*, the *extraction error*, and the *preparation error*. The delimitation error is the error generated when the geometry of sample drawn (i.e., its limits) is different from the expected geometry of sample. This error occurs when samples are contaminated by uphole dilution, recovery of core samples is incomplete, the sampling probe fails to reach the bottom of a pile, the channel of a channel sample does not have parallel sides, a scoop fails to confine a sample, etc. The extraction error occurs when the dynamic properties of some fragments cause them to be erroneously ejected from or included in the sampling tool. The bouncing of streaming material out of a sampling tool is an example as are the inadvertent levering of a large external fragment into the sampler and the manual handling of fragments too large to pass the aperture of the splitter. The preparation error occurs during the passage of sample from one location to another or from one time frame to another. It is seen as losses or contamination due to the various stages of storage and preparation. Grinding and sieving, oxidation and reduction, hydration and dehydration, etc. contribute to preparation errors. Preparation errors also include sabotage, fraud, and unintentional mistakes.

The first stage of secondary sampling is reduction of the primary sample to a manageable state. Usually this requires crushing and grinding, but for XRD texture analysis the sample will usually be sawed, etched, and polished. The particular reduction strategy will depend on the goals of the analysis and the physical and chemical properties of the primary sample, but it will always be important that little of the sample be lost. Most X-ray diffraction and fluorescence analyses of solids require that the sample be finely powdered. Grinding is very energy intensive, so it is important that no more of the primary sample be ground finely than is necessary. Minimum grinding is achieved by dividing or splitting the primary sample into representative portions, one or more of which can be appropriately ground.

1.4 SAMPLING PROCEDURES

Splitting is a crucial early step in secondary sampling. At its simplest, the primary sample is divided into two equal (in every sense) parts, and one is archived in case it is later needed. If only a small amount is needed for analysis, n stages of splitting can proceed to $1/4, 1/8, 1/16, \ldots, 1/2^n$ and the final $1 - 1/2^n$ is archived. Splitting a granular material is a dynamic process which usually involves pouring. Pouring always causes segregation, so it is vital that the segregated fractions always appear in each split in their original proportions; otherwise, an extraction error occurs.

1.4.1   Cone and Quartering

The oldest method of splitting still in common use is cone and quartering. While not the best method of splitting, it is attractive for the broad range of sample sizes that can be treated. Samples ranging from tens of kilograms to a few grams can be split by cone and quartering. In this method, a granular sample is first carefully poured vertically onto a clean surface in order to form a cone. For small samples in the laboratory, this surface can be a weighing paper. The dynamics of correct pouring will assure that coarser fragments are uniformly concentrated around the periphery of the cone and the finer ones are uniformly distributed in the center. In the second part of the method, the cone is quartered by precisely dividing it vertically into four equal sectors. This quartering must be done without generating a delimitation error. The easiest, but riskiest, way is to first cut the cone vertically into two equal halves with a sharpened rectangular plate, and without removing the plate, draw one half away from the other far enough so that there is no contact between the two halves when the plate is removed. The two halves are then simultaneously cut again by the plate at 90° to the first cut and drawn away from the remaining two quarters. It is important that the four quarters are not in contact after cutting. Finally, one or more quarters are removed for archiving and further processing.

A better method of cone and quartering is to prepare the cutter as two sharpened plates joined at right angles in their center. (See Figure 1.5.) The plates must be made of a very slick material such as stainless steel and joined so that sample does not adhere to the joint. This cutter assures that the cone will be divided properly in a single action. Some practitioners of cone and quartering prefer to flatten the cone into a cake before cutting it. This operation facilitates cutting through a very large

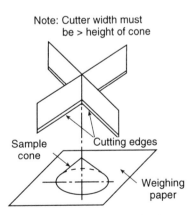

Figure 1.5. Cone and quartering layout.

cone. However, if the cone remains erect, it is easier to locate the axis of symmetry of the pile.

1.4.2 Alternate Shoveling

If the primary sample contains tens of kilograms of material, alternate shoveling is preferred to cone and quartering for reduction of sample size. In this method, as with cone and quartering, the primary sample is poured in a symmetrical pile on a clean surface. The original pile is made into two nonoverlapping piles by shoveling from the original alternately into the two new piles. It is important that the shovelsful all be the same size and that each new pile contain the same number of transfers. In this regard, it is difficult to come out exactly even, so the size of the shovel has to be small relative to the size of the pile. The method is better for large samples than for small, because the delimitation error of one shovelful can then be unimportant.

For further reduction in the size of the primary sample, one of the two new piles is chosen at random and the process repeated. In this way, the sample is reduced to a size that can be further reduced prior to final grinding.

1.4.3 Linear Japan Cake

Another shoveling or scooping method that allows probabilistic sampling is the linear Japan cake. The sample from a previous stage is spread in a linear pile from which a single scoop can be randomly extracted in one motion. The pile is constructed by pouring back and forth along its design length. Several samples can be extracted in this way. As a rule of thumb, the pile should be at least 25 times longer than the width of the scoop. Delimitation error can be avoided only if the scoop has vertical sides, sharpened cutting edges, and can hold all the sample delimited by the scoop dimensions. Any material that the scoop pushes out of the pile has to be recovered to the sample. Picard suggests that this method be used even with small samples in the laboratory balance room. The linear Japan cake is shown in Figure 1.6.

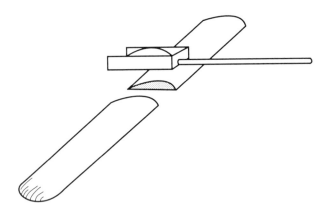

Figure 1.6. Linear Japan cake randomly sampled.

Pan width equals riffler width

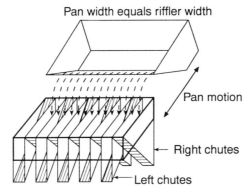

Figure 1.7. Elements of a riffle splitter. Box frame shown as transparent.

1.4.4 Riffle Splitting

The riffle splitter, sometimes called a Jones splitter, is one of the best and simplest devices for splitting. It is a box containing an even number of adjacent chutes alternately inclined in opposite directions. When the primary sample is properly poured into the top of the splitter box, half of the sample passes through one set of chutes and half through the other and into two receptacles. If the sample fraction collected in one receptacle is passed through the splitter n times, the size of the last split is $1/2^n$. Figure 1.7 diagrammatically shows the geometry of such a splitter.

The well-maintained riffle splitter will generate no significant delimitation or extraction errors if the sample passes equally through all the chutes. This condition can only be achieved by pouring the sample uniformly along the long axis of the splitter box from a scoop or a hopper extending along the full length of the splitter box. The splitter as well as its scoop or hopper should be made of polished stainless steel with all inside corners round and smooth enough to avoid trapping any of the sample.

Riffle splitters are available in various sizes in order to accommodate samples of different granularities. Pitard offers the rule of thumb that the aperture width of the chutes should be at least $(2d + 5$ mm) where d is the maximum dimension of the largest fragment. Note that the apertures of the smallest splitters have to be large enough to allow convenient cleaning.

1.4.5 Sectorial Splitting

Sectorial splitters allow the sample to be initially split into more than two parts. Several funnels are assembled together with common sides forming the radii of a circle and their third sides forming the perimeter of the same circle. Each funnel's brim describes a sector of the circle. In one type of sectorial splitter, the ensemble of funnels and their receptacles are rotated in a falling stream of the sample. In another type, the sample stream is made to sweep past each funnel. Figure 1.8 shows the essential elements of an eight-fold sectorial splitter.

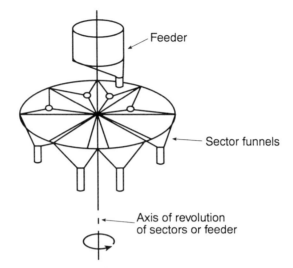

Figure 1.8. Elements of a sectorial splitter.

A disposable four-sector splitter can be folded from a square weighing paper and can be used for splitting small amounts of material. Figure 1.9 shows how it is folded. The four troughs are placed on four receptacles and the sample is slowly poured vertically onto the apex in a rotary motion. A more permanent small splitter can be made from a square polished stainless steel sheet.

1.4.6   The Holistic Approach to Sampling

It should be apparent from the above descriptions of sampling that every sampling step affects every subsequent analytical step. Often the earliest sampling steps are the least rigorous, and subsequent sampling, even when well done, takes on the aspect of damage control. It cannot be too strongly emphasized that good sampling

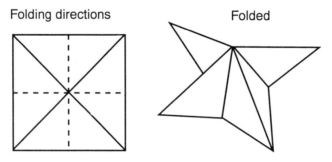

Figure 1.9. Disposable four-sector splitter.

is absolutely crucial to the meaningfulness of the ultimate analysis. If sampling is not representative in proportion to the importance of a project, then high quality analytical work is at best, decorative and at worst, dishonest.

Sampling should be viewed as a holistic process. Every step is materially linked to all other steps so that the analytical project can be related to reality, for it is through sampling that the whole analytical project is materialized. The complete sampling program should be planned in detail from sampling in the field to sampling in the balance room before the first sample from the bulk is taken. A flow sheet should be constructed that specifies the amount of sample required for each analytical step so that there will be an archived fraction at each one of them in case an analysis needs to be rerun. Of course, there should be a paper trail of every sample's progress through the sampling-preparation-analysis net.

An analysis only relates directly to the whole when the whole is totally analyzed. Every other kind of analysis is indirectly related to the whole through the sampling chain. We humans are a speculative species and we love imagining what truths an analysis might reveal. Let us never forget that sampling truth controls analytical truth.

1.5   METHODS OF COMMINUTING POWDERS FOR X-RAY DIFFRACTION

There are many methods used for grinding samples specifically for XRD. Manually grinding using a mortar and pestle (agate, corundum, mullite, etc.) has been traditionally used for routine service work. For finer and controlled particle size grinding, there are a variety of machines available commercially—vibratory rod mills (McCrone Micronizing Mill), vibratory disc mills (Tema), swing/disc mills (SPEX Shatterbox), planetary ball mills (Fritsch Pulverisette), Motorized Mortar Grinders (Brinkman), impact grinders (SPEX Mixer/Mill), and many others. All are designed to reduce the sample's particle size so as to meet the requirements for an ideal specimen.

1.5.1   Requirements for an Ideal Specimen

The requirements for an ideal XRD specimen are discussed in detail by Smith [15] and by Bish and Reynolds [16], and interested readers are referred to those papers. Smith points out that to achieve high degrees of accuracy in quantitative XRD, three conditions must be met: total randomness of the crystallite orientations, sufficient number of crystallites in the experimental specimen to meet statistical requirements, and sufficient diffraction intensity measured to satisfy counting statistics. Assuming the amount of phase present in the sample is well above the limit of detection and yields sufficient intensity, the particle size of the sample becomes all important because it is inversely related to the number of randomly oriented crystallites. Thus, reduction of the sample particle size by grinding is one of the chief considerations in sample preparation.

1.5.2 Ideal Particle Size for X-Ray Diffraction Applications

It has long been commonly accepted that for routine qualitative work, a particle size of <325 mesh (44 μm) or <400 mesh (38 μm) is satisfactory for most X-ray diffraction applications (major phase identifications, etc.). It is found that manually grinding a typical inorganic alumina silicate using a mortar and pestle readily yields a particle size of about 38 μm, which is acceptable for pattern matching and phase identification. Indeed, there are many materials (micaceous, fibrous, elastomeric, polymeric) where achieving a particle size finer than this level is very difficult, and requires specialized techniques [16]. For quantitative work, however, most authors agree that particles on the order of 10 μm or less are required, and Smith [15] points out that particles of about 1 μm size are still not sufficiently small to achieve a standard error (2.3σ) of <1% in quantitative analyses. In order to achieve particles in the range of 1–10 μm, a mechanical grinder is indispensable. There are, however, several precautions which must be considered when using a mechanical grinder.

1.6 GRINDING FOR X-RAY DIFFRACTION APPLICATIONS

Analysis of powders by XRD requires that they be extremely fine grained to achieve good signal-to-noise ratio (and avoid fluctuation in intensity), avoid spottiness and minimize preferred orientation. Reduction of powders to fine particles also ensures enough particle participation in the diffraction process. The recommended size range is around 1–5 μm [17, 18, 19] (see section 4.3.1), especially if quantification of various phases is desired. For routine qualitative evaluation of mineral components, the samples are usually ground to pass through a 325 mesh sieve (45 μm). For a detailed discussion of grinding equipment, refer to Chapter 8.

1.6.1 Mechanical Grinding: Precautions and Considerations

As a general rule, wet grinding, using a lubricant such as acetone, ethyl alcohol, methyl alcohol, isopropanol, or water is usually preferred to dry grinding [20]. Also, in both wet and dry grinding, the longer the grinding time, the smaller the particle size. There is, however, a point of diminishing returns, in that continued grinding introduces some undesirable effects without significantly reducing the particle size. For instance, particle agglomeration can occur as a result of prolonged grinding, and this effect is especially prevalent with dry grinding. As an example, we have found that increasing the dry grinding time of a zeolite A powder up to 60 minutes using a McCrone Micronizing Mill (alumina rods) caused pronounced particle agglomeration with an apparent particle size increase from 6.7 μm (three minutes grinding) to 11 μm (60 minutes grinding). (All particle sizes reported here were measured using a Leeds & Northrup MicroTrak particle size analyzer, Model 799SRA, and are defined as the mean diameter of the volume distribution of particles.) Prolonged grinding, both wet and dry, introduces line broadening either due to diminished crystallite size or the occurrence of lattice strain, or both. It may also lead to the creation of surface amorphous layers [21]. While some line broadening can be tolerated when using integrated intensities in quantitative analyses, significant line broadening may be associated with shifts in line positions for low angle reflections [16], and will almost surely be accompanied by changes in peak tails with consequent errors in background estimations.

1.6.2   Dry Grinding for X-Ray Diffraction Applications

The ultimate end point of dry grinding may sometimes be the transformation of crystalline structures into totally amorphous materials. Filio et al. [22] transformed a mixture of talc, kaolinite, and gibbsite into an amorphous state by extensive dry grinding, up to four hours. Kosanovic et al. [23] used dry grinding in a planetary ball mill with an agate vessel and tungsten carbide (WC) balls to produce amorphous materials from zeolite A (1 h), from zeolite X (1.5 h), and from mordenite (3 h). Another deleterious effect of prolonged dry grinding is the introduction of contaminants to the sample. This problem can be particularly severe if the material to be ground is very hard and abrasive, in which case sample contamination can occur rapidly. For instance, we have seen the presence of crystalline WC in samples of glass chips dry ground in a swing mill grinder (WC cups and pucks) for only eight minutes. In contrast to this result, we have dry ground zeolites for over 30 minutes under the same conditions without detecting WC contamination. Although not as common, prolonged dry grinding can also cause phase transformations in some materials. Tonejc et al. [24] showed that both boehmite and gibbsite could be transformed into α-alumina (corundum) by dry grinding in a planetary ball mill, WC balls and vial, for 2.5 hr (boehmite) and 7 hr (gibbsite). Additionally, WC contamination was severe in the ground material in both cases.

1.6.3   Wet Grinding for X-Ray Diffraction Applications

Experience has shown that wet grinding using a McCrone Micronizing Mill is the most efficient method of reducing the particle size of most inorganic materials, while avoiding many of the deleterious effects that can be associated with mechanical grinding. Using the McCrone Micronizing Mill with corundum grinding elements, as outlined in the following procedure, we have reduced the particle size of zeolite 13X molecular sieve beads to 4.9 μm mean diameter by grinding 3 g of sample with 20 ml of methanol for one minute. The beads were first coarse ground by hand to about 200 mesh. This particle size reduction was accomplished without detectable sample contamination or line broadening. For softer materials, such as hydrated aluminas, boehmite and bayerite, which are susceptible to line broadening, we have found that 30 seconds of wet grinding with methanol reduced the mean diameter particle size to < 9 μm with no line broadening. The wet grinding procedure detailed below has been shown to be effective for a variety of inorganic materials and should result in a mean particle size of about 5–10 μm. Actual results may vary depending on the relative hardness of the material being ground.

Wet Grinding Procedure for Inorganic Samples Using the McCrone Micronizing Mill

- Select the polypropylene grinding vessel and the corundum grinding elements. The sample to be ground should be at least 200 mesh. Weigh out 3.0 g of the sample to be ground and add it to the top layer of the grinding elements. Shake the jar to allow the sample to fall to the bottom.

- In a laboratory hood, add 20 ml of methanol to the jar and securely tighten the lid on the jar. Install the jar in the McCrone Mill and grind for one minute. Observe the jar during the grinding for a tight fit of the lid and no loss of lubricating agent.

- When the grinding is complete, return the jar to the hood and remove the lid, substituting a pouring lid which contains a pour hole and vent. Pour the wet slurry into a wide petri dish. Add additional methanol to the jar, shake to disperse and again pour the slurry into the petri dish. Repeat the flushing with methanol until the pourings are clear. Allow the methanol slurry to evaporate to dryness in the hood, taking appropriate measures to capture and reclaim the evaporated methanol.

- Place the dried sample cake in a laboratory oven at 100°C and dry for 30 minutes to remove any residual methanol. Remove the petri dish from the oven, allow to cool, and break up the sample cake using a spatula. Note that to totally remove all adsorbed water and traces of methanol, higher drying temperatures and/or vacuum drying may be needed.

- If the sample is to be analyzed directly, it is advisable to break up the dried sample cake further by gently regrinding using a mortar and pestle. If an internal standard is to be added to the sample followed by mechanical mixing, the dried agglomerates will be broken up by interaction with the internal standard during the mixing step.

- For mechanical mixing of the ground sample and an internal standard, about seven minutes of shaking in a glass sample vial (1 in. diameter × 1.5 in. height) using a SPEX Mixer/Mill (no mixing balls) is usually sufficient for thorough mixing.

Grinding Polymers, Plastics, Elastomers, etc. using the SPEX freezer/mill

Where polymers and plastics are available in sheet form, they can usually be analyzed as received as flat specimens, albeit with the risk of encountering preferred orientation. Frequently, however, these kinds of materials are received for analysis in irregular shapes or granular forms that require particle size reduction for optimum analysis conditions. The following procedure using the SPEX freezer/mill will reduce the particle size of a typical chip plastic to about 60% <325 mesh in four minutes of grinding under liquid nitrogen. The ground sample is suitable for XRD or XRF analyses; however, for trace XRF analyses, be aware that the potential for sample contamination exists from the steel impactor components and/or the polycarbonate vial centers.

- Cut the plastic or polymer into chips about 3 to 5 mm on a side.
- The SPEX freezer/mill accommodates three sample vials. In operation it grinds one sample while prechilling the other two. Load each of the three vials with two grams of sample. Insert the loaded vials into their respective chambers in the apparatus.
- Carefully fill the freezer/mill chamber with liquid nitrogen, close the cover, and allow it to come to thermal equilibrium.
    CAUTION: Wear a protective face shield and nonporous, thermally insulated gloves while handling liquid nitrogen!
    A five-liter Dewar flask with fitted insulated "caddie" makes a convenient vessel for adding liquid nitrogen (Thomas Scientific).
- When the freezer/mill chamber has come to thermal equilibrium conditions as evidenced by the reduction in vented vapors, grind the sample for four minutes.
- Open the chamber, remove the vial that contained the ground sample using the vial retrieval tool, place the vial on the bench, and allow it to come to ambient conditions. Alternately, the vial can be opened directly using the extractor and vial opener tools for immediate access to the ground sample.
- At this point, insert one of the prechilled grinding vials in the grinding position, add more liquid nitrogen if necessary, and repeat the grinding procedure on the second sample.

1.6.4 Types of Grinding

Grinding can be accomplished either though hand grinding or in a mechanical grinder. Hand grinding is tedious and time-consuming and is done using a pestle and mortar made out of agate, alumina, mullite, boron carbide, or tungsten carbide. When dealing with extremely small amounts of sample, hand grinding is probably the most viable option. Grinding of sensitive minerals that may be damaged or undergo alteration during size reduction (e.g., kaolinite, calcite) is usually carried out under alcohol (acetone or isopropyl alcohol) in a pestle and mortar. Grinding in liquid nitrogen may also be effective. Various mechanical grinders are available for automatic grinding including McCrone micronizing mill, Brinkman "Retsh" grinder, Pitchford grinder, and Spex shatter box. They are used both under dry and wet grinding conditions. The choice of a grinder depends on various factors, the most important ones being the sample integrity, possible chemical contamination, and degree of damage afforded to the crystallites.

Figure 1.10 shows the XRD patterns for a sandstone sample ground using three different grinders and hand grinding. The hand ground sample was continually sieved to remove fine particles and prevent excessive grinding. The effects of grinding can be monitored by observing peaks around 7 Å (due to kaolinite) and 3.24 Å and 3.18 Å (due to feldspars). The shatter box may be too severe on these "soft" minerals, depending on the grinding time. Contamination may also be somewhat dependant on

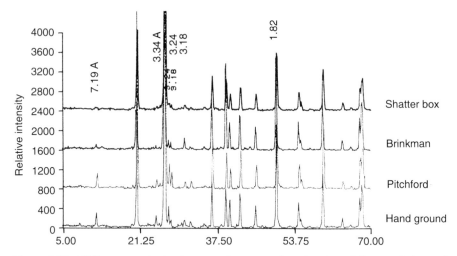

Figure 1.10. XRD patterns obtained on a sandstone sample, using three different grinders and hand grinding.

the type of grinder employed. When grinding samples which are mixtures, different physical properties may lead to significantly different results with the different phases; for example, in a clay quartz mixture the clay may protect the quartz by *lubricating* the grinder platens. Care must be taken to avoid physical phase separation (i.e., by sieving) if the goal is phase quantification.

1.6.5   Effects of Grinding

The effects of excessive grinding include lattice distortion and possible formation of an amorphous layer (Beilby layer) outside the grains. The McCrone micronizing mill has been used extensively to grind samples, especially to a very small size (1–5 μm). The effects of this mill on the amorphous character and particle size distributions of α-quartz powder were investigated by O'Connor and Chang [25]. They found that five minutes wet grinding produced the best particle size distribution and least amorphous component in these powders. Dry grinding with the micronizing mill was also quite satisfactory. Note that an *amorphous* surface layer may result from excessive grinding, but it also results as a function of time and rearrangement of the surface atoms even in such materials as quartz.

1.7   ABSORPTION PROBLEMS

1.7.1   Absorption of X-rays

All matter is made up of atoms. Each atom is in turn made up of a nucleus surrounded by electrons in discrete energy levels. The number of electrons is equal to the atomic number of the atom, when the atom is in the ground state. When a beam of X-ray photons is incident upon matter, the photon may interact with the individual atomic electrons. The absorption (attenuation) process can be described by considering a beam of X-ray photons of intensity I_0 incident on the specimen at an angle normal to the surface of the specimen. A portion of the beam will pass through the absorber, the fraction I being given by the expression:

$$I = I_0 \exp(\mu \rho x) \tag{1.10}$$

where μ is the mass attenuation coefficient of absorber for the wavelength in question, ρ the density of the specimen, x the distance travelled through the specimen. Three processes can now occur. The first of these processes is the attenuation of the transmitted beam as described above. Second, the incident beam may be scattered and emerge as coherently and incoherently scattered wavelengths. Third, fluorescent radiation may arise from the sample. It derives from equation (1.10) that a number $(I - I_0)$ of photons are lost in the absorption process. Although a significant fraction of this loss may be due to scattering, by far the greater loss is due to the photoelectric effect. Photoelectric absorption occurs at each of the energy levels of the atom; thus, the total photoelectric absorption is determined by the sum of each individual absorption within a specific shell. Where the absorber is made up of a number of different elements, as is usually the case, the total absorption is made up of the sum of the products of the individual elemental mass attenuation coefficients and the weight fractions of the respective elements. This product is referred to as the total matrix absorption. The value of the mass attenuation coefficient is a function of both the photoelectric absorption and the scattering. However, the influence of photoelectric absorption is usually large in comparison with the scattering, and to all intents and purposes, the mass attenuation coefficient is equivalent to the photoelectric absorption coefficient.

The photoelectric absorption is made up of absorption in the various atomic levels, and it depends on the atomic number. A plot of the mass attenuation coefficient as a function of wavelength contains a number of discontinuities called absorption edges, at wavelengths corresponding to the binding energies of the electrons in the various subshells. As an example, Figure 1.11 shows a plot of the mass attenuation coefficient as a function of wavelength for the element tungsten. It will be seen that as the wavelength of the incident X-ray photons become longer, the absorption increases. A number of discontinuities are also seen in the absorption curve: one K discontinuity, three L discontinuities, and five M discontinuities. Because it can be assumed that the mass attenuation coefficient is proportional to the total photoelectric absorption it is made up of absorption in the K level, plus absorption in the three L levels, and so on. If a slightly longer wavelength were chosen, and this wavelength would fall to the long wavelength side of the K absorption edge, the photons are insufficiently energetic to eject K electrons from the tungsten, and because there can be no photoelectric absorption in the level, the term for the K level drops out of the equation. Because this K term is large in comparison with the others, there will be a sudden drop in the absorption curve corresponding exactly to that wavelength at which photoelectric absorption in the K level can no longer occur. Clearly, a similar situation will occur for all other levels, giving the discontinuities in the absorption curve indicated in the

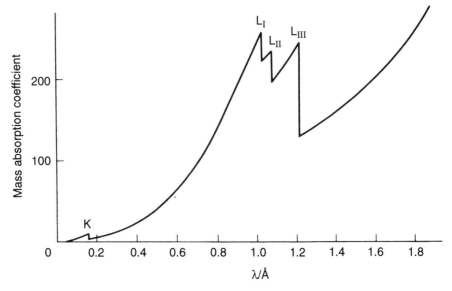

Figure 1.11. A typical absorption curve.

figure. Because each atom has a unique set of excitation potentials for the various subshells, each atom will exhibit a characteristic absorption curve. This effect is very important in quantitative X-ray spectrometry, because the intensity of a beam of characteristic photons leaving a specimen is dependent upon the relative absorption effects of the different atoms making up the specimen. This effect is called a "matrix effect" and is one of the reasons why a curve of characteristic line intensity as a function of element concentration may not be a straight line.

1.7.2 Scattering and Diffraction of X-rays

Scattering occurs when an X-ray photon interacts with the electrons of the target element. Where this interaction is elastic (i.e., no energy is lost in the collision process), the scattering is referred to as coherent (Rayleigh) scattering. Because no energy change is involved, the coherently scattered radiation will retain exactly the same wavelength as that of the incident beam. The origin of the coherently scattered wave is best described by thinking of the primary photon as an electromagnetic wave. When such a wave interacts with an electron, the electron is oscillated by the electric field of the wave and in turn radiates wavelengths of the same frequency as the incident wave. All atoms scatter X-ray photons to a lesser or greater extent, the intensity of the scattering being dependent on the energy of the incident ray and the number of loosely bound outer electrons—in other words, upon the average atomic number. It can also happen that the scattered photon gives up a small part of its energy during the collision, especially where the electron with which the photon collides is only loosely bound. In this instance the scatter is referred to as incoherent (Compton scattering).

XRD is a combination of two phenomena—coherent scattering and interference. At any point where two or more waves cross one another, they are said to interfere. Interference does not imply the obstruction of one wave train by another, but rather describes the effect of superposition of one wave upon another. The principal of superposition is that the resulting displacement at any point and at any instant may be found by adding the instantaneous displacements that would be produced at the same point by independent wave trains if each were present alone. Under certain geometric conditions, wavelengths that are exactly in phase may add to one another, and those that are exactly out of phase may cancel each other out. Under such conditions, coherently scattered photons may constructively interfere with each other, giving diffraction maxima. A crystal structure consists of a regular periodic arrangement of atoms, with layers of high atomic density existing throughout the crystal structure. Planes of high atomic density mean, in turn, planes of high electron density. Because scattering occurs between impinging X-ray photons and the loosely bound outer orbital atomic electrons, when a monochromatic beam of radiation falls onto the high atomic density layers, scattering will occur. In order to satisfy the requirement for constructive interference, it is necessary that the scattered waves originating from individual atoms, i.e., the scattering points, be in phase with one another. The geometric conditions for this condition to occur is illustrated in Figure 1.12. Here, a series of parallel rays strike a set of crystal planes at an angle θ and are scattered as previously described. Reinforcement will occur when the difference in the path lengths of the two interfering waves is equal to a whole number of wavelengths. This path length difference is equal to $BC + CD$, because $BC = BD = x$, n must equal $2x$ for reinforcement to occur, where n is an integer. It will also be seen that $x = d \times \sin \theta$, where d is the interplanar spacing; hence, the overall condition for reinforcement is as given in equation (1.11):

$$n\lambda = 2d \sin \theta \qquad (1.11)$$

This equation is known as Bragg's law. Bragg's law is important in wavelength dispersive spectrometry, where, by using a crystal of fixed $2d$, each wavelength will be diffracted at a unique diffraction (Bragg) angle. Thus, by measuring the

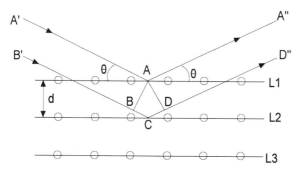

Figure 1.12. Diffraction of X-rays.

diffraction angle θ, knowledge of the d-spacing of the analyzing crystal allows the determination of the wavelength. Because there is a simple relationship between wavelength and atomic number, as given by Moseley's law [26], one can establish the atomic number(s) of the element(s) from which the wavelengths were emitted.

1.7.3   Absorption in X-ray Fluorescence

The distances travelled by X-ray photons through solid matter are not very great for the wavelengths and energies of the characteristic lines used in XRF and XRD analysis. The x term in equation (1.10) represents the distance travelled, in centimeters, by a monochromatic beam of X-ray photons, in a matrix of average mass attenuation coefficient expressed in cm^2/g. Because the densities of most solid materials are in the range of 2 to 7 g/cm^3 and values of mass attenuation coefficients are typically in the range 50 to 5,000 cm^2/g, values for x range from a few μm to several hundred μm. The path-length x is related to the depth of penetration D of a given analyte wavelength as shown in equation (1.12):

$$D = x \times \sin \theta_2 \qquad (1.12)$$

where θ_2 is the angle at which the radiation emerges form the specimen. In X-ray fluorescence this angle would be the *take-off angle* of the spectrometer (In X-ray diffraction, θ_2 is simply the value θ, i.e., one-half of the Bragg angle). Although the depth of penetration is a rather arbitrary measure, it is nevertheless useful, because it does give an indication of the thickness of sample contributing to the measured fluorescence radiation from the specimen for a given analyte wavelength or energy. In most spectrometers the value of θ_2 is between 30° and 45°, so the value of D is generally about one-half the value of the path length x.

1.7.4   Absorption in X-ray Diffraction

Because a (nominally) single wavelength is used in XRD, absorption problems are much simpler than those in XRF. Nonetheless, two effects due to absorption must always be considered.

1. *Beam Penetration.* As in XRF, the X-ray beam entering a specimen during a diffraction measurement, penetrates only to a finite depth, this depth being determined by the total mass attenuation coefficient of the specimen. If the mass attenuation coefficient is large the penetration will be small, and if the mass attenuation coefficient is small the penetration will be large. Because limitation of the beam penetration means that a smaller volume of specimen will be analyzed a high absorbing specimen will be much more likely to be influenced by particle statistical effect (see section 4.3.1) than would a low absorbing specimen.

One important difference between the beam penetration effect in XRD as compared with XRF is that, whereas in XRF the angle between the specimen and the emerging beam is fixed, in XRD this angle is variable. One result of the variable specimen surface-emergent beam angle is that the effective penetration depth D increases with increase in the diffraction angle.

2. *Matrix Effects.* Again, as in XRF, because the specimen is made up of several elements, each with its own mass attenuation coefficient, the emerging (diffracted) beam of X-radiation will be subject to the absorption influences of the matrix elements. Fortunately, because XRD is generally employed using a single wavelength, absorption effects in XRD are simpler than those in XRF.

1.8 CRYSTALLINITY AND PARTICLE EFFECTS IN X-RAY DIFFRACTION APPLICATIONS

Figure 1.13 illustrates the makeup of the analyzed specimen. Atoms join together to make molecules; a specific arrangement of molecules make up a subcell; several subcells make up a unit cell; many unit cells make up a crystallite, and many crystallites make up a specimen. The values of the measured d-spacings depend upon the dimensions of the unit cell. The intensities of the diffracted lines will also depend upon the type and arrangement of atoms making up the unit cells but also, will depend very much on the orientation of the crystallites. An "ideal" specimen is one made up of a large number of randomly oriented crystallites. Where the crystallites tend to orient in a specific direction the specimen is referred to as "preferred," or suffering from "preferred orientation." The relative intensities of lines from preferred specimens may vary by tens of percent for similar lines from a random specimen.

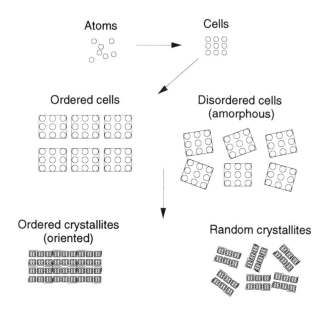

Figure 1.13. Makeup of the prepared specimen: atoms, cells and crystallites.

REFERENCES

[1] Krumbein, W. C. and Graybill, F. A. (1965), *An Introduction to Statistical Models in Geology*. McGraw-Hill Book Co., New York, 475 pp.

[2] Arkin, Herbert and Colton, R. R. (1970), *Statistical Methods*, 5th ed., Barnes & Noble College Outline Series, Barnes & Noble: New York, pp 160–161.

[3] Renault, J., McKee, C. and Barker, J. (1992), "Calibrating for X-ray analysis of trace quartz," *Adv. X-ray Anal.*, **35**, 362–373.

[4] Renault, J., McKee, C. and Barker, J. (1994), "New X-ray diffraction method of additions for crystalline silica," *Anal. Chim. Acta*, **286**, 129–133.

[5] Richardson, J. M. (1993), *A practical guide to field sampling for geological programs. Analysis of Geological Materials*, Chris Riddle, ed., Marcel Dekker: New York, pp 37–64.

[6] Wenk, H. R. (1985), "Measurement of Pole Figures," in "Preferred Orientation in Deformed Metals and Rocks" in *An Introduction to Modern Texture Analysis*, Ch. 2. Hans-Rudolf Wenk, ed. Academic Press: Orlando, FL, pp 11–47.

[7] Youden, W. J. (1985), *Experimentation and Measurement*, Rev. ed. National Science Teachers Association, Special Publications Dept., 1742 Connecticut Ave. N.W., Washington, DC 20009, 97 pp.

[8] Natrella, M. G., *Experimental Statistics*, National Bureau of Standards Handbook 91, issued August 1, 1963, NIST: Gaithersburg.

[9] Huff, Darrell (1982), *How to Lie with Statistics*, W.W. Norton & Co., New York. 142 pp.

[10] Shaw, D. M. (1961), "Manipulation errors in geochemistry: a preliminary study," *Trans. Royal Soc. Canada*, vol. LV, series III, June, 1961, 41–55.

[11] Ramsey, M. H., Thompson, M., and Hale, M. (1992), "Objective evaluation of precision requirements for geochemical analysis using robust analysis of variance," *J. Geochem. Expl.*, **44**, 23–36.

[12] Pitard, Francis F. (1993), *Pierre Gy's Sampling Theory and Sampling Practice*, CRC Press, Boca Raton, FL 488 pp.

[13] Ridley, K. J. D., Turek, A., and Riddle, C. (1976), "The variability of chemical analyses as a function of sample heterogeneity, and the implications to the analyses of rock standards," *Geochim. Cosmochim. Acta.*, **40**, 1375–1379.

[14] Plesch, R. (1981), "Runaways in X-ray spectrometry," *X-ray Spectrom.*, **10**, 8–11.

[15] Smith, D. K. (1992), "Particle statistics and whole-pattern methods in quantitative X-ray powder diffraction analysis," *Adv. X-ray Anal.*, **35**, 1–15.

[16] Bish, D. L., and Reynolds, R. C. Jr. (1989), "Sample preparation for X-ray diffraction," in *Reviews in Mineralogy Vol. 20: Modern Powder Diffraction*, D. L. Bish and J. E. Post, eds. Mineralogical Society of America, Washington, DC, pp. 73–99.

[17] Brindley, G. W. and Brown, G. (1980), "Crystal Structures of Clay Minerals and Their Identification," Mineralogical Society Monograph No. 5, Mineralogical Society, London, 495 pp.

[18] Klug H. P. and Alexander, L. E. (1974), *X-ray Diffraction Procedures*, Wiley: New York, 996 pp.

[19] Cullity B. D. (1978), *Elements of X-ray Diffraction*. 2nd Ed., Addison-Wesley Publishing Co., Menlo Park, CA.

[20] O'Connor, B. H. and Chang, W. J. (1986), "The amorphous character and particle size distributions of powders produced with the micronizing mill for quantitative X-ray powder diffractometry," *X-ray Spectrom.*, **15**, 267–270.

[21] Okada, K., Kuriki, A., Yano, S., and Otsuka, N. (1993), "Mechano-chemical effect for some Al_2O_3 powders produced by attrition milling," *J. Mat. Sci. Letts.*, **12**, 862–864.

[22] Filio, J. M., Sugiyama, K., Kasai, E., and Saito, F. (1993), "Effect of dry mixed grinding of talc, kaolinite, and gibbsite on preparation of cordierite ceramics," *J. Chem. Eng. Japan*, **26**, 565–569.

[23] Kosanovic, C., Bronic, J., Subotic, B., Smit, I., Stubicar, M., Tonejc, A., and Yamamoto, T. (1993), "Mechanochemistry of zeolites: Part 1. Amorphization of zeolites A and X and synthetic mordenite by ball milling," *Zeolites*, **13**, 261–268.

[24] Tonejc, A., Stubicar, M., Tonejc, A.M., Kosanovic, K., Subotic, B., and Smit, I. (1994), "Transformation of γ-ALOOH (boehmite) and $Al(OH)_3$ (gibbsite) to $\alpha - Al_2O_3$ (corundum) induced by high energy ball milling," *J. Mats. Sci. Letts.*, **13**, 519–520.

[25] O'Connor, B. H. and Chang, W. J. (1986), "The amorphous character and particle size distribution of powders produced with the micronizing mill for quantitative X-ray powder diffractometry," *X-ray Spectrom.*, **15**, 267–270.

[26] Moseley, H. G. J., "High frequency spectra of the elements," *Phil. Mag.*, **26** (1913), 1014–1034; and **27** (1914), 703–713.

CHAPTER 2

SPECIMEN PREPARATION PROCEDURES IN X-RAY FLUORESCENCE ANALYSIS

CONTRIBUTING AUTHORS: FERET, JENKINS

2.1 COMMON PROBLEMS IN PREPARING AND PRESENTING SPECIMENS FOR X-RAY FLUORESCENCE

2.1.1 General

Because X-ray spectrometry is essentially a comparative method of analysis, it is vital that all standards and unknowns be presented to the spectrometer in a reproducible and identical manner. Any method of specimen preparation must give specimens which are reproducible and which, for a certain calibration range, have similar physical properties including mass attenuation coefficient, density, particle size, and particle homogeneity. In addition the specimen preparation method must be rapid and cheap and must not introduce extra significant systematic errors, for example, the introduction of trace elements from contaminants in a diluent. Specimen preparation is an all-important factor in the ultimate accuracy of any X-ray determination, and many papers have been published describing a multitude of methods and recipes for sample handling.

2.1.2 Types of Samples Analyzed

In general samples fit into three main categories.

1. Samples which can be handled directly following some simple pretreatment such as pelletizing or surfacing. For example, homogeneous samples of powders, bulk metals, or liquids.
2. Samples which require significant pretreatment. For example, heterogeneous samples, samples requiring matrix dilution to overcome interelement effects, and samples exhibiting particle size effects.

A Practical Guide for the Preparation of Specimens for X-Ray Fluorescence and X-Ray Diffraction Analysis, Edited by Victor E. Buhrke, Ron Jenkins, and Deane K. Smith. ISBN 0-471-19458-1 © 1998 John Wiley & Sons, Inc.

3. Samples which require special handling treatment. For example, samples of limited size, samples requiring concentration or prior separation, and radioactive samples.

Each of these categories will be discussed in turn and where possible practical examples will be given illustrating the various sample treatments which are available.

The ideal specimen for XRF analysis is one in which the analyzed volume of specimen is representative of the total specimen, which is, itself, representative of the sample submitted for analysis. The ideal specimen should also remain stable prior to and during the actual analytical sequence.

2.2 PRESENTING THE SPECIMEN TO THE SPECTROMETER

2.2.1 Physical Form of the Specimen in X-ray Fluorescence Analysis

There are many forms of specimen suitable for XRF analysis, and the form of the sample as received will generally determine the method of pretreatment. It is convenient to refer to the material received for analysis as the sample, and that which is actually analyzed in the spectrometer as the specimen. While the direct analysis of certain materials is certainly possible, more often than not some pretreatment is required to convert the sample to the specimen. This step is referred to as specimen preparation. In general, the analyst would prefer to analyze the sample directly, because if it is taken as received, any problems arising from sample contamination that might occur during pretreatment are avoided. In practice, however, there are three major constraints that may prevent this ideal circumstance from being achieved: sample size, sample size homogeneity, and sample composition heterogeneity. Problems of sample size are frequently severe in the case of bulk materials such as metals, large pieces of rock, etc. Samples of suitable size can be cut from finished stock including glass, ceramic, or plastic materials by use of diamond tools or punches. Pieces of fiber or cloth can usually be clamped in a cell between two sheets of polyethylene terephthalate (mylar) film or alternatively mixed with a resin bonded polymer and cast into blocks.

Problems of sample composition heterogeneity will generally occur under these circumstances as well, and in the analysis of powdered materials, heterogeneity must almost always be considered. Table 2.1 lists some typical sample forms and indicates possible methods of specimen preparation. Four types of sample form are listed: bulk solids, powders, liquids, and gases. The sample as received may be either homogeneous or heterogeneous; in the latter case, it may be necessary to render the sample homogeneous before an analysis can be made. Heterogeneous bulk solids are generally the most difficult kind of sample to handle, and it may be necessary to dissolve or chemically react the material in some way to give a homogeneous preparation. Heterogeneous powders are either ground to a fine particle size and then pelletized, or fused with a glass forming material such as borax. Solid material in liquids or gases must be filtered out and the filter analyzed as a solid. Where

TABLE 2.1. Forms and Treatments of Samples for Analysis

Form	Treatment
Homogeneous bulk solids	Grind to give flat surface
Heterogeneous bulk solids	Dissolve or react to give solution or homogeneous melt
Homogeneous powders	Grind and press into a pellet
Heterogeneous powders	Grind and fuse with borax
Homogeneous liquids	Analyze directly or dilute
Homogeneous (dilute) liquids	Preconcentration
Heterogeneous liquids	Filter to remove solids
Airborne dusts (gases)	Respirate through a filter to remove solids

analyte concentrations in liquids or solutions are too high or too low, dilution or preconcentration techniques may be employed to bring the analyte concentration within an acceptable range. These various specimen preparation techniques will be discussed in detail in succeeding sections of this chapter.

2.2.2 Specimen Placement

It is important that the prepared specimen be placed correctly in the specimen holder, and that the holder plus specimen be placed correctly into the spectrometer. Referring to Figure 2.1, it will be seen that radiation leaving the X-ray tube at a and b, penetrate into the specimen to a depth c and d. The collimator *sees* the specimen at e when observed from g, and at f when observed from h. The ideal intersection points for the line pairs ac, eg and bd, fh is at the specimen surface. However, because the

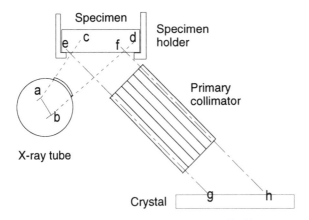

Figure 2.1. Placement of specimen and holder.

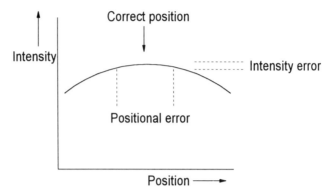

Figure 2.2. Position of the specimen holder in the spectrometer.

radiation has a finite penetration depth into the specimen, the collimator is unable to *observe* all of the volume irradiated by the X-ray tube. Hence, if the specimen surface were to be moved further and further away from the X-ray tube window, the observed intensity would steadily decrease. A similar intensity decrease would be seen if the specimen surface were brought *nearer* to the X-ray tube window. The overall effect of this phenomenon is shown in Figure 2.2.

As the position of the specimen is moved from a *too low* position to the ideal position, the intensity increases. As the specimen is moved further from the ideal position to a *too high* a position, the intensity will again decrease. Any positional error of the specimen thus translates to an intensity error. Positional errors can arise from poorly constructed specimen cups, from incorrect (rotational) positioning of the X-ray tube or, most commonly, from incorrect seating of the specimen in the cup. This latter effect may be due to the collection of particles of previously run specimens in the inside edge of the cup, or to dents or deep scratches of the cup surface.

2.2.3   Photochemical Decomposition

Special care must be taken when specimens may be influenced by the X-ray beam itself. As a simple example, in tetraethyl lead, TEL, the ligand between the lead atoms and the C_2H_5 groups is rather weak. When analyzing gasoline samples for TEL by XRF, the weak ligands rupture due to photoionization with subsequent decomposition of the TEL. As another example, if nylon specimens are being analyzed for trace metals by XRF, photoionization by the X-ray beam produces much bond rupture in the nylon and subsequent recombination of the radicals produced. The overall effect is that the molecular weight of the nylon increases sharply with radiation exposure with an associated change in physical properties (flexibility, etc.). It must also be remembered that when *damp* specimens are being X-rayed, ozone may be produced (as evidenced by its smell) which can, in turn, cause oxidation of the specimen.

2.3   HANDLING SAMPLES WITHOUT PRETREATMENT

While the direct analysis of the sample in the spectrometer would appear to offer significant advantages both in speed and avoidance of contamination in the specimen preparation process, in point of fact, direct analysis is only possible on special occasions. As an example, it is sometimes possible to compress small sample chips into a briquet by high pressure and analyze directly. Although this procedure may give a somewhat uneven surface, it is possible to correct for the uneven surface by ratioing analytical line intensities to another major sample element line [1].

By far the greatest potential problem in the direct analysis of bulk solids and powders is that of local heterogeneity. As was shown in section 1.7.3, the actual penetration of X-rays into the specimen is generally rather small and is frequently in the range of a few microns. Where the specimen is heterogeneous over the same range as the penetration depth, what the spectrometer actually analyzes may not be representative of the whole specimen. For the lower atomic number elements, penetration depths may be of the order of only a few microns. The problem in specimen preparation is then to ensure that the relatively thin surface layer actually analyzed is truly representative of the bulk of the sample. This problem may manifest itself either in terms of specimen composition heterogeneity, particle heterogeneity, or particle size inhomogeneity as illustrated in Figure 2.3. This figure represents the measurements of two elements A and B, first in a heterogeneous specimen, then in a heterogeneous particle. In the case of specimen heterogeneity, if the penetration of the X-ray beam is of the same order as the particle size, the measured intensity ratio of lines from the two phases A and B is quite different to measurements that would be obtained from a homogeneous sample. A similar effect may be seen if

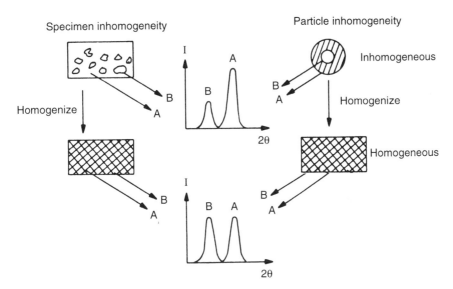

Figure 2.3. Heterogeneity in powder specimens.

individual particles are heterogeneous. Such an effect can occur, for example, in the case of partially oxidized pyrite (FeS_2) and chalcopyrite ($CuFeS_2$) minerals. If the penetrating power of the characteristic radiation is sufficiently large, it may be possible to analyze certain materials directly. This condition is particularly important for the analysis of living matter. As an example, measurements have been carried out to determine the amount of in vivo tibia lead using a ^{109}Cd radioisotope source spectrometer [2]. Such a technique allows rapid monitoring of workers with chronic lead exposure, and it has been found that this technique is acceptable for such screening.

2.4   USE OF PRESSED POWDER PELLETS

2.4.1   General

The most common method of preparing powder samples is first to grind, and then to pelletize at high pressure. While grinding is an extremely quick and effective means of reducing the particle size, there is always the potential problem of contaminating the sample during the grinding process. In practice, some materials may be sufficiently soft and homogeneous to allow direct pelletization and analysis. Such a technique is applicable to many pharmaceutical products [3], although it is often necessary to add a small amount of cellulose as a binder. In general, however, the sample is too hard for this technique to be applied, and means of reducing the particles to an acceptable size must be sought. Many studies have been carried out to quantify the magnitude of this problem—not just in XRF, but also in other spectroscopic disciplines (e.g., [4]). Probably the most commonly employed grinding device is the disk mill (Shatterbox) [see Chapter 8 for a list of equipment vendors]. This device consists of a series of concentric rings plus an inner solid disk that are shaken back and forth very vigorously. This device produces particles which are both small and of similar size. The actual sample container may be made of hardened steel, agate, or tungsten carbide. These mills are very efficient and are able to reduce powders to less than 325 mesh in a matter of minutes. Unfortunately, they can also be a source of contamination. As an example, Tuff [5] has reported a detailed study of the contamination of a quartzite cobble (99.5% quartz and 0.5% alumina) during crushing and grinding. He reported major pickup of iron, manganese, cobalt, and chromium from the steel jaw crusher.

2.4.2   Direct Pelletizing of the Specimen

Where powders are not affected by particle size limitations, the quickest and simplest method of preparation is to press them directly into pellets of constant density, with or without the additional use of a binder. In general, provided that the powder particles are less than about 50 μm in diameter (300 mesh) the sample should usually be pelletized at 15–20 tons per square inch. Where the self-bonding properties of the powder are good, low pressures of perhaps up to 2 to 5 tons per square inch can be employed or, in extreme cases, recourse made to the use of a binder. High pressure pelletizing in a die, or directly in a sample cup often results in a fracture of a pellet following the removal of pressure from the die. The fracturing is due to slight deformation of the die under pressure and can be avoided by pressing the powder in a former which will irreversibly yield under the high pressure. For example, one method of sample preparation which has been employed successfully in the analysis of slags and sinters consists of pressing the sample in lead rings or rubber "O" rings between two hardened steel platens. It is sometimes necessary to add a binder before pelletizing and the choice of the binding agent must be made with care. As well as having good self-bonding properties, the binder must be free from significant contaminant elements and must have low absorption (unless for some reason the mass absorption coefficient of the matrix needs to be increased). It must also be stable under vacuum and irradiation conditions, and it must not itself introduce significant interelement interferences. Of the large number of binding agents which have been successfully employed, probably the most useful are starch, ethyl cellulose, Lucite, polyvinyl, alcohol, and urea. Use of a binding agent invariably decreases the overall matrix absorption, and so dilution of a sample with a certain quantity of binder does not necessarily mean that the sensitivity for a certain element will drop by an equivalent amount. Frequently the addition of one or two parts by volume of a carefully chosen binder makes little or no difference to the absorption of a medium average atomic number matrix. However, the addition of a binder can lower the average atomic number of a sample quite significantly with a subsequent increase in the scattering power of the sample. Thus, addition of a binder often increases the background radiation of a sample, and this effect can become important in the determination of small quantities of elements with wavelengths shorter than about 1 Å where background is difficult to remove. As a result, it is often best to use the minimum quantity of binder mixed with the sample and to add additional binder as a backing for the sample to give it extra strength. This backing technique can also successfully be employed when the available amount of sample is too small to give a pellet of sufficient mechanical strength. Where necessary, further stability can be achieved by spraying the pellet with a 1% solution of formvar in chloroform.

2.5 LITHIUM AND SODIUM TETRABORATE FUSION METHODS

2.5.1 Advantages of Fusion

As was stated in the previous section, inhomogeneity and particle size effects can often be minimized by grinding and pelletizing at high pressure. However, all too frequently the effects cannot be completely removed because the harder compounds present in a particular matrix are not broken down. The effect is to produce systematic errors in the analysis of specific types of material typified by siliceous compounds in slags, sinters and certain minerals.

Probably the most effective means of preparing a homogeneous powder sample is the borax fusion method. This method was first proposed by Claisse in 1957 [6]; its principle involves fusion of the sample with an excess of sodium or lithium tetraborate and casting into a solid bead. Chemical reaction in the melt converts the phases present in the sample into glass-like borates, giving a homogeneous bead of controllable dimensions that is ideal for direct placement in the spectrometer. (For a detailed description of the principles of the borax fusion method, the reader is referred to references [7, 8]). While manual application of this technique is rather time consuming, a number of automated and semiautomated borax bead making machines are commercially available (e.g., [9, 10]), and these devices are able to produce multiple samples in a matter of minutes. The ratio of sample to fusion mixture is quite critical, because this will not only determine the speed and degree of completion of the chemical reaction, but also the final mass absorption coefficient of the analyzed bead and the actual dilution factor applied to the analyte element. Some control is possible over these factors, by use of fusion aids (such as iodides and peroxides) and use of high atomic number absorbers as part of the fusion mixture (barium and lanthanum salts have been employed for the purpose). Quantitative comparisons of the variables is not easy because of synergistic effects. Claisse's original method proposed a sample-to-sodium tetraborate ratio of 100:1, and in 1962 Rose et al. [11] suggested the use of lithium tetraborate in the ratio of 4:1. The same authors also suggested lanthanum oxide as a heavy absorber. Some years later Norrish and Hutton [12] concluded that for whole rock analysis a ratio of 5.4:1 was ideal, along with added lithium carbonate to lower the fusion temperature. More recently, Bower and Valentine [13] have published a critical review of these various techniques as applied to whole rock analysis and have compared the results obtained from different recipes with similar data obtained from pressed pellets. It should be mentioned that fusion dilutes the specimen and trace elements may no longer be detectable following borax fusion.

2.5.2   Choice of Fusion Mixture

The original Claisse technique [6, 14] consisted of fusing the sample with borax and casting into a solid button, but many variations of the original method have been described, [15], of which the more important are the use of lithium tetraborate [16] as an alternative to the sodium salt. Lithium tetraborate offers the advantage of having a lower average atomic number than the corresponding sodium salt and if used with lithium carbonate, with which it forms a eutectic mixture, when in the ratio of 6:1, it has a melting point lower than that of borax. It is, however, somewhat hygroscopic and prefused mixtures of lithium borate-lithium carbonate should be stored in well-stoppered bottles. This point should also be remembered when working with fused, ground, and pressed discs, because the pickup of moisture is quite a rapid process. XRD measurements have indicated the presence of Li_2CO_3, H_2O on the disc surface after they have been left for as little as 24 hours in a humid laboratory.

2.5.3   Choice of Reaction Crucible

The actual fusion reaction can be performed at 800–1000°C containing the fusion mixture in a crucible made of, for example, platinum, nickel, or silica. Each of these materials has certain advantages to offer but all suffer from the disadvantage that the melt tends to wet the sides of the dish and it is impossible to effect complete recovery of the fused mixture. Use of graphite crucibles overcomes this problem to some extent but probably the neatest way of avoiding the difficulty is to use a crucible of platinum + 3% gold. This alloy is hardly wetted at all by borate fusion mixtures, wastage is avoided, and cleaning is made much easier. Although the source of heat is usually a standard laboratory muffle furnace or even simply a Meeker burner, slightly more sophisticated methods have been tried, of which the most promising is the high frequency heating furnace. All of the bead techniques suffer from the limitation that the necessary annealing process may take several hours to complete, although this time can often be reduced by preparing very thin beads [17] or by special handling of the bead [18]. The time lost can, however, be almost completely avoided by grinding the bead immediately after fusion and pelletizing the resultant powder [19]. Other fusion techniques employ sodium carbonate [20], sodium and potassium bisulfate [21] with or without added sodium fluoride [22] and ammonium metaphosphate [23] as the reacting melts, followed by pelletizing the melt or dissolving in water and analyzing the resultant solution. A glass, formed by fusion with borax, may partially recrystallize during the cooling process, and it is necessary that any microscale heterogeneities be avoided. This homogeneity may be checked by taking an XRD pattern of the powdered bead where no diffraction lines should be observed. Many of the problems and failures encountered in the preparation of fused beads are due to the lack of appreciation of the chemical processes which occur during the fusion process. Essentially the purpose of the fusion is twofold, first to completely react the compounds present in the sample with the fusion mixture to give total solution and second to cool the fused melt to form a solid glass (i.e., completely noncrystalline). Incompleteness of either of these two stages will result in an inhomogeneous sample which, in addition to being bad in itself, can also lead to cracking of the glass bead during the cooling stage. For example, fusion with sodium or lithium borates will yield glassy borates of the elements in the sample but the number of stages in the reaction may be considerable. For instance, a divalent metal oxide MO may produce any number of reaction products:

$$Na_2B_4O_7 \rightarrow 2NaBO_2(\text{metaborate}) + B_2O_3$$

$$NaBO_2 + MO \rightarrow NaMBO_3(\text{orthoborate})$$

$$B_2O_3 + MO \rightarrow M(BO_2)_2(\text{metal metaborate})$$

$$M(BO_2)_2 + 2NaBO_2 \rightarrow Na_2M(BO_2)_4(\text{complex borate})$$

The final distribution of reaction products depends to a large extent on reaction temperature and initial sample to borax weight ratio. It is, however, important to avoid reaction products which lie outside the glassy region of the system; such problems can often be avoided simply by reference to the appropriate phase diagram—many of which are available in the literature, particularly for the lithium borate system [24].

2.5.4 Controlling the Fusion Reaction

Movement to the appropriate part of the phase diagram can always be achieved quite simply by varying the initial weight ratio of sample to borax to boric acid, thus achieving some control over the reaction products. As an example, in the series of reactions indicated above, the production of metal orthoborate can be made the preferred reaction by ensuring an excess of B_2O_3 in the reaction mixture; in practice this excess would be done by using a mixture of borax and boric acid as the reaction mixture rather than borax alone. It is also important to realize that sodium tetraborate is the weak acid salt of a strong base and as such may be rather unreactive towards very basic materials. Lithium borate being the weak acid salt of a weak base may be far more suitable for this type of sample. Where reaction proceeds rather slowly, the process may be accelerated by use of oxidizing conditions formed by use of oxidizing agents such as potassium nitrate or barium peroxide in the reaction mixture. Again it should be remembered that use of graphite crucibles always tends to give reducing conditions and may be rather unsuitable for certain types of material such as easily reducible oxides. As with liquid solutions, all of the solid techniques suffer from the fundamental disadvantage of high dilution and the increase of scattered background. The dilution required to overcome a matrix effect can sometimes be minimized by use of smaller quantities of fusion mixture along with high absorbing compounds such as lanthanum oxide. Matrix absorption can thus be stabilized for standards and samples without significantly increasing the scattering power of the matrix. A great advantage of the solid-solution method over the liquid-solution method is that cell windows are not required, and vacuum conditions can still be employed.

2.6 LIQUIDS AND SOLUTIONS

2.6.1 General

Where the concentration of an element in a sample is too low to allow analysis by one of the methods already described, workup techniques have to be employed in order to bring the concentration within the detection range of the spectrometer. Concentration methods can be employed where sufficiently large quantities of sample are available and the efficiency of the workup technique depends to a very large extent on the ingenuity of the operator. Techniques which have been employed are necessarily rather diverse in character. For example, gases which are contaminated with solid particles can be treated very simply by drawing the gas through a filter disc followed by direct analysis of the disk. The technique has also been successfully employed for the analysis of trace amounts of zinc and lead in air. In this instance actual amounts of less than a microgram could still be detected on the discs with a normal spectrometer. Low concentrations in solution can be concentrated with ion exchange resins [25] and eluted and evaporated onto filter paper discs. Alternatively, the ion exchange resin itself can be used as a sample support either by pelletizing [26, 27, 28] the exchange resin or by use of ion exchange filter papers [29]. This latter method has proven to be extremely useful for the determination of metals in very dilute solutions because it is possible to select ion exchange resins which are reasonably specific for one type of ion. Concentration is easily brought about by shaking the solution with a known amount of resin for several minutes, followed by filtering, drying, and pelletizing. For example, 0.1 ppm of gold in solution can be determined by this method with a total sample preparation time of less than 15 minutes. Concentration can sometimes be affected simply by evaporating the solution straight onto confined spot filter paper [30, 31, 32], and this technique has been employed with great success in the field of clinical chemistry. An alternative method which has also given excellent results in this field [33] entails direct evaporation of the solution onto planchettes.

2.6.2 Preconcentration Methods

Conventional X-ray spectrometers allow measurement of all elements in the periodic table down to and including fluorine ($Z = 9$) or to boron ($Z = 4$) in very modern systems. The dynamic range of the XRF method is large, covering about six orders of magnitude from the low ppm range to 100%. The interelement (matrix) effects are predictable and relatively easy to correct for, particularly in the case of a reasonably modern computer controlled system. Matrix effects are minimal in the case of so-called thin-film samples, that is, a sample of thickness in the tens of micron range. The detection limits obtainable directly on the unprepared sample are generally barely sufficient for the analysis of trace elements in natural waters. It is, therefore, common practice to use a preconcentration step. The sample resulting from preconcentration is usually in the form of a thin sample and as such, meets the thin film criterion. Consequently, the conversion of XRF raw data to element concentration is a rather simple procedure.

X-ray spectrometers are generally not transportable and as a result almost all analyses are performed in the laboratory. This procedure in turn requires the transportation of the samples for analysis from the site of interest to the X-ray machine. The actual volume of water taken for analysis is generally rather large, not just because natural waters are usually plentiful and sample errors can be reduced easily by taking large aliquots, but, of equal importance, the very low concentration levels of the elements of interest generally require the preconcentration of a large volume of sample to allow a measurable signal to be obtained. To this end it is common practice to preconcentrate, and this step can sometimes be conveniently done at the sampling site.

The sensitivity of the X-ray spectrometer varies by about six orders of magnitude over the measurable element range, when expressed in terms of rate of change in response per rate of change in concentration. The minimum detectable concentration limit is inversely proportional to the sensitivity m (i.e., counts per second per percent) the spectrometer and directly proportional to the square root of the background response (R_b in c/s) at the analytical wavelength. The lower limit of detection (LLD) given in percentages is given by the expression:

$$LLD = \frac{3}{m} \times \sqrt{\frac{R_b}{t_b}} \qquad (2.1)$$

where t_b is the counting time. For a fixed analysis time the detection limit is proportional to $m/\sqrt{R_b}$ and this function is taken as a "figure of merit" for trace analysis. The value of m is determined mainly by the power loading of the source, the efficiency of the spectrometer for the appropriate wavelength, and the fluorescent yield of the excited wavelength. The value of R_b is determined mainly by the scattering characteristics of the sample matrix and the intensity-wavelength distribution of the excitation source.

In general, the background is low when the sensitivity is low, and these factors result in an overall variation in the detection limit of the technique of about two orders of magnitude. The most sensitive elements are generally the transition elements where direct measurements down to one part per million are possible in analysis times of about 100 seconds. The least sensitive elements are the lower atomic numbers boron ($Z = 4$) through silicon ($Z = 14$) where detection limits are around 0.01%, which is generally insufficient for the analysis of natural waters, so preconcentration techniques have to be employed.

2.6.3 Direct Methods

The detection limits directly achievable by the XRF generally lie in the low ppm range, or in absolute terms, a nominal detection limit of 0.1 to 0.01 μg on a thin-film sample. In favorable cases, such as the transition elements, this limit may go to about 0.1 ppm, and in the most unfavorable case typified by the lower atomic number elements, to about 50 ppm. One of the special problems encountered in the direct analysis of water samples stems from the need to support the specimen under examination. Most conventional spectrometers irradiate the sample from below, and the support film both attenuates the signal from the longer wavelength characteristic lines as well as introducing a significant "blank." Absorption by air also becomes an important factor for the measurement of wavelengths longer than about 2 Å, and the need to work in a helium atmosphere introduces further attenuation of the longer wavelength signals. Thus, in almost all cases, some preconcentration technique is applied to the water sample before analysis (e.g., [34, 35]).

2.6.4 Concentration by Evaporation

While the direct analysis of a solution offers an ideal sample for analysis, all too often the concentrations of the analyte elements are too low to give an adequate signal above background. In these cases preconcentration techniques must be used to bring the analyte concentration within the sensitivity range of the spectrometer. In principle, one could easily accomplish this step simply by evaporation. In order to achieve detection limits at the low ppb level, this would mean the evaporation of about 100 ml of water. Although it is not theoretically necessary to take the sample completely to dryness, it is much more convenient to do so because, by this means, one can obtain a handleable specimen. Unfortunately, taking the sample to dryness does cause experimental problems due to fractional crystallization, splashing of the sample as it reaches dryness, and so on. For these practical reasons, preconcentration by evaporation has not found a great deal of application. On the other hand, the application of evaporation preconcentration techniques does show some promise when combined with special techniques for reducing the relatively high inherent background observed in classical S-ray fluorescence methods. The method of TRXRF (total reflection X-ray fluorescence) utilizes the total reflection of X-rays from a highly polished surface. Aiginger and his coworkers [36] have achieved sensitivities down to the ppb level by evaporation of small (about 5 μl) samples of water onto a very flat quartz-glass plates. A thin layer of insulin was used to give a good distribution of the evaporated sample across the surface of the optical flat, and an energy dispersive spectrometer was used to measure and analyze the characteristic treated polyurethane foam [37], plus a variety of other techniques well known to the analytical chemist.

2.6.5 Use of Ion-Exchange Resins

By far the most studied and apparently the most useful method of preconcentration is the use of ion-exchange resins. The major advantage of most ion-exchange methods is that the functional group is immobilized onto a solid substrate providing the potential to batch extract ions from solution. The sample itself can be either the actual exchanged resin or a separate sample containing the appropriate ions re-eluted from the resin. The success of the method depends to a large extent on the recovery efficiency of the resin which in turn is determined by the affinity of the ion-exchange material for the ions in question and the stability of complexes present in solution. Preconcentration factors of up to 4×10^4 can be achieved from suitable ion-exchange material using around 100 mg of resin [38].

2.7 HANDLING VERY SMALL SAMPLES

The irradiation area in a typical X-ray spectrometer is of the order of 5 cm^2, and the penetration depth of an average wavelength about 20 μm. These dimensions mean that even though 20 g of sample may be placed in the spectrometer, the analyzed volume is still only of the order of 50 mg. The smallest sample that will give a measurable signal above background is at least three orders of magnitude less than this mass, so the lowest analyzable sample is of the order of 0.05 mg, provided that the sample is spread over the full irradiation area of the spectrometer sample cup. Where this configuration is impracticable, the smallest analyzable value is increased by roughly the total irradiation area of the cup (about 5 cm^2) over the actual area of the sample that the primary X-ray beam sees.

2.7.1 Size of the Specimen in X-ray Fluorescence Analysis

Most laboratory type spectrometers place constraints on the size and shape of the analyzed specimen. In general, the aperture into which the specimen must fit is in the form of a cylinder, typically 25–48 mm in diameter and 10–30 mm in height. Although the specimen that is placed in the spectrometer may be rather large, because

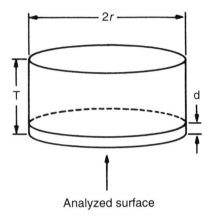

Analyzed surface

Figure 2.4. Volume of specimen actually analyzed.

of the limited penetration depth of characteristic X-ray photons, the actual mass of specimen analyzed is quite small. This condition is shown in Figure 2.4. In the illustration, the specimen is represented as a disk of thickness T, density ρ, and diameter $2r$. As was discussed in section 1.7.3, an estimation can be made of the approximate path-length x travelled by characteristic X-ray photons, by assuming a certain fraction of radiation is absorbed [see equation (1.10)]. It was also shown that the penetration depth D is related to the path length and the spectrometer takeoff angle θ_2, as given in equation (1.12). The mass of the sample analyzed, m_a, is given by

$$m_a = \pi \times r^2 \times d\rho \tag{2.2}$$

Combination of equations (1.10) and (2.2) gives a value for $d\rho$ equal to

$$d\rho = \frac{4.6 \times \sin \theta_2}{\mu} \tag{2.3}$$

Substituting for d in equation 2.2 gives

$$m_a = \frac{4.6 \times \sin \theta_2}{\mu} \times \pi \times r^2 \tag{2.4}$$

Using a value of 32° for θ_2 and 1.5 cm for r, the approximate value for m_a is given as

$$m_a = \frac{17.5}{\mu} \quad \text{(grams)} \tag{2.5}$$

A second important area of materials analysis involving small samples involves the investigation of thin films. Although the technique is limited to rather small areas, typically a few square millimeters, it does provide useful information about bulk composition of surface films. In the analysis of even large masses of material, it can be easily shown that the actual quantity analyzed is very small. Because penetration depths are of the order of a few tens of microns, and because the irradiation area is typically a few square centimeters, the actual weight of sample analyzed is only of the order of a few milligrams. Thus, when analyzing small amounts of material, it is far better to spread the sample out over the irradiated area than to make a special limited area sample holder. In fact, excellent success can be achieved even on mg samples. A good example of this is found in the analysis of air pollutants collected on filter paper.

2.7.2   Direct Analysis of Limited Quantities of Sample

Where only very limited quantities of material are available for analysis, considerable ingenuity may be required on the part of the analyst in reproducibly presenting a suitable specimen to the spectrometer in a manner which will provide the maximum X-ray intensity from the elements to be analyzed. A major problem in the quantitative analysis of small quantities is that there is great difficulty in weighing accurately a sample whose total weight is only of the order of a few milligrams. Where a sample weight can be obtained with a sufficient accuracy, fusion techniques can be applied with some success particularly where the elements to be estimated are of relatively high atomic number. In the case of medium atomic number elements, sufficient sensitivity can still be obtained even after very high dilutions, perhaps up to 1:10,000 [39]. Probably a far easier way of handling weighable quantities of sample is to dissolve in a suitable solvent and evaporate onto filter paper disks. Care should be taken that always the same surface area be wetted; this requirement can be achieved by applying a wax ring of fixed diameter on the paper and by using a fixed volume of solution [40]. This method is certainly potentially more accurate because internal standards can be added very precisely to the disks by spotting known amounts of dilute standard solutions with the aid of micropipettes. There is in fact a very great incentive in spreading the sample thinly because the relative efficiency of production of useful X-rays falls off very rapidly with the increase in depth of the sample. This effect can be demonstrated with reference to Figure 2.5 taken from [41]. If a small crystal of lead sulfate weighing about 10 milligrams were placed directly in a sample cup and irradiated it would be found that, whereas only a very small fraction of the total sample would effectively produce S $K\alpha$ radiation, practically the whole sample would contribute to the production of Pb $L\alpha$ radiation. This difference is simply due to the path-lengths of S $K\alpha$ and Pb $L\alpha$ radiation in $PbSO_4$. Pb $L\alpha$ with a mass attenuation coefficient of about 100 cm^2/g has a path-length of approximately 70 μm. However, as the absorption of the primary exciting radiation cannot be neglected in this case, and because the take-off angle is approximately 35° in most spectrometers, only some 20 μm of the total thickness contributes to the production of Pb $L\alpha$ radiation which leaves the sample. However, S $K\alpha$, with a mass attenuation coefficient of over 1000 cm^2/g has a path-length of only 6 μm; hence, only the radiation produced in the first few microns of the total depth of the sample can escape. By spreading the sample very thinly, an average depth of about 2 μm can be obtained for a sample

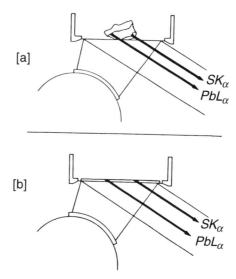

Figure 2.5. Because the path-length of S $K\alpha$ is relatively short, only a fraction of the lump of sample can contribute to the useful radiation (*a*). However, by spreading the same amount of sample very thinly (*b*) the whole of the sample can contribute of the measured intensity.

area of 6 cm^2; all of the sample would now effectively contribute to both S $K\alpha$ and Pb $L\alpha$.

A further incentive for spreading a limited sample into a thin film and rotating it over the primary beam is that the actual positioning of a single particle of powder in a sample cup is critical both from the point of view of intensity and goniometer setting. The first of these points arises because the distribution of X-radiation over the irradiated area of the cup is far from uniform. Figure 2.1 illustrates the reason for the variation in X-ray intensity. Owing to the fact that the distance (*a–c*) is significantly shorter than (*b–d*) the intensity distribution falls off going from (*c–d*). It will also be seen that the majority of the radiation arriving at *g* starts from *e* and that from *f* finishes up at *h*. This variation can give rise to peak shift effects when a specimen is used which is significantly smaller than the area defined by the mask of the X-ray tube window between the extreme edges of the sample cup. The effect can be further complicated by the fact that different areas of the analyzing crystal diffract radiation from different portions of the sample. Both of these effects can be completely removed by rotating the sample during analysis.

The relatively large solid angle of acceptance of the detector helps to minimize this effect, but angular variations of up to 0.5° 2θ are still possible. The spreading of the sample into a thin film represents an extremely attractive means of analysis for a number of other reasons, not the least of which is that matrix effects are often negligible. The limit as far as sample preparation is concerned is that the relative intensity of an analytical line is related both to the total weight and the self-absorption of the sample. One possible means of correcting for this limitation is to measure the attenuation by the sample of an intensity from a sample giving a similar wavelength to that being studied. This method is particularly useful where trace quantities of an element are being measured in a major constituent whose concentration can be assumed to be constant.

2.8   RADIOACTIVE SAMPLES

The handling and pretreatment of radioactive samples presents two major problems additional to those already discussed. First, the containment of a sample is all important because the potential health hazard can be considerable and it is vital that all of the precautions normally associated with the handling of radioactive materials be strictly adhered to. Secondly, all radiation which can enter the detector, other than that arising as deliberately excited X-radiation from the sample, must be kept at an absolute minimum. Although the majority of published work on X-ray analysis of radioactive materials deals with diffraction rather than spectrometry, the problems of containment and shielding are similar. However, because it is invariably necessary that procedures based on X-ray spectrometry be rapid, in order to compete with alternative methods of analysis, the extra effort involved in preparing samples in suitable containers, combined with the relatively high attenuation of the longer wavelengths by the necessarily thicker than normal cell windows, has tended to limit the number of published applications. The safe containment of radioactive materials invariably requires the use of a double container consisting perhaps of a primary PVC covering on the sample plus a second container which can comprise the sample cell [42]. Heat-sealed PVC sachets can also be used for the primary containment of samples consisting of borax beads or pellets. γ-active materials usually present less of a handling problem because the ingestion hazard is not so important. In this instance single containment will often suffice. The magnitude of the additional shielding required to prevent the stray radiation from entering the detector depends to a large extent upon the type and the magnitude of the activity of the sample [42]. Stray radiation entering the detector from directions other than that defined by the geometry of the spectrometer can be almost completely removed by means of suitably placed lead shielding. The real problem is to minimize scattered radiation which enters the detector along the same optical path as the X-radiation—the major source of this interference being due to radiation from the sample passing through the primary collimator and being scattered by the analyzing crystal. Should this scatter consist of γ-radiation, its effect can be almost completely removed by careful application of pulse height selection, because the energy of γ-radiation is usually an order of magnitude greater than the X-radiation being measured.

2.9   SPECIAL PROBLEMS WITH LOW ATOMIC NUMBER ELEMENTS

In XRF analysis low ($Z < 22$) and ultralow atomic number ($Z < 11$) elements are especially difficult to deal with, and the degree of difficulty increases as the atomic number decreases. Many factors contribute and each may have a different impact on measured intensities. On the part of the instrumentation the most important are:

1. *X-ray tube* (which does not excite long wavelengths efficiently) [43].

   - Choice of the tube (end window tube excites light elements more efficiently than a side window tube) [44, 45].
   - Choice of the excitation parameters (low kV and high mA are beneficial) [44, 45].
   - Tube target (for example, the Sc target excites light elements better than the Rh tube using a side window tube) [44, 45, 46].
   - Thickness of the X-ray tube window (the thinner the better).

2. *Crystal.* Generally, multilayers provide more intensity than organic or inorganic crystals[47].

3. *Radiation path.* Transmission of the analyte radiation decreases with vacuum quality. For example, transmission of Mg $K\alpha$ ($Z = 12$) is almost 100% when the pressure is below 0.1 mm Hg, and approximately 85% when it is 2 mm Hg. In addition, helium transmits only 80% at 10 Å (Mg $K\alpha$) [43].

4. *Detector.* Flow-proportional detector quantum counting efficiency decreases at the long wavelengths. The transmittance at 10 Å of 6 μm Mylar and 25 μm Be is only 26.3% and 19.9%, respectively [48].

As far as the specimen itself is concerned, two main specimen preparation methods need to be considered: dilution (fusion, dry, solution) and direct analysis. Dilution inevitably lowers measured intensities for light elements as the specimen-to-flux, specimen-to-binder, or specimen-to-solvent ratio increases. Consequently, sensitivity (expressed in Kcts/percentages) suffers. On the other hand, dilution by the way of fusion (or solution) overcomes the inhomogeneity effects (particle size, shape, density, mineralogical), and results in a homogeneous material. Despite the loss of intensity due to fusion, better estimated accuracy is often obtained by virtue of improved specimen homogeneity. For example, determining Na_2O in aluminas 1σ of 100 ppm was obtained when fusion was employed, as opposed to 210 ppm for briquetting [49]. To minimize the intensity loss in fusion, a low specimen-to-flux ratio is proposed [50, 51, 52]. However, it usually prolongs the fusion time, requires higher fusion temperatures, and may result in an incomplete fusion. On the other hand, addition of a heavy absorber [53] (to stabilize the matrix and minimize the interelement effects) lowers sensitivities of light elements even more than dilution alone. Rather than using a heavy absorber, it is preferable to correct the interelement effects mathematically, employing a suitable equation (model) for a homogeneous specimen obtained in fusion. Also, Li-borates rather than Na-borates constitute the best choice of flux for the low atomic number element analysis. To compensate the intensity loss due to fusion a large disk diameter was proposed [49] (a large spectrometer mask must be selected for this purpose). The diameter increase from 30 to 40 mm resulted in a 20–30% intensity gain for the elements analyzed in alumina. For many powders having low cohesion, which need to be prepared as briquettes, the binder addition affects sensitivities caused by dilution. To minimize the sensitivity loss the binder addition should be as small as possible, usually 5–20% of the total specimen weight [43]. Obviously, the binder's mass attenuation coefficient should be low. Consequently, an organic binder, not a metallic binder, is the best choice for light elements.

A serious limitation in the low atomic number element analysis occurs for a liquid specimen. In many applications a SPEX X-Cell [54] is used, which consists of a cup and snap-on ring covered by a thin film which serves as the specimen window. The film (Polycarbonate, Mylar, Polypropylene, Kapton) effectively contains the liquid while allowing uniform transmission of the X-rays to and from the sample. This film practically stops characteristic radiation of light elements. For example, the radiation transmission of the Mg K is only 20% when 3511 Kapton is used. For other films the transmission is worse, or much worse.

As the atomic number of the analyte decreases, its wavelength increases. Consequently, the major factors affecting radiation of light elements are as follows.

1. The infinite thickness decreases. As a result of the reduced infinite thickness, only the top layer of the specimen contributes to the analysis. Therefore, it must be representative of the bulk sample. In general, when dealing with a powder material, it is considered homogeneous if

$$[L(\mu_{av} - \mu)] \ll 1 \tag{2.6}$$

where

$$L = \text{average particle size,}$$

$$\mu_{av} = \text{average mass attenuation coefficient of the specimen,}$$

$$\mu = \text{mass attenuation coefficient of the analyte [55].}$$

In the specimen preparation for a low atomic number element analysis, it must be assured that the infinite thickness is larger than the average particle size. Then, the intensity of the secondary radiation will not be affected significantly by mineralogical and particle size effects. In order to improve the surface representation any coarse-grained or rough-finished specimen should be rotated during analysis. In alloy surface preparation, special care must be taken so that the integrity of the analytical surface is preserved. For example, if certain low atomic number constituents are missing in the surface (as compared with the bulk specimen structure), their measured intensity will be uncontrollably lower than expected. To minimize intensity loss of low atomic number elements, the surface must also be flat and free of surface roughness (see carbon determination in cast iron and steel). Shortly before the analysis, clean the analytical surface with a paper tissue (free of any elements to be measured) saturated with alcohol. Handle the specimen by the edges to avoid finger prints which alter intensities of especially low atomic number elements (e.g., sodium).

For a low atomic number specimen if the critical thickness is exceeded, then all samples and standards must have the same thickness. In case of plastics,

TABLE 2.2. Spectral Interferences: Low and Very Low Atomic Number Elements

		Spectral Interferences			
Line	W'length (Å)	Interference	W'length(Å)	Solution	Note
B $K\alpha$	67.0	O $K\alpha$ (3)	70.9	spectral correction	1
C $K\alpha$	44.0	Fe $L\alpha$ (2)	40.8	spectral correction	1
O $K\alpha$	23.7	Na $K\alpha$ (2)	23.8	spectral correction	1
F $K\alpha$	18.3	Rh $L\alpha$ (4)	18.4	spectral correction	2
		Fe $L\alpha$, $L\beta$	17.6,17.3	spectral correction	1
Na $K\alpha$	11.91	Zn $L\beta$	11.98	spectral correction	
Mg $K\alpha$	9.89	Ca $K\alpha$ (3)	10.08	spectral correction	1
				or TLAP + fine collim	
		Sn $L\beta$ (3)	9.92	spectral correction	1
Al $K\alpha$	8.34	Sc $K\beta$ (3)	8.34	spectral correction	3
		Br $L\alpha$	8.38	spectral correction	4
P $K\alpha$	6.155	Cu $K\alpha$ (4)	6.167	spectral correction	1
K $K\alpha$	3.774	Cd $L\beta$	3.738	spectral correction	1

(1) If the interference is caused by a major or minor element; (2) If the rhodium tube is used; (3) If the scandium tube is used; (4) if LiBr is used as a nonwetting agent.

which are easily penetrated by radiation, care must be taken that the sample cup does not contribute to the measured radiation.

2. The fluorescent yield decreases. For example, the fluorescent yield is:

15% for Ca $K\alpha$ ($Z = 20$)
0.9% for Na $K\alpha$ ($Z = 11$)
0.338% for F $K\alpha$ ($Z = 9$)
0.185% for O $K\alpha$ ($Z = 8$)
0.09% for N $K\alpha$ ($Z = 7$)
0.038% for C $K\alpha$ ($Z = 6$)
0.0125% for B $K\alpha$ ($Z = 5$)

3. The matrix absorption increases.
4. The chemical effect becomes more important. (Dependence of radiation wavelength with analyte chemical state increases.)
5. Elastic and nonelastic scatter increases.
6. Certain line overlaps are serious and must be corrected (see Table 2.2).

REFERENCES

[1] Tokuda, T., et al. (1985), "Improvement of sample preparation method for X-ray fluorescence spectrometry in iron and stell analysis," *R & D, Res. Dev. (Kobe Steel Ltd.)* **35**, 8–11.

[2] Somervaille, L.J., Chettle, D.R., and Scott, M.C. (1985), "In vivo measurements of lead in bone using X-ray fluorescence," *Phys. Med. Biol.*, **30**, 929–943.

[3] Sanner, G. and Usbeck, H. (1985), "Application of X-ray fluorescence analysis in the determining of chlorine (sulfur) and fluorine," *Pharmazie*, **40**, 544–548.

[4] Thompson, G. and Bankston, D. C. (1970), "Sample contamination from grinding and sieving determined by emission spectrometry," *Appl. Spectrosc.*, **24**, 210–220.

[5] Tuff, M. A. (1985), "Contamination of silicate rock samples due to crushing and grinding," *Adv. X-ray Anal.*, **29**, 565–571.

[6] Claisse, F. (1957), "Accurate X-ray fluorescence analysis without internal standards," *Norelco Reporter*, **4**, 3–7.

[7] by Jenkins, R., Gould, R. W., and Gedcke, D. A. (1981), *Quantitative Spectrometry*, Marcel Dekker: New York, Chapter 7, "Specimen preparation."

[8] Tertian, R. and Claisse, F. (1982), *Principles of Quantitative X-ray Fluorescence Analysis*, Wiley–Heyden: London.

[9] Claisse Fusion Stirrer Device, SPEX Industries, Metuchen, NJ.

[10] Willay, G. (1986), "*Perl–X* and its derivatives in mineral analysis," *Cah. Inf. Tech./Rev. Metall.*, **83**, 159–173.

[11] Rose, H., Adler, I., and Flanagan, F. J. (1960), U.S. Geol. Survey Prof. Pap. 450–B, 80 pp.

[12] Norrish, K. and Hutton, J. (1969), "X-ray spectrographic method for the analysis of a wide range of geological samples," *Geochim. Cosmochim. Acta*, **33**, 431–453.

[13] Bower, N. W. and Valentine, G. (1986), "Critical comparison of sample preparation methods for major and trace element determination using XRF," *X-ray Spectrom.*, **15**, 73–78.

[14] Claisse, F. and Samson, C. (1961), "Heterogeneity effects in X-ray analysis," *Adv. X-ray Anal.*, **5**, 335–354.

[15] Blanquet, P., *L'analyse par Spectrographie et Diffraction de rayons X*, (Madrid, 1962), Philips, Eindhoven.

[16] *A.R.L. Spectrographer's News Letter*, 1954, **7**, 1–3.

[17] Norrish, K. and Hutton, J. T., 1964, C.I.S.R.O. Division of Soils, Adelaide, Divisional Report 3/64.

[18] Carr–Brion, K. G. (1964), "A rapid method for casting fused beads for use in XRFA," *Analyst*, **89**, 556–557.

[19] Anderman, G. and Allen, J. D. (1960), "The evaluation and improvement of X-ray emission analysis of raw mix and finished cements," *Adv. X-ray Anal.*, **4**, 414–432.

[20] Campbell, W. J. and Thatcher, J. W. (1958), "Determination of calcium in wolframite concentrates by fluorescent X-ray spectrography," *Adv. X-ray Anal.*, **2**, 313–332.

[21] Cullen, T. J. (1960), "Potassium pyrosulfate fusion; determination of copper mattes and slags," *Anal. Chem.*, **32**, 516–517.

[22] Cullen, T. J. (1962), "Addition of NaF to potassium pyrosulfate fusions for X-ray spectrometric analysis of siliceous samples," *Anal. Chem.*, **34**, 862.

[23] Rinaldi, F. (1966), *Scientific and Analytical Bulletin*, No. 5, Philips: Eindhoven.

[24] Dubrovo, S. K. (1964), *Vitreous Lithium Silicates*, Nauka: Leningrad (English translation available).

[25] Grubb, W. T. and Zemany, R. D. (1955), "X-ray emission spectrography with ion exchange membrane," *Nature*, **176**, 221.

[26] Minns, R. E. (1964), *Proceedings of Exeter Conference on Limitations of Detection in Spectrochemical Chemical Analysis*, Hilger and Watts: London.

[27] Van Niekerk, J. N. and De Wet, J. F., (1960) "Trace analysis by X-ray fluorescence using ion exchange resins," *Nature*, **186**, 380–381.

[28] Van Niekerk, J. N., De Wet, J. F., and Wybenga, F.T. (1961), "Trace analysis by combination of ion exchange and X-ray fluorescence—determination of uranium in sulfate effluents," *Anal. Chem.*, **33**, 213–215.

[29] Campbell, W. H., Spano, E. F., and Green, T. E. (1966), "Micro and trace analysis by a combination of ion–exchange resin–loaded papers and X-ray spectrography," *Anal. Chem.*, **38**, 987–996.

[30] Natelson, S. and Sheid, B. (1960), "X-ray spectroscopy in the clinical laboratory. II. Chlorine and sulfur—automatic analysis of ultramicro samples," *Clin. Chem.*, **6**, 299–313.

[31] Natelson, S. and Sheid, B. (1960), "X-ray spectroscopy in the clinical laboratory. III. Sulfur distribution in electrophoretic protein fractions of human serum," *Clin. Chem.*, **6**, 314–326.

[32] Natelson, S. and Sheid, B. (1961), "X-ray spectroscopy in the clinical laboratory. IV. Phosphorus—total blood iron as a measure of hemoglobin content," *Clin. Chem.*, **7**, 115–129.

[33] Lund, P. K. and Mathies, J. C. (1960), "X-ray spectrometry in biology and medicine, parts I, II and III," *Norelco Reporter*, **7**, 127–139.

[34] Leyden, D., et al. "Comparison of Trace Element Enrichment for XRF Determination," *Pergamon Ser. Environ. Sci. 1980*, **3**, (Anal. Tech. Environ. Chem.), 469–76.

[35] Smits, J., Nelissen, J., and Van Grieken, R. (1979), "Comparison of Preconcentration Procedures for Trace Metals in Natural Waters," *Anal. Chim. Acta*, **111**(1), 215–26.

[36] Wobrauschek, P. and Aiginger, H. (1975), "Total Reflection X-ray Fluorescence Spectrometric Determination of Elements in Nanogram Amounts," *Anal. Chem.*, **47**, 852–855.

[37] Torok, S., Braun, T., Van Dyke, P., and Van Grieken, R. (1986), "Heterogeneity effects in direct XRF analysis of traces of heavy element pre–concentrated on polyurethane foam sorbents," *X-ray Spectrom.*, **15**, 7–11.

[38] Leyden, D.E., Patterson, Th.A., and Alberts, J.J. (1975), "Preconcentration and XRF determination of Cu, Ni and Zn in sea water," *Anal. Chem.*, **47**, 733.

[39] Jenkins, R. (1962), "Applications of X-ray fluorescence analysis in the oil industry," *J. Inst. Pet.*, **49**, 246–256.

[40] Natelson, S. and Bender, S.L., (1959), "X-ray fluoresence (spectroscopy) as a tool for the analysis of submicrogram quantities of elements in biological systems," *Microchemical Journal*, **3**, 19–34.

[41] Jenkins, R. and de Vries, J. L. (1975), *Practical X-ray Spectrometry*, 2nd ed., Springer: New York, p 158.

[42] Bremner, W. B. (1964), "Analysing radioactive materials by X-ray fluorescence," *Proceedings of 4th M.E.L. Conference*, (Sheffield), 65–67, Philips: Eindhoven.

[43] Bertin, E. P. (1979), *Principles and Practice of X-ray Spectrometric Analysis*, Plenum Press: New York.

[44] Gurvich, Y. M. (1984), "Comparison of various X-ray tube types for XRF analysis," *American Laboratory*, **16**, 103–108.

[45] Feret, F. (1986), "Effect of Sc and Rh X-ray tube parameters on the analysis of light elements," *Canadian J. Spectrosc.*, **31**, 15–21.

[46] Kikkert, J. (1984), Research and Development, February, pp 132–135.

[47] Nicolosi, J. A., Groven, J. P, Merlo, D., and Jenkins, R. (1986), "Layered synthetic microstructures for long wavelength spectrometry," *Optical Engineering*, **25**, 964–969.

[48] Jecht, U. and Petersohn, U. (1967), "Counters and analyser crystals for the X-ray fluorescence analysis of low atomic number elements," *Siemens-Z.*, **41**, 59–63.

[49] Feret, F. (1990), "Alumina characterization by XRF," *Adv. X-ray Anal.*, **33**, 685–690.

[50] Andermann, G. and Allen, J. D. (1961), "X-ray emission analysis of finished cements," *Anal. Chem.*, **33**, 1695–1699.

[51] Feret, F. (1982), "Minimum flux fusion processing of iron ores for XRF analysis," *X-ray Spectrom.*, **11**, 128–134.

[52] Longobucco, R. J. (1962), "X-ray spectrometric determination of major and minor constituents of ceramic materials," *Anal. Chem.*, **34**, 1263–1267.

[53] Rose, H. J. Jr., Adler, I., and Flanagan, F. J. (1963), "X-ray fluorescence analysis of light elements in rocks and minerals," *Appl. Spectrosc.*, **17**, 81–85.

[54] Obenauf, R. H. et al. (1991), *SPEX Handbook of Sample Preparation and Handling*, SPEX Industries, Inc., Metuchen, NJ.

[55] Brindley, G. W. (1945), "The effect of grain or particle size on X-ray reflections from mixed powders and alloys, considered in relation to the determination of crystalline substances by X-ray methods," *Phil. Mag.*, **36**, 347–369.

CHAPTER 3

SPECIMEN PREPARATION IN X-RAY FLUORESCENCE

CONTRIBUTING AUTHORS: BUHRKE, CREASY, CROKE, FERET, JENKINS, KANARE, KOCMAN

3.1 WHOLE ROCK ANALYSIS

3.1.1 General

Whole rock analysis presents a variety of challenges to the analyst. Heterogeneous, naturally occurring geological materials must be obtained, processed, and presented to the spectrometer in such a way that a few milligrams within the critical thickness of the analytical surface of the specimen adequately represents as much as many tons of real-world material. Samples from the field might comprise granular sands, coarsely crushed rocks, or whole rock cores. Samples might be chunks, dry free-flowing powders, or wet sludgy muds. In the field of whole rock analysis, the analyst is indeed challenged to prepare the most useful specimen from every sample. Standard operating procedures for sample size reduction, blending, and subsampling frequently require additional judgment depending on the nature of the specific sample at hand.

For several decades, XRF analysts would type match unknown and calibration specimens to account for particle and mineralogical effects. Fusion specimen preparations eliminate such effects while grinding and pressing powder specimens can only minimize such effects. The analyst must be aware of this critical difference. While a broad range of rocks and minerals can be analyzed with acceptable accuracy using a single, wide-range calibration curve based on fused specimens, accuracy will not be as high for the same specimens finely ground and pressed into powder briquettes. (See, for example, [1, 2, 3, 4].) Powder pressing, however, preserves volatile elements and phases and maintains the highest possible concentration of analytes in the specimens.

Handling apparent contaminants can be problematic. For example, imagine a centimeter-sized piece of limestone among several hundred pieces of similar-sized crushed silicate rock we wish to analyze. Such apparent contamination is not infrequent: common carriers such as rail cars, tank trucks, and barges haul many types of industrial materials. Does this single piece of limestone in this small sample represent

A Practical Guide for the Preparation of Specimens for X-Ray Fluorescence and X-Ray Diffraction Analysis, Edited by Victor E. Buhrke, Ron Jenkins, and Deane K. Smith. ISBN 0-471-19458-1 © 1998 John Wiley & Sons, Inc.

the concentration of limestone in the entire lot of material? If you follow your laboratory's standard sample processing methods, would the limestone be crushed and distributed among the other sample particles so that any small aliquot would contain the correct proportion of the limestone? Was the limestone added inadvertently and should it be removed? The laboratory must have a policy on how to handle such apparent contaminants. If the odd particle likely represents a proportionate amount from the original lot, then it should be included and the sample must be processed so as to distribute the contaminant throughout the analytical sample. (If you begin by simply splitting the sample as received, there is a 50% probability that you will discard the contaminant in the first step of processing!) If the odd particle is known or likely to be a one-time contaminant (e.g., a coal seam in a long rock core), it is acceptable to remove and retain the particle while recording its mass and circumstances under which it was found. In such a situation, a footnote in the analytical report describing the contaminant and its removal from the bulk sample is appropriate. Apparent contaminants should only be removed after consultation with the submitter or client for whom the analyses will be performed.

Theoretical and practical aspects of particle size effects on XRF intensities were thoroughly evaluated by Claisse and Samson [5] and more recently for the special case of cubic particles by Bonetto and Riveros [6]. The need for specimen particle homogeneity on a micrometer scale becomes clear when we realize that the critical thickness for Na $K\alpha$ in a silicate rock matrix is about 2 μm. To get the mean particle size to one-fifth the critical thickness (an oft-quoted rule of thumb) means grinding to a mean particle size of about 0.5 μm. Grinding this fine is a practical impossibility for routine laboratory preparation of XRF specimens. The practical limits appear to be 95% finer than 10μm with a mass median diameter of about 3 μm.

3.1.2   Specimen Preparation Procedures

Obtaining Samples and Reducing Field Samples to Testing Size   This subject, as applied to XRD, has been discussed in section 1.5 of this book. Other excellent sources of information are the discussion by Allen (Chapter 1 in [7] and two ASTM standards [8, 9]). Theoretical aspects of representative sampling have been discussed by Ingamells [10, 11, 12, 13, 14] and by Gy [15, 16]. In practice, it is usually sufficient to reduce a portion of a field sample to pass a 150 μm (No. 100) U.S. Standard Sieve at which point it is ready for subsampling for specimen preparation. The following method has been found to work well.

Whole rock cores or large chunks are hammered or crushed with the aid of a pneumatic or hydraulic crusher to produce pieces several centimeters in size. These pieces are further crushed using a jaw crusher or dual-roller crusher. If the crushing elements are carefully aligned and set so the minimum opening is approximately one millimeter, the crushed material should mostly pass a 2.4 mm (No. 8) U.S. Standard Sieve. (The crusher opening will determine the second-largest dimension of the crushed particles.) This crushed rock can be split using a straight chute (Jones) riffler, spinning riffler, or coning and quartering techniques described in [7, 9]. The amount of material of a given particle size deemed representative is given in [8].

Do not discard material that does not pass through—you may selectively remove the same constituent!

A suitable portion of crushed material should be spread into pans for drying; the material of construction of the pans should be chosen to avoid the presence of analytes. Smooth, thin-wall aluminum or polypropylene pans have been found to work well for many rock types. The sample should be spread in as thin a layer as possible and dried at an appropriate temperature to remove free moisture before attempting finer grinding. For example, calcareous or siliceous rocks can be dried in shallow aluminum pans at 105°C in a forced air oven for two hours. Rocks containing significant clay portions might require drying overnight to get to constant weight. Samples with hydrate water must be dried at a temperature that will produce a constant weight product of known stoichiometry. For example, crushed gypsum samples should be dried at 40–50°C for as brief a time as possible, to avoid dehydration to hemihydrate (plaster).

After drying, the crushed rock is passed through a disk pulverizer to reduce all particles to pass a 150 μm sieve. If special care is taken to align and maintain the disk pulverizer, a single pass of material can result in nearly all product passing the 150 μm sieve. Disk plates should be shimmed and adjusted so the rotating plate runs in a plane with deviation less than 50 μm (0.002 in.) and the stationary plate is parallel to the rotating plate with a similar tolerance. Feed the crushed material slowly and steadily to avoid clogging the plates. Be careful that no metal fragments are present in the crushed sample as these fragments will ruin the plates instantly. The disk pulverizer can be cleaned between analyses by passing about 50 g oven-dry coarse silica sand through the pulverizer. A vacuum cleaner with a stiff brush attachment works well to remove most of the dust between samples or after cleaning. Aluminum oxide disk pulverizer plates can also be wiped with a rag dipped in 10% hydrochloric acid to remove stubborn, acid-soluble material that sometimes clings tenaciously to the plates. Wipe the plates with a damp rag to remove all traces of acid and then dry to remove moisture before further use. Use adequate ventilation and respiratory protection at all times when grinding samples.

The sample is now dry and fine enough to pass a No. 100 sieve. The sample should be placed in a labelled, clean, and dry container such as a glass screw-top jar for analysis. In the analytical preparation area, a portion of the sample is weighed and specimens are prepared either as pressed powder briquettes or fused disks.

Pressed Powder Briquet Preparation   Thoroughly remix the sample in its jar by rotating in a figure-eight motion with two hands. Weigh 7.000 ± 0.001 g of sample into a weighing boat by taking several separate gram-size portions using a spatula or scoop, being careful to take representative material from throughout the sample in the jar. Place the weighed sample into the body of a swing mill for fine grinding. Swing mills, also called ring-and-puck mills, are available from several vendors and are made of various materials including hardened steel, tungsten carbide, aluminum oxide, and zirconium oxide. The choice of mill depends on cost and possible contaminants that might be introduced into the specimen during fine grinding from the walls of the mill. Tungsten carbide mills, for example, are made with a cobalt binder that can add 5 mg/kg Co to a 7 g silica sand sample during four minutes grinding [17]. Tuff [18] found even greater contamination (Co: 35 ppm, W: 422 ppm) as did Goto and Ohno [19]; Hickson et al. [20] found contamination from several types of grinding media including ceramic, tungsten carbide, and steel mills.

Add two small drops (approximately 40 μl) propylene glycol on top of the powder sample in the mill as a grinding aid. Cover the mill and grind 4 min at 1000 rpm. If your mill runs at a different speed, adjust the time to obtain about 4,000 rotations of the mill. This length of grinding time will produce a finely pulverized specimen with 90–95% finer than 10 μm and mass median diameter approximately 3 μm. This particle size distribution will be obtained using steel and tungsten carbide mills after grinding for four minutes, for a wide variety of minerals. However, mills made from zirconia or alumina may require longer grinding times due to the lighter mass of their grinding media.

After grinding, remove the mill cover and add 0.5 g binder such as commercially available XRF specimen preparation tablets. Replace the mill cover and grind an additional 30 s to distribute the binder into the specimen. This two-step grinding process has been found to produce highly homogeneous, rugged specimens. If the binder is added at the start of grinding, agglomerations will occur that are visible on the surface of the finished briquet.

Binders such as waxes or methyl cellulose help to hold the specimen particles firmly together after pressing, minimizing degradation and dusting of the analytical surface and thereby prolonging the useful life of the briquet and reducing instrument contamination. Any soft, waxy substance that acts as a binder by mechanically sticking particles together will inhibit very fine grinding and homogenization of the specimen particles. Using a grinding aid followed by adding the binder in a second step assures that you will obtain the most complete fine grinding possible in the mill. However, grinding times longer than four min (in steel or tungsten carbide mills) do not reduce the maximum particle size and do increase the proportion of finest particles, exhausting the grinding aid and leading to agglomeration. Such agglomeration will be evident as caking and sticking of specimen on the walls of the grinding media (ring, puck, and mill body). Excess binder will also produce caking and sticking of material that must be scraped off the mill. The correct amount of grinding aid will produce a free-flowing, pulverized sample that brushes easily out of the mill leaving no residue. The correct amount of binder, generally 5–10% by weight of the sample, will produce an extremely durable specimen surface without caking in the mill.

Brush the finely ground specimen into a labelled, 31 mm aluminum sample cap and press at 345 MPa (50,000 psi). Hold the pressure for one minute and release the pressure over a period of one additional minute. Eject the specimen and examine for cleanliness and homogeneity of the analytical surface. Store covered until ready for analysis.

Some commercially available presses do not require an aluminum sample cup. The amount of binder might need to be increased slightly (10–12% by weight of sample) for these types of specimens.

Pressed powder specimens can be used over many months if the minerals do not react with atmospheric moisture, oxygen, or carbon dioxide. Specimens should be stored in tightly covered containers with their analytical surfaces protected from handling. The surface can be cleaned by gently brushing or blowing with filtered compressed air immediately before analysis. The use of grinding aid and binder dilutes the sample, so calibration standards and unknown analytical samples must be prepared according to the same procedure.

References [21, 22, 23] provide some examples of alternative approaches to powder specimen preparation.

Fused Bead Preparation   Many recipes for fusions are given in this book. For a wide range of geological materials, stable, homogeneous glass beads can be produced using lithium tetraborate as flux with a flux-to-sample ratio of 5:1. Careful temperature control at $1000 \pm 25°C$ is necessary to minimize evaporation of alkalis and sulfate. Fusions can be accomplished in 3–10 minutes depending on the solubility of the sample minerals in the hot flux. Sample and flux are weighed to a constant ratio on their ignited basis. The ignited sample is mixed with flux and fused. Heating the sample in air before combining with flux serves to oxidize reduced elements and burn off carbonaceous substances that will not fuse in the molten lithium tetraborate.

Thoroughly remix the sample in its jar by rotating in a figure-eight motion with two hands. Determine the loss on ignition at $1000°C$ on a several-gram portion of the sample. Take several separate gram-size portions using a spatula or scoop, being careful to take representative material from throughout the sample in the jar. Cool the ignited material in a desiccator. Some siliceous samples will have sintered during heating and will need to be gently reground by hand using a mortar and pestle to pass a No. 100 U.S. Standard Sieve before mixing with flux. If the ignited sample contains particles coarser than 150 μm, these particles will not dissolve completely during the fusion, leaving inclusions in the bead that act as local stress concentrations, causing the bead to crack or shatter. Unoxidized, carbon-rich particles will also lead to beads that crack or shatter upon cooling. Unignited carbonate minerals will rapidly decompose in the fusion mixture, releasing CO_2 and resulting in a miniature volcano that requires cleanup and starting over. The complete ignition and grinding of the ignited sample provides the best opportunity for rapid, successful fusion.

For a 7 g bead (sufficient to fill a 31 mm diameter mold), weigh enough flux to give 5.833 ± 0.001 g, taking into account its loss on ignition. Weigh 1.167 ± 0.001 g ignited sample and mix thoroughly in a small container (not in the platinum crucible—avoid scratching the crucible inner surface) using a noncontaminating stirrer. Check that the mixture is a uniform color and that particles of the sample are completely distributed in the flux. Very fine samples that agglomerate can be dispersed in the flux by gentle mixing with a mortar and pestle followed by smearing the mixture on a piece of clean paper with a broad spatula to check for homogeneity.

Place the flux and sample mixture into a clean 95% platinum–5% gold crucible and add two drops of 50% (w/v) LiBr (aq.) as antiwetting agent. Load the crucible into the fusion apparatus and heat gently over an oxidizing propane or propylene flame for several minutes to remove moisture and drive off air trapped among the particles. Increase the heat to achieve $1000 \pm 25°C$ and hold at this temperature until the mixture is molten. Stir for five minutes followed by casting the specimen into its 95% platinum–5% gold mold. Various brands of fusion apparatus have different rates of mixing. The most vigorous machines can effect complete dissolution of most minerals in lithium tetraborate at 1000°C in a few minutes. Machines that rock gently, or fusions performed in a muffle furnace with occasional mixing by hand, require longer fusion times. The shortest possible fusion is desirable to minimize evaporation of elements such as potassium and sulfur, and to minimize evaporation of the antiwetting agent. The analyst must be aware that some minerals will decompose and lose some elements during heating; the analysis performed on the fused bead might not represent the composition of the original sample if volatile phases are present.

Fused beads can be analyzed without further surface preparation if the surface is flat. Flatness and cleanliness can be restored to old beads by gently lapping on a 30 μm diamond-impregnated, metal-bonded grinding wheel. Water used as a lubricant does not seem to dissolve water-soluble ions during the few seconds grinding time.

3.1.3   Special Problems and Hints

Some samples are difficult to fuse into stable glasses at a flux-sample ratio of 5:1. For example, calcium sulfates often crystallize and require sodium tetraborate flux or addition of glass-forming elements to the lithium tetraborate flux. Alunite, $KAl_3(SO_4)_2(OH)_6$, is notorious for evaporating even during brief fusions at 1000°C. Sulfide ores can require special flux mixtures to retain the sulfur. Baker [24] determined that sulfate can best be retained in fusions by using a flux mixture of one part lithium metaborate plus two parts lithium tetraborate (1:10 sample-flux ratio). For sulfides, 19.6% lithium nitrate plus 80.4% lithium tetraborate (1:10 sample-flux ratio) will quantitatively retain sulfide sulfur except in high grade oil shales which require a higher ratio of tetraborate to nitrate.

Iron-rich samples can be quite viscous and difficult to pour unless additional antiwetting agent is added. High-intensity light, such as a fiber-optic microscope illuminator, is useful to examine dark, iron-rich solidified disks for homogeneity. Sometimes spots of partly dissolved, poorly mixed sample particles will be observed. Longer or more vigorous fusion might be necessary to completely dissolve and homogenize such samples.

Samples ground in a tungsten carbide swing mill, rather than an aluminum oxide pulverizer, sometimes lead to bright blue fused disks. The color is an indication that some cobalt binder from the tungsten carbide grinding media got into the sample. Such a contaminated sample should be discarded and the original sample reprocessed. The mill should be inspected for cracks and chips and repaired or replaced as necessary. Aluminum contamination can come from aluminum oxide pulverizer plates that chip or flake, especially around the edges. The solution to this problem is to carefully chamfer the edges of the plates using a cylindrical diamond grinding bit in a rotary tool driver.

Silica-rich fusion samples can create several problems. Zones of immiscible silicate-borate melts can form within a fused disk, making inhomogeneous specimens. Sometimes small residual silicate particles remain in the centers of these zones which can be observed under a stereo microscope. The solution to this problem is to be sure the sample particles are fine enough and well dispersed in the flux; fusion time might have to be increased to permit complete dissolution and homogenization of the melt. High-purity silica sands can be difficult to bind in briquettes. Even when finely ground, the silica particles do not hold together with less than approximately 10–15% binder.

Powdered carbonate rock samples must be weighed and fused quickly after cooling following the loss on ignition. The free lime produced by heating the samples is extremely hygroscopic; storage in a desiccator, even over magnesium perchlorate, will not keep the samples from picking up a few percent moisture in several hours. If such samples are stored more than a few hours, they should be reignited at 900°C and cooled immediately before weighing for fusion.

3.2 LIMES, DOLOMITIC LIMES, AND FERROLIMES

3.2.1 General Discussion

Lime (CaO) and ferrolime (CaO containing as much as 10% Fe_2O_3) are used in *pig iron* and steel production as additives during the reduction of iron ore into metal. Lime forms a *slag*, in which unwanted contaminants such as silicon, phosphorus, and sulfur are concentrated and removed from the molten charge. Hydrated lime, $Ca(OH)_2$ is one of the oldest building materials. It is used in mortar and plasters (mixture of hydrated lime and sand) and in agriculture for soil conditioning. Free iron must be removed or oxidized. Dolomitic lime (CaO.MgO) results from calcination of dolomite. Pulp and paper industry mills operating in "Kraft pulping cycle" produce a variety of by-products originating from chemical reactions involving lime, sodium carbonate, sodium hydroxide, and sodium sulfide and sulfate. These products are known as reburned lime, dregs, grits, and smelts and contain impurities such as silicon, magnesium, aluminum, iron, manganese, potassium, strontium, and phosphorus in various forms and concentrations. Some of these by-products, recycled in the process, have a tendency to accumulate in pipes or tanks. Their analysis and identification is, therefore important during the cleanups and tune-ups of the recovery systems.

Lime analysis usually involves the determination of twelve elements which are expressed as weight percentages of CaO, MgO, Fe_2O_3, MnO, Cr_2O_3, SiO_2, Al_2O_3, K_2O, SO_3, P_2O_5, SrO, and Na_2O. Calibration standards consist of a variety of calcined limestones, dolomites, and their mixtures. If necessary, calibration ranges on iron and silica can be extended by "spiking" the standards with pure oxides of the elements (for fusions only).

The specimen preparation method, described by Kocman [25], is suitable for the analysis of a variety of limes, ferrolimes, hydrated limes, limestones, and other lime-like samples used in pulp and paper, steel, and construction materials industry. Istone [26] describes a similar method used as a problem solving tool in the lime-kiln recovery process of a Kraft pulp mill and uses also the method for evaluation of filler distribution levels. These methods could also be applied to carbonates, such as calcite, aragonite, dolomite, and magnesite, provided that sulfur values are not important to the analyst. Sulfur, present as elemental sulfur, sulfide, or sulfite, may be partially lost from the sample due to the calcination (ashing) and high fusion temperatures.

3.2.2 Sampling

Representative samples are obtained using sampling procedures described in section 1.4 of this book. Approximately 200 g of production sample (lime, hydrated lime, or limestone) would be the minimum amount accepted by a laboratory. This quantity would be easy to obtain in an industrial plant environment, where hundreds of tons of lime are produced daily. "Grab samples" taken from railroad cars, bags, conveyor belts, or silos should be at least 100 g to 500 g size, in order to "represent" the bulk. Deposits, scales, and smelts are usually available in much smaller quantities and the analyst may be forced to work with samples as small as 2 to 10 grams. Important: when samples of freshly burned or "reburned" lime (CaO) are taken in production, it is imperative to collect these materials into predried, airtight glass or plastic jars. The jar should be filled fully with the sample (with no space left) in order to prevent carbonation of lime (formation of $CaCO_3$) by carbon dioxide present in the air.

3.2.3 Specimen Preparation—Fused Glass Disk

Apparatus This list includes the following: drying oven, grinding equipment which does not contaminate the sample i.e., puck or ball mill with grinding vessel lined with alumina surface, 1000°C muffle furnace, automatic fluxer or similar equipment for fusion of the glass disks, 95% platinum–5% gold-alloy crucibles and molds for the fluxer, alumina or platinum crucibles for "loss on ignition," airtight jars or vials for specimen storage. Equipment suppliers are listed in Chapter 8.

Drying and Grinding Predry the collected sample for 2–6 hours at 105°C depending on the moisture content. Lime and ferrolime samples can take longer to dry, while carbonates such as limestone or dolomite may contain very little moisture. Grind the predried sample to -74 μm (-200 mesh). Use a "puck" or ball mill lined with corundum (Al_2O_3) or hardened steel to avoid contamination. When limestones, dolomites, or magnesites are ground, large pieces of the rock coming from a quarry are air dried at a room temperature and passed through a "jaw crusher." The resulting gravel size (approximately 25 mm) rock is then ground in a disk mill, which further reduces the particle size to about 150 μm (100 mesh). In general, breaking a smaller rock or sample with a hammer and grinding it in a puck or ball mill is sufficient for calcination of the sample and for fusion.

Calcination (Ashing) of the Sample About 4–6 g of dry and pulverized sample is calcined in an alumina or platinum crucible in a muffle furnace at 1000°C for 1 hr. The sample is cooled in a desiccator and the *loss on ignition* (LOI) is calculated from the weight difference, before and after calcination. Samples of by-product or reburned limes which contain more than 1% of sodium may adhere to, or react with the glaze of ordinary china crucibles. In this case it is recommended to use platinum crucibles for calcination. LOI is merely a number representing the total sum of weight loss due to residual moisture, combustible carbon or sulfur (other than sulfate), chemically bound water from hydrated lime, and carbon dioxide resulting from the decomposition of carbonates at high temperature. These individual losses occur at different temperatures and can be quantitatively recorded and examined by thermal analysis (DTA-TGA) performed separately on the sample. Loss can generally be reduced by oxidation.

Preparation of Fused Disk

1. Weigh 1.000 g of freshly calcined sample directly into the fusion crucible (made from 95% platinum–5% gold alloy).
2. Add 5.00 g of dried (400°C) lithium tetraborate and 0.30 g of lithium fluoride.
3. If necessary, add approximately 10–20 mg of lithium bromide (LiBr) as a nonsticking or release agent. (LiBr is vaporized during the fusion and would not affect the results.)
4. Mix well with a small spatula without spilling the sample from the crucible.
5. Fuse in a *fluxer* for at least 4 min in the flame with a 1 min stationary preheat before casting. Samples containing more than 2% of quartz as an impurity may require longer fusion times as quartz dissolves slowly in the flux.
6. The resulting disk is released from the mold, properly labelled with a round label, and stored in a small paper or plastic envelope until it is presented to the spectrometer for analysis. For longer storage or preservation of calibration disks, use of a *dry box* or desiccator is mandatory. Humidity present in air will dull the mirror-shiny surface of the disk and make it unusable for future analysis.

IMPORTANT: Fuse only freshly calcined samples. If the sample is stored overnight or even for a few hours in a desiccator or in open air, etc., recalcine the sample again at 1000°C for 30 minutes in the furnace, prior to weighing and disk fusion. Lime samples have a tendency to rehydrate and recarbonate rather fast when left in contact with air.

3.2.4 Results

The results are presented in the form of a computer printout of the weight percentages of the individual elements expressed as oxides and their sum. The sum will be theoretically 100%, provided that rehydration or recarbonation of the lime samples is minimized and that no other nondetermined elements are present. Results for carbonates are recalculated back to the original dry sample with respect to their carbon dioxide content or LOI.

3.2.5 Discussion

The described specimen preparation method is suitable for analysis of a large variety of lime and ferrolime samples produced in industry. Lime fuses exceptionally well with lithium tetraborate and produces perfect, transparent disks which are used for multielement XRF analysis without further finishing. The addition of a release agent (LiBr) is recommended in order to avoid sticking of high sodium content ($>1\%$ Na_2O) limes to the mold. Sticking can also be caused by trace quantities of copper (>10 ppm Cu) in the sample.

3.3 LIMESTONES, DOLOMITES, AND MAGNESITES

3.3.1 General Discussion

There are basically two different specimen preparation methods suitable for the XRF analysis of limestone and dolomite rocks. The pressed powder method described by Kocman [27] uses finely ground rock mixed with a small amount of organic binder, pressed into disks using aluminum cups as a support. The second method, published by King et al. [28], employs calcination (ashing) of the rock to its oxides, which are subsequently fused with a lithium tetraborate flux and cast into a disk. The pressed powder method is more popular in industry, because it is quicker and allows for a precise determination of sulfur (lost almost entirely during the heating of the sample above $600°C$) and a number of trace elements. Determination of sulfur in limestones and dolomites is very important for the process and environmental control of sulfur dioxide emissions in the production of cements and limes. Pressed powder disks are also more suitable for use with both energy dispersive and wavelength dispersive XRF spectrometers. The fusion method, when used as described in section 3.2.3 is slightly more precise for the major elements, calcium and magnesium, but due to the dilution with flux, lacks the required sensitivity for the determination of trace elements. It is, therefore, less suitable for use with energy dispersive systems.

3.3.2 The Pressed Powder Method

Summary   Finely ground, approximately -325 mesh (45 μm), dried sample is slurried with a solution of polymerized methylacrylate in acetone. After drying under an infrared heat lamp, the resulting cake is crushed in an agate mortar and pressed into a pellet, which is analyzed by XRF spectrometry. The concentration of carbon dioxide is determined on a separate sample as a loss on ignition between $500°C$ and $1000°C$.

Elements Determined   All the elements determined are expressed as their corresponding weight percent oxides. Commercially available reference materials (standards) usually cover the calibration ranges shown in Table 3.1. See Chapter 9.

TABLE 3.1. The Most Common Calibration Ranges Used in the Industry and Standard Deviations Based on the Available Reference Materials (Standards)

Element	Calibration Range %	Std. Dev. %
CaO	30–56	0.3
MgO	0–22	0.3
SiO_2	0–20	0.1
Fe_2O_3	0–1.2	0.01
MnO	0–1	0.02
TiO_2	0–1	0.01
Al_2O_3	0–4.5	0.05
P_2O_5	0–1	0.005
SO_3	0–0.5	0.005
K_2O	0–1.2	0.01
Na_2O	0–0.5	0.03
SrO	0–1.5	0.01

Sample Grinding Calcite or dolomite rock is crushed in a "jaw-crusher" and passed through a disk mill (alumina or tungsten carbide plates) to obtain a preground, approximately −100 mesh (−150 μm) sample. A representative portion of this sample (approximately 25 g) is dried at 105°C for 3–6 hr and subsequently ball-milled or swing-milled (mill with a puck) to approximately −325 mesh (45 μm). Screening of the sample is not necessary as the grinding times are determined empirically on a typical sample from the variations in intensity of the major element peaks of Ca and Mg as described in detail by Wheeler [29].

IMPORTANT: The sample is properly ground and assumed free from particle size effects when the calcium and magnesium intensities remain constant (no peak intensity change with prolonged grinding). Typical grinding times used were 5 minutes for a "puck" swing-mill and 10 min when using a tungsten carbide ball-mill on a 25 g portion of disk-mill preground calcite or dolomite rock.

Binder Solution (10% Methyl Methacrylate) Dissolve slowly 100 g of polymerized methyl methacrylate (Kodak P-4942 methyl-methacrylate powder or clean pieces of Plexiglas) in 900 ml of acetone and mix well on a magnetic stirring plate. Such a stock solution of binder is kept in a well-closed polyethylene bottle for further use.

Preparation of a Pellet For the preparation of a standard 32 mm (1.25 in.) diameter pellet, follow the procedure outlined below.

1. Weigh 8.00 g of the prepared sample into a clean 60 mm diameter disposable aluminum dish (Fisher cat. # 08-732 or similar).
2. Add to the sample exactly 5 ml of 10% methacrylate-acetone solution with a disposable syringe. Mix well with a stainless steel spatula.

3. Dry the dish contents under an infrared heating lamp. The slurry dries completely in about 3–5 minutes, when mixed occasionally in the dish.

4. If the 5 ml volume of the binder is insufficient to produce a homogeneous mixture, add a few milliliters of acetone to the sample.

5. The resulting dry cake is crushed and quickly ground by hand in a large agate mortar to about -100 mesh (150 μm) and subsequently pressed into a 32 mm diameter aluminum backing cup (SPEX #3619, Chemplex #505, or similar) using a pressure of 25 tons on the ram of the hydraulic press.

IMPORTANT: The ratio of the pulverized sample to the volume of the prepared binder solution must be kept constant for the preparation of calibration standards and for the unknown samples. Any deviation from the chosen ratio will affect the accuracy of the analytical results. Therefore, if necessary, extra acetone (and *not* the binder solution) is added to the weighed sample for improved wetting and better homogenization of the mixture. The extra acetone evaporates from the sample during the infrared drying cycle and will not affect the analysis.

Surface Quality of the Prepared Disk   The resulting disk must have a flat and smooth surface, without any visible scratches or defects. The homogeneity of the surface is represented by a continuous white or grey shade of the surface. If the surface shows spots or areas of slightly different color, this usually indicates insufficient homogeneity of the limestone-binder mixture. The disk thickness should be greater than 3 mm to assure its dimensional stability and correct determination of higher atomic number elements (iron, strontium, etc.).

3.3.3   Determination of Carbon Dioxide, CO_2

Carbon dioxide is determined on a separate representative (approximately 5 g) sample as a loss on ignition between 500°C and 1000°C. The sample is weighed into a porcelain crucible and placed for 30 minutes in a furnace heated to 500°C, removed, cooled, and weighed. This heat treatment will remove any remaining moisture, organics, sulfur, or chemically bound water from clays or other minerals, present as impurities. The sample is then heated in a 1000°C furnace for another 30 minutes, cooled in a desiccator and weighed within two hours (to prevent sample recarbonation by carbon dioxide from the atmosphere). The second weight difference at 1000°C corresponds to evolved carbon dioxide. More precise determination of impurities and carbon dioxide in the sample is obtained by thermal analysis (DTA-TGA) scan to 1000°C or by determination of carbon dioxide by gas evolution techniques.

3.3.4   Calibration Standards

Calibrations for the individual elements (Table 3.1) can be prepared from about twenty commercially available, certified reference carbonates (standards). Standards BCS-368 and BCS-393 are certified dolomite and calcite rocks from Great Britain, standards NBS 1C and NBS 88A are from NIST (formerly NBS) in the United

States. Standards DO 1-1 and DO 2-1 are certified reference samples of limestone and dolomite from the Institut de Recherches de la Siderurgie in France. The IPT certified samples originate from Instituto de Pesquisas Technologicas in Brazil. Two limestone and dolomite samples JLs-1 and JDo-1 were certified by the Geological Survey of Japan. Three GFS samples can be purchased from G.F. Smith Co. in Columbus, Ohio. Some recent certified samples originate from China (GBW 7214 and 7215, calcite and GBW 7216 and 7217 dolomite). There are also synthetic limestone series of reference samples available from China for the determination of trace elements, namely GBW 07712 to GBW 07720. See Chapter 9.

3.3.5 Results

The results are presented as a computer printout of the weight percentages of the individual oxides and their sum by the spectrometer. In order to bring this total to 99+%, the previously determined concentration of carbon dioxide is added to the results.

3.4 NATURAL AND BY-PRODUCT GYPSUM

3.4.1 General Discussion

Gypsum is known as one of the oldest building materials, dating back to the age of the pharaohs. In the Middle Ages, whole cities were built from natural gypsum, and some of them have been preserved in Morocco. Gypsum is essentially calcium sulfate dihydrate ($CaSO_4.2H_2O$) with some minor impurities such as quartz, calcite, and dolomite. Today, gypsum is used for the manufacture of wallboard (Gyproc, Sheetrock, etc.), joint fillers, quickset fillers, decorative plasters, gypsum bricks, gypsum based flooring cements, and many others. Gypsum is often added to cements as a setting retardant. Gypsum is frequently used in agriculture for conditioning of soils or to supply nutritional sulfur and catalytic support of fertilizers. Very pure gypsum is used in dentistry and sculpting. In the last decade, by-product gypsum, formed by desulfurization of smokestack flue gases (the so-called FGD gypsum produced at coal-fired electric power plants), titano-gypsum (by-product in titanium-dioxide production) and phospho-gypsum (by-product of phosphate fertilizer production) has started to replace natural gypsum rock, mainly in wallboard production. Thus, analysis of gypsum is quite important in industry and research for determination of impurities, product quality control, and by-product gypsum landfill requirements.

3.4.2 Remarks

A rapid and precise specimen preparation technique suitable for multielement analysis (up to 14 elements) of gypsum, hemihydrate, anhydrite, and gypsum-related products is described. Samples of gypsum are calcined at 1000°C and the resulting anhydrite ($CaSO_4$) is fused with sodium tetraborate into disks. The choice of sodium tetraborate as a flux is critical as rapid decomposition of anhydrite to CaO and SO_3 occurs in lithium based fluxes at normal fusion temperatures (800–1100°C). An alternative sample preparation method using pressed gypsum or anhydrite is also described. The latter is more suitable for the use with X-ray energy dispersive systems and in process or quality control applications, where speed is the driving factor.

3.4.3 Elements Determined

The major and minor elements (expressed as wt% of oxides) determined in gypsum, hemihydrate, anhydrite, or by-product gypsum are usually CaO, SO_3, MgO, Al_2O_3, SiO_2, Fe_2O_3, K_2O, P_2O_5, SrO, and TiO_2. Together with water and carbon dioxide (both determined by other analytical methods) these cover most of the impurities present in gypsum, such as calcite, dolomite, quartz, silicates, various clays, metal oxides and hydroxides, etc. Chromium, vanadium, and fluorine are usually determined as additional elements in titano-gypsum or phospho-gypsum samples analysis.

3.4.4 Dehydration of Gypsum

Dehydration of gypsum to hemihydrate and anhydrite can be described by the following equations:

$$CaSO_4 \cdot 2H_2O \rightarrow CaSO_4\ 0.5H_2O + 1.5\ H_2O \tag{3.1}$$

$$2\ CaSO_4 \cdot 0.5H_2O \rightarrow 2CaSO_4 + H_2O \tag{3.2}$$

While equation (3.1) is completely reversible and explains the dehydration of gypsum to hemihydrate (plaster of paris) and vice versa, equation (3.2) is irreversible above 360°C. Consequently, if a specimen containing gypsum is heated to 900–1000°C and anhydrite is formed, it will not rehydrate back to hemihydrate or gypsum when cooled to room temperature and exposed to moisture in the air.

3.4.5 Drying of Gypsum

Because the first dehydration of gypsum to hemihydrate starts at 60 to 70°C, all gypsum samples *must be dried only at maximum temperature of* 45 ± 5°C. Under no circumstances should the temperature exceed 50°C. Sufficient drying is achieved in 10–12 hrs (overnight). Overdrying or drying at the usual laboratory temperature of 105°C would lead to partial dehydration of the gypsum and give incorrect analytical results.

3.4.6 Grinding

Most natural and by-product gypsums are easily ground to 100–200 mesh (150–75 μm), except for samples containing high levels of quartz impurity (3–10% SiO_2). Any type of swing mill employing a puck in a barrel, made from fused alumina (Al_2O_3) or tungsten carbide (WC) is suitable. Short grinding times of 2–3 minutes are recommended in order to prevent overheating of the gypsum. Ball mills are less suitable due to longer grinding times, sticking, and heat generation. Samples with high quartz impurity may require repetitive regrinding for short periods of time in order to avoid overheating of the sample.

Fusion Techniques   Fusion of calcined (preashed) samples with lithium tetraborate or lithium metaborate was successfully used in the past for the precise analysis of silicates, minerals, cements, refractories, and some ores. The method is highly recommended for gypsum samples, although with some restrictions on the use of lithium based fluxes. Kocman [30] reported partial decomposition of anhydrite in lithium-based fluxes. The observed decomposition resulted in visible bubbles, porosity, and cracking of the cast disk. Stronger adherence of the disks to the standard (95% platinum–5% gold alloy) was also observed. Subsequent investigation of the thermal stability of the melt revealed serious decomposition of anhydrite, $CaSO_4$, in lithium-based fluxes, such as lithium tetraborate ($Li_2B_4O_7$) and lithium fluoride (LiF). The cause was most probably the formation of lithium sulfate (Li_2SO_4) with a relatively low boiling point of 845°C. Decomposition of anhydrite in lithium based fluxes increases drastically with increased fusion temperatures above 900°C (pronounced weight loss of the melt). The decomposition seriously affects the sample-to-flux ratio and leads to incorrect analytical results. For this reason the use of lithium-based fluxes is not recommended for analysis of gypsum.

The Use of Sodium Tetraborate $Na_2B_4O_7$ (Borax) for Fusion   Anhydrite, $CaSO_4$, is very stable in molten sodium tetraborate, $Na_2B_4O_7$ (borax flux), even at high temperatures and prolonged fusion times. Disks fused from 0.5 g of calcined gypsum (anhydrite) and 6 g of sodium tetraborate produced clear and transparent disks with perfect surfaces without any visible decomposition. The use of high sample to flux ratio (1:12) is dictated by the lower solubility of anhydrite, $CaSO_4$, in molten borax. The use of a sodium based flux means that the analyst has to sacrifice the determination of sodium in the sample (typically below 200 ppm sodium in natural gypsum). If necessary, sodium can be determined in gypsum by atomic absorption spectrophotometry. The advantages of using sodium tetraborate flux outweigh the loss of sodium as a determined element. The use of releasing agents such as lithium bromide (LiBr) was eliminated completely. The disks fused from sodium tetraborate flux release easily from molds each time, without any sign of sticking, cracking, or crystallization.

Details of the procedure described below, as well as preparation of synthetic gypsum calibration standards are provided by Kocman [30].

Preparation of a Fused Disk

1. Determine the loss on ignition in a muffle furnace at 1000°C (1 h) using a pulverized (-150 mesh) sample dried at 45°C. The resulting $CaSO_4$ is stable up to 1200°C and no sulfur losses will occur.
2. Weigh 0.500 g of the freshly calcined (ashed) sample into a Claisse crucible (made from 95% platinum–5% gold alloy).

3. Add 6.00 g of dense sodium tetraborate flux (borax, $Na_2B_4O_7$). For best results use Spectroflux #200 flux or Claisse Corp. high density #0630 borax. The borax flux must be predried for several hours at $350°C$ and kept in an airtight container.

4. Mix the weighed sample and flux thoroughly with a stainless steel microspatula.

5. Fuse for 1 minute on fluxer (stationary heat) to melt the mix and swirl in the flame for 4 minutes.

6. Wait 1 minute (no swirling) to increase the temperature of the melt and pour into a 32 mm mold.

7. Cool the disks undisturbed for about 2–3 minutes. Further cooling can be accomplished by blowing air on the bottom of the mold. Three to six disks can be produced and cooled simultaneously in about 15 minutes. The minimum thickness of the disk should be 4 mm.

Storage of Fused Disks   Disks fused with sodium tetraborate flux are more prone to moisture damage than their lithium tetraborate counterparts. This reactivity is especially true in the summer months when humidity even in an air-conditioned room is quite high (about 50%). The disks are properly labeled with round stickers and stored in a desiccator or dry box for further use. Special care must be taken with fused calibration disks (spiked) which may be used every 2–3 years for major recalibration of the XRF spectrometer. Their preparation is labor intensive and, if correctly stored, these disks may last with the preserved original gloss (or shine) for at least 10–15 years.

Preparation of a Pressed Pellet   As gypsum is quite soft and easily pulverized to -200 mesh (75 μm), pellets can be pressed directly (without the use of binder) from a $45°C$ dried sample. The following simple procedure should be followed for the preparation of a 32 mm diameter (1.25 in.) pellet.

1. Weigh approximately 6 g of dried, pulverized (75 μm, -200 mesh) gypsum sample into a plastic or aluminum disposable dish.

2. Transfer the sample into a die provided with an alumina backing cup (SPEX or Chemplex type).

3. Press at 25 tons pressure on the ram for 30 seconds.

4. Slowly release the pressure and unload the pellet.

The resulting pellets are of very high quality, with extremely flat and shiny surfaces. Such pellets may not, however, survive even the low vacuum of the X-ray fluorescence spectrometer, which is required for the determination of elements below atomic number 20. The hydrogen-bonded water in gypsum is partially removed from the surface of the pellet and may cause a slow, but steady change in chemical composition of the analyzed surface. The decomposition under vacuum affects the peak intensities, mainly for the major elements—calcium and sulfur. However, if an energy dispersive or wavelength dispersive spectrometer is equipped with a helium flushing system and the surface layers of the disk are not overheated by the absorption of the primary X-ray beam, the pressed powder method can be used with great success for quick quantitative evaluation of major and trace elements, as required in a plant or quality control environment. The pressed powder method described above is therefore more suitable for a process control or quick analysis (for CaO and SO_3 and impurities) where sample preparation speed is an important factor and ashing and fusion of the sample is impractical. When helium is used as a spectrometer chamber medium, pressed gypsum can be analyzed with high precision. Helium is also employed when trace elements are determined directly on the pressed disks. The highest analytical precision is obtained by using fusion of the calcined sample with sodium tetraborate (borax).

3.4.7 Results

Results obtained on pressed pellets show directly the concentrations of individual elements expressed as weight percent oxide. Addition of the loss on ignition (LOI) value or better the values obtained for the chemically bound water and carbon dioxide (determined by other techniques) complement the analysis. In the case of a fused disk, the results are reported by the spectrometer on water, carbon dioxide free basis. These values should be recalculated with respect to LOI to the original dried sample. Thermoanalytical data (DTA and TGA) provide more details on the proportion of water, organic impurity, carbon dioxide, etc. present in the sample. The use of this technique is recommended for more complex analyses.

3.4.8 Reference Materials

Four certified natural gypsum rock standards (GYP-A, GYP-B, GYP-C and GYP-D) and three recently (1992) certified by-product gypsum (FGD-1, FGD-2 and TIG-1) are available from Domtar Inc. Research Centre in Senneville, Quebec, Canada [31] or from designated suppliers (see list in Chapter 9). These standards have a status of "Canadian Certified Reference Materials" and cover a wide range of elemental concentrations suitable for calibration of X-ray fluorescence spectrometers. These reference materials are commercially available from suppliers (see section 9.4.3, "Sources for XRF Standards"). Concentration ranges can be easily extended to higher elemental concentrations by addition of precalculated quantities of additives (oxides of the elements) to pure anhydrite $CaSO_4$. For more details on the sample preparation procedure as well as spiking of anhydrite fusions and preparation of additional synthetic standards, the reader is referred to a section by Kocman in ASTM publication STP 861 [30].

3.4.9 Discussion

When fusions of sulfur-bearing minerals are performed at high temperatures (in our case 1000°C) a question of sulfur loss and thermal stability often arises. West and Sutton [32] reported the first sign of dissociation of $CaSO_4$ into CaO and SO_3 above

1225°C. These results were verified by Kocman [30], who found anhydrite stable up to 1200°C. The only other sulfur loss may originate from pyrite, FeS_2, which is a rare impurity found in natural gypsum. The pyrite sulfur will also be lost during a wet chemical analysis of gypsum when the gypsum sample is dissolved in hydrochloric acid.

3.5 GLASS SANDS

3.5.1 General Discussion

The method described is suitable for quartz sands and glass sands containing over 95% SiO_2. These methods are generally used in glass bottle manufacturing and the glass industry. Six elements, as oxides, which are associated with impurities in the sand are Al_2O_3, K_2O, CaO, TiO_2, Fe_2O_3, and Cr_2O_3. These six elements comprise usually 0.2% to a maximum of 5% of the sample. The rest of the sand consists of pure quartz, α-SiO_2. XRF analysis [33] for these six elements is done on a disk prepared by fusion of the ignited sand sample with lithium tetraborate flux. Iron and chromium are the most important impurities in glass sands because their presence strongly affects the coloration of the end product. Iron and chromium, if present even at ppm levels, will produce green or brown glass. In extra pure glass sands, chromium and iron are usually present in the low (0.2–20 ppm) range. The lower limit of detection (LLD) of these elements will depend on the X-ray tube target (usually Rh or Mo) and the type of spectrometer used. In general, the LLD is around 2–3 ppm for iron or chromium in a 32 mm diameter fused glass disk. Detection limits below 1 ppm can be achieved by using pressed pellets and "end window" XRF tubes. Pellets are prepared by thoroughly mixing finely ground sand with a wax type binder and pressing the mix at a high pressure into a supporting cup.

For more details on impurities in glass sands, the reader is referred to the article issued on the subject by the Committee of the Society of Glass Technology [34].

3.5.2 Selection of a Representative Sample

A representative sand sample is chosen from a covered railroad hopper car, truck, or pile by methods described in section 1.4, "How to Obtain a Representative Sample from the Bulk." A minimum quantity of about 30 g is required to allow for replicate analyses and additional tests.

Specimen Preparation—Fused Glass Disk

Apparatus This includes a drying oven, 1000°C muffle furnace, grinding equipment which does not contaminate the sample, i.e., a puck or ball mill with zirconia or alumina liner is favored over hardened steel or tungsten carbide vessels, which are unsuitable due to possible contamination with iron or cobalt. Also needed are automatic fluxer or similar equipment for fusion of disks, 95% platinum–5% gold alloy crucibles and molds for the fluxer, ordinary platinum crucibles for loss on ignition, precleaned plastic vials.

Drying and Grinding   Dry the sample at 105° C ± 5°C for 2–4 hours depending on moisture content. Grind approximately 10–20 g of representative sample to −44 μm (−325 mesh). Do not screen the sample to avoid possible contamination. Establish correct grinding times and particle sizes on your equipment (puck mill, ball mill) by pregrinding a test portion of the bulk sample or other similar sand.

Loss on Ignition   Weigh approximately 3–4 g of preground sand directly into a clean platinum crucible. The crucible was previously heated to 1000°C for 10–15 minutes and cooled in a desiccator. Determine the loss on ignition at 1000°C for 1 hr. Store the calcined sample for further use in a small plastic vial.

IMPORTANT: Due to very small losses on ignition in these types of samples, use utmost care when weighing a sample before and after ignition. In general, ignited samples should be weighed right after they cool down to room temperature. Some impure sands may actually show weight increase after heating to 1000°C. This increase is caused by the oxidation of FeO to Fe_2O_3. In this case, the loss on ignition is considered to be zero.

3.5.3   Fusion with Lithium Tetraborate and Lithium Fluoride

1. For a 32 mm diameter disk, accurately weigh out 1.000 g of ignited sample, 5.000 g of lithium tetraborate, and 0.3 g of lithium fluoride. The individual portions are weighed directly into a Claisse crucible (95% platinum–5% gold), which is placed into the fluxer apparatus.

    IMPORTANT: Use lithium tetraborate ($Li_2B_4O_7$) and lithium fluoride (LiF) fluxes of extra high purity, which have been analyzed and which have impurities listed on the bottle or in the flux certificate. Tetraborate fluxes suitable for this method are supplied by Claisse Corp. [35] and EM Science Corporation [36].

2. Add 20 mg of solid LiBr (using a microspatula) as a non-sticking or release agent to the flux. Lithium bromide decomposes in the hot flux; therefore a well-ventilated area around the fluxer is recommended.

3. Fuse on flame with a 1 minute preheat and at least 5 minutes of swirling at 800–1000°C (propane-air flame) in order to allow sand particles to dissolve completely in the flux. One minute of stationary heating is recommended just prior to casting, in order to reheat the flux mixture to a maximum temperature.

4. Cast into a 95% platinum–5% gold mold (32mm diameter) and cool to room temperature (some fluxers have a builtin cooling fan).

5. Empty the mold onto a clean sheet of paper or lint-free tissue, label with a round sticker and store in a small paper or plastic bag.

Sample Storage   Fused lithium tetraborate disks should be stored for future use in a desiccator or dry cabinets. Atmospheric moisture, especially during the summer, dulls the surface of the disk. Pressed pellets can be stored in open air for several months. If long-term storage is considered, a dry environment should be used.

Problems

Disk Cracking   In most cases, cracking of the disk when cooled is caused by poor grinding of the sample. The larger sand particles (mostly quartz) remain undissolved in the flux and form crystallization centers, which cause cracking. The disk shows a starlike pattern of cracks.

Remedy: Extend your grinding times. Longer grinding should decrease the particle size of the sample. Prolong fusion on flame when using an automated fluxer. Times set for fusion and mixing on flame may be too short. You can remelt the disk from the cracked pieces and recast it.

Disk Sticking to the Mold, Unable to Release Disk   This problem is common with silicate samples containing copper in ppm concentrations. As glass sands contain very little copper (below 1 ppm copper) disk sticking should not be a problem.

Remedy: Addition of a few milligrams of LiBr to the flux (as recommended above) should eliminate sticking and the disk could be easily released from the mold.

Specimen Preparation—Pressed Powder Pellet   An alternative specimen preparation method can be used when iron and chromium are present in the sand below 10 ppm levels. The method uses a small amount of binder mixed with the pulverized sand. The resulting mixture of sand-binder is pressed into a pellet at a high pressure (15 to 20 tons).

Apparatus   This list includes a hydraulic press capable of up to 25 tons ram pressure, motorized mixing equipment, aluminum disposable dishes, plastic vials.

1. For a 32 mm diameter pellet, accurately weigh 6.000 g of pulverized glass sand into a 50 ml plastic vial. Add 1.000 g of Hoechst Wax as a binder. Close the vial with a cap.
2. Mix thoroughly for 15 minutes by shaking on a mechanical shaker or similar equipment.
3. Weigh approximately 8.0 g of the prepared mixture into a press die. Use aluminum backing cup for dimensional stability of the pellet. Press at 20 tons pressure on the ram. Keep under high pressure for at least 10 seconds and slowly release. Use the same pressure, time, and release rate for all specimens and standards.
4. The resulting pellet must have a flat and smooth surface.
5. Label the pellet with a round sticker and store it in a small paper envelope or plastic bag for further use.
6. The pressed pellet is presented to the spectrometer for analysis.

3.5.4   Discussion

Both methods described are suitable for the analysis of sands and glass sands. The lithium tetraborate fusion method is more accurate for the determination of minor elements in sands. This effect is mainly due to the dilution with the flux and elimination of the surface and mineral effects. Pressed pellets are used when low detection limits are desirable especially on chromium and iron. Note that the use of 40 mm molds for fusion eliminates the edge rounding within the 30 mm diameter of most spectrometer sample cups. This size would, of course, require slightly larger sample and flux weights.

3.6 CEMENTS, CLINKER, AND RAW MATERIALS

3.6.1 Types and Compositions of Cements

Hydraulic cements consist of inorganic compounds that set and develop strength by chemical reaction with water by formation of hydrates, and are capable of doing so underwater [37]. Portland cement is the generic name for a particular class of hydraulic cement. Portland cement is produced by pulverizing portland cement clinker with calcium sulfate. The clinker is made by burning a mixture of finely ground raw materials (traditionally limestone, clay, and sand) at high temperature in a large rotary kiln. Analysis of raw materials, clinker, and finished portland cement is commonly performed using XRF spectrometry. The concentrations of the elements shown in Table 3.2, commonly expressed as oxides, are routinely determined for manufacturing process control and for conformance with specification. These concentration ranges are typical for ordinary portland cements.

The raw materials used to make clinker nowadays include many industrial wastes such as spent catalysts, mine tailings, fly ash, blast furnace slag, foundry sands, etc. Specimens of these materials, traditional geological raw materials, and blended hydraulic cements can be prepared for analysis following the methods given in Section 3.1.2 for whole rock analysis.

TABLE 3.2. Typical Concentration Ranges for Portland Cement

Analyte	wt %
SiO_2	20–23
Al_2O_3	3–6
Fe_2O_3	2–4
CaO	61–67
MgO	0.5–4
SO_3	2–4.5
Na_2O	0.1–0.5
K_2O	0.1–1.3
TiO_2	0.2–0.3
P_2O_5	0.1–0.25
SrO	0.05–0.3
Mn_2O_3	0.05–0.3
Loss on ignition	1–2

Other types of hydraulic cements that lend themselves to analysis by XRF spectrometry include calcium aluminate cements; portland-pozzolan cements containing fly ash, blast furnace slag, or natural pozzolans; masonry cements containing portland cement clinker and finely ground limestone; expansive cements containing calcium aluminosulfates; and others. Specifications for hydraulic cements are found in ASTM Volume 4.01 which comprise C 91 Masonry Cement, C 150 Portland Cement, and C 595 Blended Cement. More than 90% of all hydraulic cements sold are portland cement. ASTM C 150 specifies several types for various purposes (general construction, low heat, sulfate resistance) denoted by roman numerals I–V. Even though these cements have different percentages of clinker minerals, specimens of all these types can be prepared and analyzed by a common method.

Specifications for oil well cements can be found in American Petroleum Institute (API) Specification 10. Hundreds of proprietary formulations exist for cement-based mortars, grouts, leveling and patching compounds, most of which can be analyzed by XRF spectrometry using simple specimen preparation techniques described below. The particle size distribution of hydraulic cements typically covers the range 0.5–100 μm with mass median diameters around 15 μm. Hydraulic cements are easily ground to high fineness for briquet specimens because they consist mostly of minerals with Mohr's hardness 4–5 and primary grain sizes ranging from less than 1 μm to around 40 μm. The chemical composition of hydraulic calcium silicate cement minerals permits rapid fusion using lithium tetraborate at less than 1000°C.

3.6.2 Pressed Powder Briquet Preparation

Cement should be sampled in accordance with methods described in ASTM C 183-95, "Standard Practice for Sampling and the Amount of Testing of Hydraulic Cement." Clinker and raw materials should be sampled and reduced to powders passing a No. 100 U. S. Standard Sieve according to the principles described in Section 3.1.2 of this book.

Thoroughly remix the sample in its jar by rotating in a figure-eight motion with two hands. Weigh 7.000 ± 0.001 g of sample into a weighing boat by taking several separate gram-size portions using a spatula or scoop, being careful to take representative material from throughout the sample in the jar. Place the weighed sample into the body of a swing mill for fine grinding. Swing mills, also called ring-and-puck mills, are available from several vendors and are made of various materials including hardened steel, tungsten carbide, aluminum oxide, and zirconium oxide. The choice of mill depends on cost and possible contaminants that might be introduced into the specimen during fine grinding from the walls of the mill. (See section 3.1.1 for a discussion of possible contaminants from grinding.)

Add two small drops (approximately 40 μl) of propylene glycol on top of the powder sample in the mill as a grinding aid. Cover the mill and grind 4 minutes at 1000 rpm. If your mill runs at a different speed, adjust the time to obtain about 4,000 rotations of the mill. This length of grinding time will produce a finely pulverized specimen with 90–95% finer than 10 μm and mass median diameter approximately 3 μm.

After grinding, remove the mill cover and add 0.5 g binder such as commercially available XRF sample preparation tablets. Replace the mill cover and grind an additional 30 s to distribute the binder into the specimen. This two-step grinding process has been found to produce highly homogeneous, rugged specimens. If the binder is added at the start of grinding, agglomerations will occur that are visible on the surface of the finished briquet.

Binders such as waxes or methylcellulose help to hold the specimen particles firmly together after pressing, minimizing degradation and dusting of the analytical surface and thereby prolonging the useful life of the briquet. Any soft, waxy substance that acts as a binder by mechanically sticking particles together will inhibit very fine grinding and homogenization of the specimen particles. Using a grinding aid followed by adding the binder in a second step assures that you will obtain the most complete fine grinding possible in the mill. However, grinding times longer than 4 min (in steel or tungsten carbide mills) do not reduce the maximum particle size and do increase the proportion of finest particles, exhausting the grinding aid and leading to agglomeration. Such agglomeration will be evident as caking and sticking of specimen on the walls of the grinding media (ring, puck, and mill body). Excess binder will also produce caking and sticking of material that must be scraped off the mill. The correct amount of grinding aid will produce a free-flowing, pulverized sample that brushes easily out of the mill leaving no residue. The correct amount of binder, generally 5–10% by weight of the sample, will produce an extremely durable specimen surface without caking in the mill.

Brush the finely ground specimen into a labelled, 31 mm aluminum sample cap and press at 345 MPa (50,000 psi). Hold the pressure for one minute and release the pressure over a period of one additional minute. Eject the specimen and examine for cleanliness and homogeneity of the analytical surface. Store covered until ready for analysis. Some commercially available presses do not require an aluminum sample cup. The amount of binder might need to be increased slightly (10–12% by weight of sample) for these types of specimens. Pressed powder cement specimens should be analyzed within a few hours after preparation. The surface of the cement briquet will react with atmospheric moisture and carbon dioxide, ruining the analytical surface within a few days. The use of grinding aid and binder dilutes the sample, so calibration standards and unknown analytical specimens must be prepared according to exactly the same procedure.

3.6.3 Fused Bead Preparation

Portland cements and clinker are quickly and easily fused using lithium tetraborate to produce stable, homogeneous glass beads at a flux-to-sample ratio of 2:1. This low dilution permits precise analysis for sodium, the lowest atomic number element of routine interest in cements. Careful temperature control at $1000 \pm 25°C$ is necessary to minimize evaporation of potassium and sulfur. Fusions can be accomplished at this temperature in as little as five minutes although ten minutes is typical. Sample and flux are weighed to a constant ratio corrected for their loss on ignition. Unlike geological materials, cement or clinker do not need to be heated in air prior to fusion since virtually all elements are present at their highest common oxidation state. Thoroughly remix the sample in its jar by rotating in a figure-eight motion with two hands. Determine the loss on ignition at $1000°C$ on a separate one-gram portion of the sample. For the fusion, take several separate portions using a spatula or scoop, being careful to take representative material from throughout the sample in the jar. For a 7 g bead (sufficient to fill a 31 mm diameter mold), weigh enough flux to give 4.667 ± 0.001 g taking into account its loss on ignition. Weigh enough cement to give 2.333 ± 0.001 g on its ignited basis and mix thoroughly in a small container using a noncontaminating stirrer. Check that the mixture is a uniform gray color and that particles of the sample are completely distributed in the flux.

Place the flux and sample mixture into a clean 95% platinum–5% gold crucible and add two drops 50% LiBr (aq.) as antiwetting agent. Load the crucible into the fusion apparatus and heat gently over an oxidizing propane or propylene flame for several minutes to remove moisture and drive off air trapped among the particles. Increase the heat to achieve $1000 \pm 25°C$ and hold at this temperature until the mixture is molten. Stir for five minutes followed by casting the specimen into its 95% platinum–5% gold mold. Various brands of fusion apparatus have different rates of mixing. The most vigorous machines can effect complete dissolution of most minerals in lithium tetraborate at $1000°C$ in a few minutes. Machines that rock gently, or fusions performed in a muffle furnace with occasional mixing by hand, require longer fusion times.

Fused beads can be analyzed without further surface preparation if the surface is flat. Flatness and cleanliness can be restored to old beads by gently lapping on a 30 μm diamond-impregnated, metal bonded grinding wheel. Water used as a lubricant does not seem to dissolve water-soluble ions during the few seconds grinding time.

3.6.4 ASTM C 114 Qualification

ASTM C 114, "Standard Test Methods for Chemical Analysis of Hydraulic Cement" includes classical gravimetric, volumetric, and colorimetric "wet" chemical analytical techniques. These "Reference Test Methods" generally are too tedious and require such a high level of analytical skill that they are not used for routine purposes such as quality control. Instead, ASTM C 114 Section 3.3 "Performance Requirements for Rapid Test Methods" permits the use of any analytical test method which can be shown to meet specified performance criteria for precision and accuracy. For the analysis of portland cements, XRF spectrometry laboratories routinely have their methods qualified to these requirements.

The general procedure is to prepare two sets of seven cement specimens on nonconsecutive days from available standard reference material (SRM) cements. The specimens are prepared and analyzed according to the laboratory's routine procedures. The results of the analyses are compared to the C 114 qualification criteria: for each analyte, the difference between the duplicates is a measure of the laboratory's precision, and the difference between the average of the duplicate values and the SRM certified value is a measure of the laboratory's accuracy. The maximum permissible variation in results is given in ASTM C 114 Table 1.

3.7 PREPARATION OF METALS

3.7.1 Sampling

In the case of metals, as with other materials, direct analytical responsibility of every analyst begins at the sampling stage. Questions that need to be asked at this stage are many. For example, is the entire procedure correct? What is the depth from which the sample should be sampled? If there is a choice of sample mold, which one should be chosen and why? Is the shape, size, and sample thickness correct? When is the specimen and its surface acceptable and when is it not? How much quality in surface preparation can be sacrificed to gain speed? Is the sample representative? Does the small amount of the specimen contributing in the analysis represent tons of metal? Usually, metal is sampled with a ladle and poured in a mold, in which it solidifies (casting). Casting is the most universally employed analytical sample preparation procedure, although procedures involving direct mold immersion in the molten metal are also employed. As a result of casting a solid specimen of a particular shape, micro and macrostructure is obtained depending on the mold type or casting process. When an alloy solidifies, the solid that forms generally has a different composition than the liquid from which it is cast. Moreover, this solid has a certain structure.

3.7.2 Solubility of Elements

To evaluate the effect of the internal sample structure on the fluorescent radiation intensity of elements one needs to consider the elements' solubility. Any solution is composed of two parts: a solute and a solvent. The amount of solute that may be dissolved by the solvent is generally a function of temperature. With decreasing temperature the solubility limits usually decrease. Matrix constituents that are not soluble in the liquid state have a natural tendency to float or sink, depending on their density. Therefore, it is important to swirl the melt (if possible) before sampling to make it homogeneous.

   The majority of the binary intermetallic phases made by aluminum and various metallic elements form a liquid solution (are completely miscible) in the liquid state. The solubility limits for selected elements [38, 39, 40] at temperatures above the melting point of aluminum are listed in Table 3.3. In the solid state, elements such as calcium, iron, phosporus, and tin are nearly insoluble in aluminum. Silicon has a low solid solubility. No element is known to have complete solubility with aluminum in the solid state. Of all elements, zinc has the greatest solid solubility in aluminum.

   Table 3.4 presents solid solubility of elements in iron. There is a considerable difference in the solubility of many elements in rapidly quenched γ-iron (austenite) and in α-iron (ferrite). Obtaining fully austenitic steels requires large amounts of stabilizing elements (carbon, nitrogen, nickel, and manganese), as compared with those elements that stabilize ferrite [41]. Austenite is relatively hard, while ferrite is a soft low-strength phase. Solid solubility of several elements depends on the level of carbon content. Elements such as manganese, cobalt, and nickel have unlimited solubility in γ-iron, whereas vanadium and chromium have unlimited solubility in

TABLE 3.3. Solubility of Elements (Percentages) in Aluminum

Element	Liquid Solubility	Solid Solubility
Na	0.18	less than 0.1
Mg	35	15
Si	13	1.7
P	less than 0.1	less than 0.1
Ca	less than 0.1	less than 0.1
Fe	less than 0.1	less than 0.1
Cu	33	5.5
Zn	95	83
Ag	72	56
Pb	1.5	0.15

α-iron. Unlimited solubility means that the atoms of the solute substitute for atoms of the solvent in the lattice structure of the solvent. Carbon solubility, for example, is high in gamma iron and very low (maximum solubility of 0.025%) in α-iron. In low carbon alloys, carbon appears in the form of cementite (Fe_3C). Theoretically, 100% cementite contains 6.7% of carbon, and it forms eutectic with γ-iron at 4.3% C.

Low solubility elements appear in the form of hard metal matrix constituents, harm alloy homogeneity, and behave differently from the base metal during surface preparation, particularly of soft alloys. In order to minimize the influence of low solid solubility of elements, casting processes promoting formation of fine grain metallographic structure of the specimen are favored.

TABLE 3.4. Solubility of Elements (Percentages) in Iron

Element	γ-Iron	α-Iron
C	2*	0.025
Al	1[c]	36
Si	2–9[c]	19
P	0.5	3
S	less than 0.1	less than 0.1
Ti	0.8	6
V	1–4[c]	unlimited
Cr	13	unlimited
Mn	unlimited	3
Co	unlimited	75
Ni	unlimited	10
Mo	3–8[c]	38

*May vary with certain limits.
[c] depends on the carbon content.

3.7.3 Alloy Metallographic Structure

A metal alloy metallographic structure consists of certain major components, such as a metal matrix, constituents, inclusions, and structural defects [41, 42, 43, 44]. The constituents and inclusions are compounds rich in alloying and impurity elements, respectively. Their volume fraction is related to the alloy's overall chemistry and solid solubility of the alloying and residual elements. When the solid solubility of an element is low, it means that constituents and inclusions are rich in this element. Consequently, they (and not the matrix) contribute predominantly to the measured intensity.

Let's consider, as an example, a selected microstructure of 4343 aluminum alloy (Figure 3.1a, 1000x magnification). This Alcan standard originates from a continuous casting process, and contains 7.6% silicon, 0.53% iron, 0.98% zinc, and other elements at a low concentration level. Large dark spots are dendrites composed of a number of cells. Space between dendrites is filled by silicon-rich constituents (lighter spots, dispersed; Figures 3.1 a, b). White areas concentrated at the dendrite's borders are rich in other alloying elements. The alloying atoms (for example, iron atoms) form a separate crystal with aluminum atoms (inter metallic compound or constituent), which has the composition Fe_xAl_y. If this specimen surface was etched to reveal grains (macrostructure, which is visible without magnification), the average

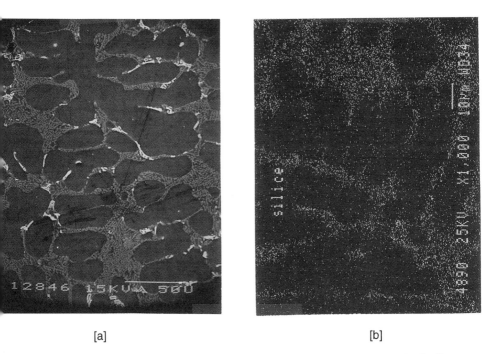

[a]   [b]

Figure 3.1. Microstructure of 4343 aluminum alloy (a) distribution of dendrites. (b) distribution of Si.

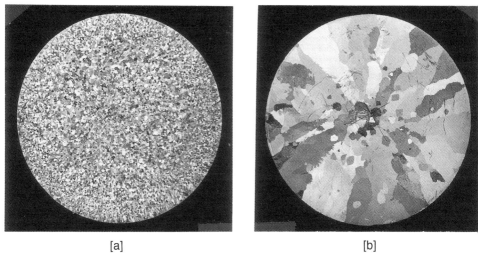

[a]   [b]

Figure 3.2. Distribution of grain size of an aluminum alloy 4343 specimen solidified at different cooling rates. (*a*) Continuous casting process. (*b*) Horizontal disk.

grain size would be around 200–250 μm (Figure 3.2*a*). The grain size also confirms that the alloy was solidified fast. If an alloy solidifies slowly in the mold, dendrites grow and so do the alloy grains. The same 4343 alloy obtained as a typical horizontal disk (10 mm thick, 56 mm in diameter), usually has an average grain size of approximately 1400 μm, and corresponds to the specimen in Figure 3.2*b*. This macrostructure also suggests that the grains have grown from the exterior surface (contact with the mold) towards center.

In the case of cast iron, when most white irons (below carbon = 4.3%) are cooled from the liquid state, the first solid formed at around 1280°C is austenite (with carbon equivalent to about 2%) in the shape of dendrites [44]. The remaining liquid becomes richer in carbon and its solidification temperature decreases. As cooling continues, the dendrites grow until the solidification is completed at 1130°C, which is called the eutectic temperature. The space between dendrites is filled by a eutectic structure comprised of austenite and cementite. Because most constituents tend to segregate at the grain borders, a sample macrostructure composed of large grains is not favorable for the analysis as it promotes heterogeneous structure.

3.7.4 Parameters Affecting the Analytical Surface Finish

The analytical surface finish is affected by a number of parameters [45]. Major parameters can be divided into three groups: surface preparation method, macro and microgeometry of the analytical surface, and subsurface representation of a base material structure. Such data can be conveniently illustrated in the form of a dispersion analysis diagram [45]. The major surface preparation methods are machining (milling, turning), mechanical grinding and polishing, and etching. They are presented in more detail in the sections following.

Subsurface representation of a base material structure involves the constituents matrix ratio (ratio of the constituents' volume in the top analytical surface layer to the constituents' volume in the base material). This ratio is related to the surface preparation conditions, chip formation micromechanics, and micrometallurgy, as well as the physicomechanical properties of the constituents (hardness, brittleness) and the matrix. The damaged layer is composed of the deformed layer (bottom) and fragmented layer (top). In addition, one can also distinguish effects such as: smearing, collapsing of gas and shrinkage cavities, striation, material translation parallel to the machined surface, and distortion of the constituents (change of their shape and location).

The macro and microgeometry of the analytical surface can be described by four major parameters: surface roughness, topography of metal matrix and constituents, surface flatness, grooves and caps orientation (spiral, linear). All of the above mentioned parameters affect the XRF analytical surface finish, which in turn alters the measured intensities of the analytes.

In general, the surface integrity depends a lot on the technique used for specimen surface preparation. Consideration should be given to the deformed layer occurring beneath the fragmented layer, in addition to the surface effects observed directly by microscopic methods. The physical and chemical properties of both the fragmented layer (top surface) and the deformed layer (beneath) differ from the bulk material.

3.7.5 Effective Thickness and Volume

XRF analysis of most solid samples, including metals, involves a relatively thin surface layer of the investigated specimen. Therefore, it should be representative of the bulk sample structure. It is not difficult to imagine that this condition may not always be met. After surface preparation the specimen top layer may not be representative of the bulk sample. For example, some of the constituents may no longer be there. Generally, the degree of surface finish for the measurement of analyte-line radiation is determined by the *effective thickness* of the specimen layer and its internal structure. In the effective layer 99.9% of the measured analyte line intensity is generated. The effective thickness corresponds to the depth from which the specific radiation can emerge and can be calculated for each matrix and tube-analyte radiation using the Togel formula [46]. Employing the Togel formula, the effective thickness can be calculated for selected elements in the aluminum matrix. Assuming that iron and zinc are predominantly excited by the tube continuum radiation, and calcium, phosphorus, silicon, magnesium, and sodium are predominantly excited by scandium K alpha radiation, effective thicknesses range from 2 to 20 μm for the lower atomic numbers to several hundreds of microns for the higher atomic numbers. The most difficult element to analyze in the aluminum matrix is silicon for which the effective thickness is about 3 μm. Assuming that the irradiated surface was 30 mm in diameter, the effective volume (EV) was also calculated (see 2.7.1). For silicon the EV is 2 mm^3, for iron it is about 92 mm^3, and for zinc the EV is larger than 180 mm^3.

By contrast, the EV in the optical emission analysis is approximately 0.5 mm^3 (based on 2 sparks and assuming a 6–8 mm diameter and 14–18 μm depth of the melting zone), and it is constant for most elements.

Similar calculations for the effective thickness of elements in the XRF analysis of iron alloys can also be made. Based on the use of the rhodium tube, it was assumed that light elements such as sulfur and lighter are preferentially excited by the tube L lines, whereas other elements are preferentially excited by the tube continuum. The calculations show that while most elements from vanadium through molybdenum have effective thicknesses between 50–130 μm, the value for nickel is relatively low (about 20 μm), because its characteristic radiation is strongly absorbed by the iron matrix. The data for carbon looks as if it shows an effective thickness of < 1 μm suggesting, perhaps, that it would be impossible to analyze. The effective thickness for this element involves no more than a few atomic layers.

3.7.6   Hardness

Among important characteristics of a metal specimen to consider regarding surface preparation is its *hardness* (resistance to cutting and abrasion) [42]. Hardness favors selection of certain techniques and eliminates others as unsuitable. For instance, in the surface preparation of hard metals, such as cast iron, selection of lathing or machining would result in too slow, too costly (mostly due to accelerated use of expensive cutters), and ineffective procedures. On the other hand, grinding stones or abrasive papers, employed in surface preparation of soft metals such as aluminum or tin, would inevitably lead to the smearing effect and quick clogging of the cutting surface. Therefore, selection of the most suitable surface preparation technique depends on individual characteristics of an alloy or a group of alloys.

A majority of soft metals are simultaneously nonferrous alloys of elements such as, for instance, magnesium, aluminum, copper, tin, or lead. One could assume that surface preparation problems encountered for these alloys are similar. In order to demonstrate the effect of different specimen preparation techniques on the XRF results, an aluminum matrix was chosen.

3.8   ALUMINUM AND ITS ALLOYS

3.8.1   Surface Preparation Techniques

Because silicon has the lowest effective volume of all the common alloying elements in aluminum and low solid solubility, a 4343 aluminum alloy which contains high silicon could be considered as an interesting choice for an XRF study. To minimize doubts regarding homogeneity, an Alcan certified standard (direct chilled cast) was used corresponding to the 4343 aluminum alloy family. The original optical emission standard was scalped to 50 mm in order to fit into the Philips PW 1427/50 sample holder used during the subsequent XRF analysis employing the Sc tube.

Eleven different surface preparation procedures encountered in aluminum preparation for analytical or metallurgical purposes were selected for consideration [45].

1. Lathe turning by a high speed steel cutter under dry conditions (HSL).
2. Milling by a head with 7 boron carbide cutters (BC M).
3. Milling by a head with 7 tungsten carbide cutters (WC M).
4. Standard mechanical polishing involving 120–600 grit grinding papers, followed by the final MgO stage (MgO-S).
5. Standard wet grinding with 120 grit waterproof corundum paper (120 CP).
6. Lathe turning by a diamond tip cutter under dry conditions (DL).
7. Standard mechanical polishing involving 120–600 grit diamond papers, followed by final 1 μm diamond paste stage (MP).
8. Standard mechanical grinding involving 120–600 grit diamond papers (600-DP).
9. Milling by a head with a finishing diamond cutter under isopropyl alcohol lubricated conditions (DM).
10. Diamond micromilling under lubricated conditions (D MIC).
11. Lathe ultraturning by a single crystal diamond toolbit under Varsol lubricated conditions (DU).

The lathe turning is usually done using a conventional device, such as manufactured by South Bend Lathe Inc. Milling is carried out on a variety of milling machines, for example LAGUN, model FUTV-1400 using spindle speed 1500 rpm. Diamond micromilling can be performed on a device such as Polycut-E made by Reichert Jung, or a semiautomatic sample milling machine made by Herzog, model HAF/2. Specimen preparation involving standard metallographical polishing follows that presented in full detail by Samuels [47]. The last technique (diamond ultraturning; DU) is unconventional and is not used on a routine basis by the industry. At the present time the DU technique is being employed in finishing surfaces of laser optics components and computer disks using devices, such as Ultra Precision 2000 from Rank Pneumo Inc. This sophisticated process is the most advanced of all surfacing techniques, but is slow. It results in an extremely thin deformation zone, ideal flatness, gives a surface of a very high reflectivity (mirror-like) without any significant artifacts.

3.8.2 X-ray Fluorescence Tests

The Si $K\alpha$ and the intensities were measured for the eleven surface preparation techniques listed in Figure 3.3. The intensities are represented in Figure 3.3 by vertical bars [45]. As is clearly seen, the lowest Si $K\alpha$ intensity was obtained for the HSL case, whereas the intensity for the DU case was higher by a factor of 2.4. For five cases: MgO-S, 1200CP, DL, MP, and 600DP, a similar Si $K\alpha$ intensity level was observed. It is clear that a variation of the depth of the damaged layer and its integrity (associated with a particular sample preparation technique) in combination with the limited depth of the silicon analytical layer, can explain the results. The volume fraction of the torn out and/or fragmented constituents in the top analytical layer is the predominant factor. Further information is provided by the tube scattered Sc $K\alpha$

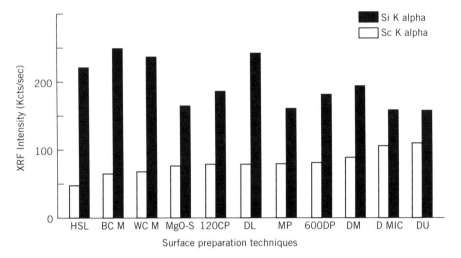

Figure 3.3. Effect of surface preparation in XRF analysis of 4343 aluminum alloy.

intensity. Because the major contribution to this intensity is made by the top surface layer, this intensity is a measure of the degree of surface roughness. It describes the total effect resulting from the metal matrix as well as from pulled out and fragmented constituents. Lower Sc $K\alpha$ intensity is obtained for a smooth surface rather than for a rough one. This loss of intensity is due to the fact that less radiation reaches the detector when it is scattered elastically in a particular direction (which does not obey the Bragg equation) than when it is scattered in all possible directions. The analytical surface roughness and the degree of the surface damage do not affect the measured intensities of all of the elements to the same extent. Silicon yields the most variation in intensity but other elements are slightly affected.

3.8.3 Discussion of Selected Surface Preparation Techniques

The thickness of the damaged layer is in the range of about 20 μm for the HSL technique and about 0.1 μm for the UL technique. Given the depth of the XRF analytical layer of about 3 μm for silicon, it becomes clear how important is the choice of the sample preparation technique. Using the high speed steel cutter lathe machining under dry conditions causes the most damage and destroys surface integrity (Figure 3.4). Detailed scanning electron microscope (SEM) observations confirm the presence of very deep, step-like features in the material, densely distributed on the caps' areas in an irregular manner. Large microvolumes of the material are totally fractured and removed from their original locations. Significant number of the constituents are removed from the surface layer.

Using diamond lathe machining under dry conditions (DL), coarse surface damage is still visible in SEM observations. During the machining, a part of the matrix is still removed together with the constituents. A significant number of the constituents

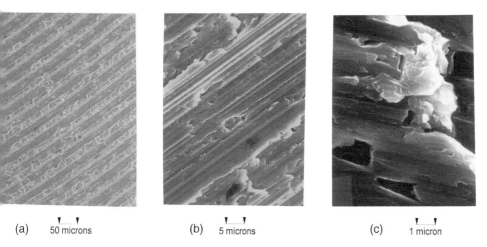

(a)   50 microns   (b)   5 microns   (c)   1 micron

Figure 3.4. AA4343; surface topography after high speed tool turning under dry conditions (*a, b, c*: different magnifications).

is missing on the caps' tops. Some constituents located on the caps' slopes begin to be cut across by the diamond cutter. When using diamond ultra turning (DU), constituents are intersected across and their shape and distribution is clearly visible (Figure 3.5). The surface is flat without any distortion (grooves and caps). Some small constituents may still be torn out from the surface.

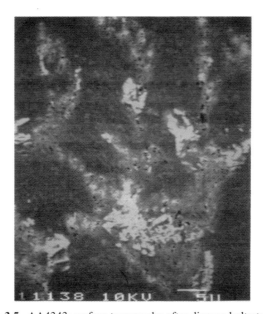

Figure 3.5. AA4343; surface topography after diamond ultraturning.

Diamond ultraturning and diamond micromilling under lubricated conditions cause the least damage and could be considered as reference techniques. However, these techniques may not be used widely in routine industrial practice because of high equipment cost. Moreover, these procedures are relatively slow (up to a few minutes per sample). Instead, diamond ultra milling and turning under lubricated conditions are more practical solutions. The highest quality surface finish is really necessary only when analyzing samples containing high Si (above 1.5%). There is no particular advantage of ultramilling over diamond milling in the case of samples with low Si.

The ideal cut produces a surface which is representative of the bulk sample. The cross section opens the interior structure without disturbing the relative position of the structural features. The constituents and inclusions are intersected and not pulled out, or fragmented. Consequently, the highest elemental intensity is obtained.

The most practical device for the routine XRF analysis of soft matrix alloys is such that can provide a required degree of the analytical surface finish and work automatically without operator assistance (Herzog HAF/2 milling machine). The milling machine should have at least four finger holders to assure good grasp and stability. It should have a cutting blade to cut off approximately a 2 mm slice from the top of the sample (if the sample is thick). Use of a diamond compact blade, instead of a circular saw blade, results in a precisely cut surface, and reduces the amount of machining required at the first stage. The cutting head should be equipped with several cutters of adjustable height and with a diamond tip finish cutter. In order to get the best quality finish, the milling machine should provide passes (cuts) of 0.0025 mm (0.001 inch) or less.

The recommended cutting conditions on a milling device are as follows.

- Final cutter: a single crystal diamond toolbit.
- Lubricant: Varsol.
- Depth of "rough" machining 0.025 mm at 20.3 mm/min.
- Depth of finish: 0.005–0.002 mm at 5.1 mm/min.
- Spindle speed: 1000 rev/min.

3.9   TITANIUM

Most workers sample titanium by taking drillings from the final ingot or slab. In this case, if XRF is to be used for the determination of metallic elements, the sample must first be melted into a button using some sort of remelt furnace. If solid sample is obtained directly from the furnace, one can proceed directly to the preparation steps. A number of factors are common to the preparation of all metals. See section 3.3.10 on steel for a discussion of these factors. Titanium samples can be surfaced with 60 to 600 grit silicon carbide (ASTM E 539) or zircon–alumina. In cases where samples are taken directly from the furnace, they are generally not in a convenient shape to be surfaced on a sander. A lathe is a very efficient tool to prepare these samples and will, if properly used, produce an acceptable surface to be analyzed directly. After a flat surface has been obtained, a final cut should be made using a slow cross feed to obtain a smooth surface.

3.10 CAST IRON

3.10.1 General Discussion

Of the four basic types of cast iron, white iron and grey iron are important in this consideration. White iron and grey iron derive their names from the appearances of their respective fracture surfaces; white iron exhibits a white, crystalline fracture surface, and grey iron exhibits a grey fracture surface [48, 42, 41, 44]. White iron is formed when the carbon in solution in the molten iron does not form graphite on solidification but remains combined with the iron, in the form of massive carbides Fe_3C. The formation of white iron is promoted by (a) low carbon content; (b) low silicon content and presence of carbide-forming alloying elements, such as chromium, molybdenum, vanadium; and (c) rapid solidification, which limits the distance over which carbon atoms can diffuse during solidification. Grey iron is formed when the composition of the iron and the cooling rate at solidification are such that a substantial portion of the carbon content separates out of the liquid to form flakes of graphite. The size and distribution of the graphite flakes strongly influence specimen homogeneity. Grey iron is not adequate for spectral analysis of carbon, and when other elements such as phosphorus and sulfur are analyzed the results have limited precision.

In order to avoid formation of the grey iron, the molten cast iron is cast into a mold that provides rapid solidification. A coin-like specimen is usually obtained, connected by a thinner and narrower feeding section with the head (Figure 3.6). The head always solidifies slower than the feeding section and specimen and forms the grey iron structure. The thickness of the solidified disk is typically 5–8 mm, and

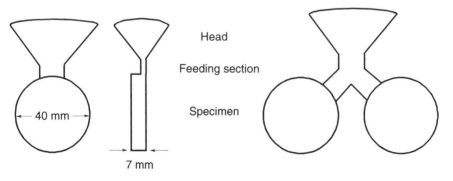

Figure 3.6. Typical shapes of cast iron samples.

the diameter is 30–50 mm. If the specimen thickness exceeds 8 mm, the structural composition might be such that white iron is formed in regions adjacent to the mold walls where the ratio of solidification is high. The grey iron might be formed in the inside regions of the disk, where freezing is slower. In other words, only the outermost layer (of approximately 1 mm) of the disks would be white. A layer thicker than 1 mm is quite often removed during the machining stage, particularly in case of visible surface imperfections. This may easily lead to analysis of grey instead of white iron.

3.10.2   Specimen Preparation Procedure

Equipment   Useful equipment includes a ladle bowl of approximately 8 kg iron capacity, a specimen mold, made of steel, a bench vice: any 4 in. model, or larger, hammer: medium size, specimen grinder equipped with a grinding wheel, bench top polishing machine (if carbon is to be analyzed), and a desiccator: with silica gel or activated alumina drying agent.

Sampling   Carry out sampling in a manner that (1) provides a specimen uncontaminated by slag, and (2) ensures safety of the operator. Take a specimen from a hot melt (below the slag layer) with a ladle bowl of approximately 1.1 l capacity or by an immersion device. Cast the hot metal in the mold in which it solidifies as a coin-like specimen, 40 mm in diameter and 7 mm thick. The disk sample must be sound and free from graphite. Secure the disk sample in the bench vice, and break off the head of the sample with the hammer.

Routine Specimen Preparation   Although the preparation procedures may vary considerably from foundry to foundry, it is general practice to employ wet-grinding procedures for the laboratory preparation of specimen prior to the analysis. In routine cast iron preparation for either XRF (no carbon) or optical emission spectroscopy analysis, the coin sample is usually ground with a 37 grit, Al_2O_3 stone wheel on a horizontal grinder. In this device the sample is fixed on a magnetic chuck with a slot in its center.

With the grinder, first remove the casting skin, using cuts of up to 25 μm in depth. Supply coolant. Then reduce the depth of cut to 13 μm or less. After several strokes of the grinding wheel, stop the machine and cut off the coolant supply. Visually check whether the sample surface is free from major cracks or holes. Then, start the machine again to execute the final 2–3 dry strokes (without the coolant) of the grinding wheel. In the dry grinding, penetration of the contaminated coolant into the specimen surface grooves is prevented. Inspect the sample surface again to make sure it is free from major defects such as large pinholes of diameter exceeding 3 mm. Cracks are tolerated, provided that the specimen rotates during analysis. If the sample surface is unsuitable, further metal should be removed or the specimen should be reversed and prepared on the other side. Shortly before the analysis, clean the analytical surface with a paper tissue saturated with alcohol. Handle the specimen by the edges to avoid fingerprints. Under no circumstances should a damp or wet specimen be used. If the specimen is to be analyzed at a later time, store it in the desiccator. Prepare the reference materials the same way as the routine specimens.

Special Problems and Hints   In general, the reference materials should have the same degree of homogeneity, considering the size, composition, and distribution of constituents, and other microstructural features. Besides homogeneity, perhaps the next most important consideration is to be sure that their complete composition is known with an accuracy which is the same as or better than that which is desired in analysis. Selected cast iron standards available commercially are as follows.

- BAS standards SS656-SS660 and SS661-SS665 (sold by Brammer Standard Company, Inc., Houston, TX 77069).
- H.R.T. standards OF series, (sold by H.R.T. Labortechnik GMBH, Dieselstr. 6, D-8011 Kirchheim bei München).
- XHPC (sold by MBH Analytical Ltd., Holland House, Queens Road, Barnet, Herts EN5 4DJ, England).

In the wet grinding procedures cooling water without additives is used in some laboratories. However, use of lubricants helps the grinding process in that the grinding wheel lasts longer. Even a solution of soap and water can be used. Oil-based coolants are unsuitable because they produce an oily swarf and a greasy specimen surface.

The grinding wheel is satisfactory in preparing cast iron for both XRF and OES analysis. Further preparation involving fine polishing is only necessary if carbon is to be analyzed by XRF. Such preparation would be to eliminate the so-called surface roughness effect which affects C analysis significantly. The characteristic radiation of C originates from the top surface layer which is less than 1 μm thick. Unfortunately, this additional operation takes minutes, which prevents XRF from being used where time is a factor. The analysis of elements becomes more difficult as atomic number decreases.

3.11   STEEL

In steel, as in all materials, there are a number of factors which are very important in the preparation of the analytical surface. These factors must be addressed, even before the instrument is calibrated, because the reference materials used to calibrate the instrument must be surfaced in the same manner as the samples.

Contamination from the grinding medium is a major contributor to errors. It must be determined, before the calibration is started, which elements are of importance and verified that the surface preparation procedure will not contaminate the surface with those elements. It may be, for example, that if both aluminum and silicon are important, separate surface preparations may be necessary for them, one which does not contaminate with silicon (alumina or zircon–alumina) and one which does not contaminate with aluminum (silicon carbide). If the determination of calcium in steel is required, it is absolutely essential that the grinding material be free of calcium.

This requires the analysis of the grinding materials because calcium is usually not very homogeneous in steel so the determination by evaluating the surface is difficult. For the analysis of steels, 60 to 240 grit alumina, zircon–alumina, or silicon carbide have been shown to be adequate. For the determination of calcium, calcium free 240 grit silicon carbide has been shown to produce the best results, (ASTM E1085). Some workers believe that a metallurgical polish is necessary for all X-ray analyses. It has been shown that, except perhaps for the most critical analyses, this polish does not improve the results, and a metallurgical polish will not guarantee that no contamination has occurred. Residual material can remain from intermediate steps. The only way to assure that this residue has not occurred is to repolish and reanalyze and compare the first and second results. Contamination can occur from other sources than the grinding materials. Calcium can be picked up from various paper products that may be used to wipe the sample. Other contaminants can be picked up from the environment, as well. It is important to assure that the surfaces are kept clean after preparation.

Proper surface preparation equipment is also necessary. While belt sanders are commonly used and can produce adequate results, they tend to wear in the center of the platen, and this wear will produce a surface on the specimen that is not flat. A disk sander is recommended because the grinding paper is fastened to the surface of the grinder, therefore no wear occurs. It is also recommended that small (therefore, inexpensive) disks be used and a new disk be used for each sample for the final surfacing. This change is to prevent sample-to-sample contamination. Also, when a disk is used, it wears and the effective grit size changes. The use of a new disk for each sample ensures that the same grit size is being used on each sample. Only after the appropriate surface preparation procedure has been determined, along with all of the other instrument parameters, is one ready to calibrate the instrument.

In order to maintain reliable analyses, consistency in surface preparation, as well as the other phases of the analytical procedure is essential. Once a surface procedure has been selected, it must be used for all reference materials and subsequent samples.

3.11.1   Evaluation of Grinding Materials

Virtually all surfacing techniques will contaminate the surface with some element(s). In order to provide accurate analyses, this contamination must be kept to a minimum or reduced to elements which are not of concern. There are two basic techniques to minimize contamination. The first is to analyze the grinding material directly, and the second is to observe the results of the surface preparation technique.

Evaluation by Direct Analysis   Samples of the grinding material were analyzed by energy dispersive X-ray spectroscopy. The major constituents are those which are most likely to contaminate the specimen. Table 3.5 shows examples of these analyses for four common types of grinding belts. The first column gives data for an alumina grinding belt and as expected, the major constituent is aluminum. However, sodium is also a major constituent, and there are traces of titanium, calcium, and iron. The potential for contaminating with aluminum and sodium is high, so this belt

TABLE 3.5. Results of XRF Analyses of Different Types of Grinding Belts

Element	Alumina	Type of Belt Silicon Carbide	Zircon/ Alumina	Silicon Carbide (240 grit)
Fluorine	minor	—	trace	—
Sodium	major	trace	minor	—
Magnesium	minor	trace	—	—
Aluminum	major	—	major	—
Silicon	minor	major	—	major
Zirconium	trace	—	minor	—
Potassium	—	—	major	—
Sulfur	minor	—	—	trace
Calcium	minor	major	major	—
Titanium	minor	trace	major	—
Iron	trace	—	—	—
Barium	—	—	—	trace

should be avoided if either of those elements are of interest. The second column gives data from a silicon carbide belt and reveals major concentrations of silicon and calcium. The third column gives data from a zirconium stabilized alumina belt showing major concentrations of aluminum, potassium, calcium, titanium, and lesser levels of zirconium and sodium. Finally, the fourth and last column gives data from a 240 grit silicon carbide. This material is free of calcium, and it is necessary to use this if one is interested in determining calcium in steel.

If this technique is employed, it is necessary to analyze the grinding materials which are being considered for use because different manufacturers of grinding belts or disks may use different sources of abrasives.

Evaluation by Surface Analysis   Evaluation by surface analysis does utilize the material of interest to the analyst and, therefore, may give better information. This technique is also more expedient because it does not require the use of additional equipment. Materials are surfaced using different grinding materials and analyzed in a routine manner. The results are tabulated and contaminations are easily identified. Tables 3.6 and 3.7 give examples of such an evaluation. For the sake of brevity, only those elements which promote contamination are shown, but in an actual evaluation all elements of interest should be tabulated.

These tables show only summaries, but in practice, this procedure should be run two or three times for each sample, going through the range of grinding materials serially. From this it will be seen that, when contamination occurs, it is very inconsistent. As can be seen, there is a certain low and repeatable level for each element which is the contamination free surface. As would be expected, silicon carbide contaminates surfaces with silicon, alumina with aluminum, and zircon-alumina with aluminum and zirconium. Metal-bonded diamond wheels contaminated the surfaces with alu-

TABLE 3.6. Evaluation of Surface Contamination–Test 1

Grinding Material	Sample 1			
	Al	Si	Zr	Ni
Diamond paste	0.016	0.234	—	0.292
Resin bonded diamond	0.025	0.245	—	0.294
Metal bonded diamond	0.070	0.290	—	0.316
SiC 80 grit	0.016	0.444	—	0.297
SiC 120 grit	0.016	0.385	—	0.301
Alumina 60 grit	0.070	0.234	—	0.296
Zircon-alumina 60 grit	0.085	0.234	0.003	0.292
Zircon-alumina 120 grit	0.061	0.234	0.006	0.292

TABLE 3.7. Evaluation of Surface Contamination–Test 2

Grinding Material	Sample 2			
	Al	Si	Zr	Ni
Diamond paste	0.019	0.285	0.051	0.256
Resin bonded diamond	0.065	0.288	0.050	0.258
Metal bonded diamond	0.081	0.315	0.051	0.290
SiC 80 grit	0.019	0.364	0.050	0.259
SiC 120 grit	0.020	0.587	0.052	0.260
Alumina 60 grit	0.028	0.288	0.051	0.256
Zircon-alumina 60 grit	0.121	0.284	0.066	0.260
Zircon-alumina 120 grit	0.053	0.285	0.053	0.260

minum, silicon, and nickel. This contamination shows the importance of tabulating all elements, because sometimes unexpected results appear.

It is very important to note that the level of contamination is far from consistent from sample to sample, as well as from run to run on the same sample. For this reason, it is not appropriate to assume that one can surface the samples and the reference materials with the same grit material and expect that the contamination effects will be negated. One must provide a contamination free surface (from the elements of interest) to the instrument, even if this means surfacing the specimen more than once to complete the analysis. For example, surface the specimen with silicon carbide to run most elements and resurface with alumina or zircon-alumina to determine silicon. This procedure is very important if one is to obtain the most accurate analyses.

3.11.2   Special Problems and Hints

Especially in the analysis of metals, each surface preparation technique inevitably leaves a surface that differs from that left by any other technique. The nature of that surface greatly affects the analysis and may be a source of serious error.

Depending on the type of metal matrix and choice of elements, some aspects in specimen preparation may be more important than others. In general, the XRF method is sensitive to the analytical surface integrity; the extent of that sensitivity depends on the atomic number of the element to be analyzed. For example, in XRF analysis of aluminum alloys, silicon is the most critical element, and magnesium the lightest that can be practically determined (sodium cannot be determined with required accuracy and precision). Fine surface finish which is necessary in analysis of the light or low solubility elements, may not be required for other elements. Clearly, the lower the element atomic number and its solubility, the finer the surface finish required. For example, in the analysis of steel a relatively coarse specimen surface composed of caps and deep grooves is adequate for most elements. But in the analysis of carbon (atomic number 6) the specimen surface must be carefully polished for the best accuracy.

One of the important effects that occurs in the interaction of the incident X-rays and the sample surface is radiation scattering. The magnitude of the surface scattering power is related to the degree of surface irregularity (*the roughness effect*), flatness, and the effective atomic number of the matrix. Both elastically (Rayleigh) and nonelastically (Compton), scattered radiation corresponding to the most intensive lines of the tube anode material can be measured. The roughness effect also has serious consequences on the secondary (characteristic) radiation, and is explained in Figure 3.7 (specimen surface configuration represented by cups and grooves in the vertical cross-section) [49]. Because the path-length varies with the degree of shielding, the required surface finish will depend both upon the wavelength of the measured radiation and the composition of the sample. Thus, XRF can be used to measure the chemical composition of the metal and to simultaneously determine the surface roughness effect, a capability other analytical methods do not possess.

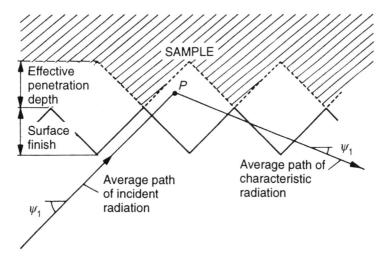

Figure 3.7. Effect of shielding due to surface finish.

Surface preparation techniques involving etching should be avoided, unless dissolution rate of the specimen matrix and constituents is similar. The purpose of etching is to make visible the many structural characteristics of the metal or alloy. The process is such that the various parts of the surface microstructure are clearly differentiated, which is not the analytical objective.

The specimen grain size depends on the mold selection and should be small to assure its homogeneity. Generally, molds in which metal solidifies quickly provide fine specimen macrostructure, as opposed to molds in which metal solidifies slowly. It is possible to diminish the specimen grain size by adding a grain refiner to the mold cavity before pouring the liquid metal, but this procedure has not found a widespread application.

In any surface preparation technique employed, special care must be taken to ensure that no additional surface contamination is introduced. This care is particularly important in the case of a relatively soft metal such as aluminum, where particles of the grinding agent may penetrate to the sample surface during surface finishing. All traces of lubricant, cutting fluid, finger marks and so on must be removed by cleaning with isopropyl alcohol prior to examination. Surface composed of cups and grooves is much more difficult to clean before analysis than a fine surface. In any case it should always be cleaned before analysis with a paper tissue saturated with alcohol. Because the paper itself can introduce contamination, it is good policy to finish the paper wiping process by flushing the sample surface with alcohol, then blow with clean compressed air. The surface of several metals is affected by corrosion which progresses with time. Therefore, the best results are always obtained when analyzing a specimen without much delay after the surface preparation.

3.12 ALUMINUM-BEARING MATERIALS

3.12.1 General Discussion of the Problem

Aluminum-bearing materials constitute a group of raw materials vital for the aluminum industry [50]–[54]. They are also called Bayer process materials, or bauxite related. Bauxite, a mixture of hydrated aluminum oxide minerals, is the primary raw material used in alumina manufacture. Other materials of concern, such as clay, red mud, and sand, are waste products. Clay is obtained during bauxite washing, whereas red mud and sand originate in the process of bauxite digestion. Concentration ranges of the eight major constituents in the above mentioned materials are given in Table 3.8. The constituents are Al_2O_3, SiO_2, Fe_2O_3, TiO_2, CaO, Na_2O, and F. Other important constituents determined separately are combined water and organic matter, represented by loss of mass (LOM).

As seen in Table 3.8 the constituents occur in very wide concentration ranges. Some constituents like Na_2O and fluorine are practically not present in the bauxite and clay matrix, but they are very important in the mud and sand matrix. Additional constituents that are of interest are represented by oxides of elements such as phosphorus, chromium, zirconium, zinc, manganese, vanadium, potassium, sul-

TABLE 3.8. Percent Concentration Ranges of Eight Major Constituents in Aluminum-bearing Materials

Constituent	Bauxite	Clay	Red Mud and Sand
Al_2O_3	30–60	20–60	5–25
SiO_2	0.5–40	20–60	1–35
Fe_2O_3	0.5–40	0.5–15	30–60
TiO_2	1–7	0–5	1–20
CaO	0–2	0–5	0–20
Na_2O	—	0–2	0–15
F	—	—	0–10
LOM	15–32	10–20	20–35

fur, magnesium, strontium and gallium. Altogether, there could be as many as 19 constituents.

3.12.2 Specimen Preparation Procedure

In the past, most specimen preparation techniques for quick, routine X-ray analysis of the aluminum-bearing materials were based on briquetting procedures. Bauxite grinding (together with a binder) followed by briquetting could still be used for exploration purposes, if samples representing one mineralogical deposit are to be analyzed. In this case, typically a 16.0 g sample portion is ground with a 4.0 g binder in a ring-and-puck grinder for 2 min. The finely ground material is pressed into a hydraulic press using a forming pressure of 40 t for 5 s to yield a briquette 40 mm in diameter.

However, fusion techniques are generally required in the analysis of mineralogically heterogeneous deposits for which uniform grinding is almost impossible. Fusion involving dilution of powdered bauxite with a flux eliminates virtually all inhomogeneity with respect to mineralogical composition, density and particle size, and gives a ready-to-analyze sample in a relatively short time. In the method which has been used over the years in a few Alcan laboratories the preparation steps to execute are as follows [52, 54]: Grind the sample to pass through a 100 mesh sieve, and dry it at 105°C overnight. Then, take a representative portion of 5 g and determine the LOI value by calcining it at 500°C for 1 hr and subsequently at 1000°C for 2 hr. Into a 25 ml plastic vial weigh 1 g of the calcined sample, and 6 g of the analytical flux composed of 50% $Li_2B_4O_7$ and 50% $LiBO_2$. Homogenize the flux and the sample before fusion by vigorously shaking the vial for a few seconds. Transfer the content of the vial to a Pt/Au crucible, add 3 drops of the LiBr solution (prepared by dissolving 30.0 g LiBr in 100 ml distilled water), and fuse using a fluxer. Employ the fluxer parameters recommended by the manufacturer.

3.12.3 Special Problems and Hints

Both specimen and reference materials must be prepared the same way. Because the specimen is calcined before the XRF analysis, the element content in the natural matrix is calculated based on the LOI value. Certain red muds and sands very rich in Fe_2O_3 require sample-to-flux ratio higher than that used for bauxites. Usually, the ratio 1:8 is sufficient to produce a homogeneous bead which would not contain undissolved particles, and would not crack. The LiBr nonwetting agent used in the fusion procedure contains bromine which causes a spectral line overlap. The Br $L\alpha$ at 8.375 Å overlaps Al $K\alpha$ at 8.339 Å. In order to minimize the effect of the overlap it is advised to use a pipette for adding the solution with the nonwetting agent in the crucible. Thus, the contribution by Br $L\alpha$ is constant and small compared with the size of the aluminum peak. It is also possible to use LiI as the nonwetting agent, but I $L\alpha$ interferes with Mn $K\alpha$. In the case of the briquetting specimen preparation technique the use of a binder is mandatory. Neither bauxite nor red mud have a self-binding property, so ground powder cannot be pressed into a mechanically stable briquette without it, even if an aluminum cup is used as a support.

3.13 ALUMINA

3.13.1 General Discussion

Alumina is produced in the Bayer process in which crushed bauxite is digested with hot sodium hydroxide solution. One of the products is aluminum trihydrate which is then calcined [55]. Depending on the degree of calcination, metallurgical or ceramic grade aluminas are produced. They differ with respect to the α-alumina content, grindability, particle shape and size, impurity content, etc. [56, 57]. Two basic specimen preparation techniques are used for elemental analysis of aluminas by XRF [58]. One is based on briquetted powders, another involves fused samples.

3.13.2 Specimen Preparation Procedure for Aluminas

Briquetting Specimen Preparation   In the method using briquetted powders, grind a 20.0 g sample and 5.0 g of an organic binder together in a rotary puck mill for 1 min. The binder, called SX, is composed of nine parts (by mass) of styrene copolymer and one part of wax [59]. The components are available in powder form consisting of grains of 10 μm size and have to be well mixed prior to analysis.

The briquetting method provides very reliable results for metallurgical grade aluminas [58]. However, ceramic grade aluminas require a longer grinding time and a separate analytical program. Moreover, the grain size is already small (in the range 1–10 μm), and further grinding is not effective. The longer grinding time of abrasive material such as ceramic alumina causes its inevitable contamination, especially in the analysis of sodium, where the performance of the briquetting method is demonstrably better for metallurgical aluminas than for ceramic aluminas. Sodium, which mostly appears as $NaAl_{11}O_{17}$, is predominantly distributed inside grains, so that grain morphology affects the results. To eliminate these inconveniences, get rid of the particle size distribution effect, and eliminate pitfalls with calibration standards. A fusion sample preparation technique should be used.

3.13.3 Fusion Specimen Preparation for Aluminas

Weigh a 4.0 g alumina sample and 9.0 g of flux in a 50 ml plastic vial. The 4.0 g sample mass ensures a representative sampling of trace elements. The flux is composed of 90% and 10% LiF. The addition of LiF helps reduce the fusion time and temperature, as opposed to pure $Li_2B_4O_7$. Flux marketed by Claisse Corporation (ultrapure) was found to be the most adequate with respect to the impurities content [58]. Homogenize the flux and the sample in a 50 ml plastic vial, then transfer the vial content to a Pt/Au crucible, add a nonwetting agent, and fuse using an automatic fluxer. Cast the molten mass in a mold of 40 mm diameter.

3.13.4 Special Problems and Hints

The sample must be dried in the drying oven at $110°C$ for 3 hr. Alumina hydrate must be first calcined at $500°$ and then at $1000°C$ for 1 hr. The binder SX used in the briquetting technique is not hygroscopic and has good self-binding properties. The background arising from the pure binder pressed into a pellet is very low for all elements except sodium. In the analysis of alumina containing sodium at a low concentration level use another binder of higher purity, such as polyethylene (more expensive than the SX). The binder performs three major functions. It serves as a lubricant, reducing contamination and facilitating cleaning of the components of the rotary mill; it isolates very fine from large grains and thus helps the grinding action; finally, it acts as a binder in the briquetting operation. The fusion technique can be applied to all types of aluminas because grain size, shape, and hardness are no longer factors to overcome. In general, the fusion technique compared with briquetting offers wider calibration ranges, "absolute" calibration based on fewer synthetic standards [58], analysis of several additional elements with the help of synthetic standards, and better estimated accuracy for most elements, except sulfur. Unfortunately, an uncontrollable amount of sulfur is always lost during fusion due to volatilization. In order to retain sulfur in the solid solution, the flux composition needs to be modified. A strong oxidizing agent such as $LiNO_3$ added to the flux should convert sulfur into SO_3, which is soluble, as compared to volatile SO_2. However, no study was carried out to determine the required amount of $LiNO_3$. To accommodate the quantities of materials as specified, the crucible must be at least 30 mm in volume. The glass disk of 40 mm diameter is of good mechanical strength, and results in a 20–30% intensity gain in comparison with a 30 mm disk, for most of the elements concerned. In contrast to the fusion technique, the briquetting technique is better for sulfur and sodium and is more economical.

3.14 BATH ELECTROLYTE

3.14.1 General Discussion of the Problem

Aluminum metal is produced from pure alumina (Al_2O_3) by an electrolytic process. In order to bring the melting point below $1000°C$, the alumina is dissolved in fused cryolite (Na_3AlF_6). To reduce the melting point further and thus minimize power consumption and fluoride emission during the electrolytic process, the bath usually contains 5–8% CaF_2, and recently magnesium and lithium salts have been added to the melt in some smelters. Generally, two distinct approaches have been used in the industry to prepare samples for bath analysis [60]–[64]. One involves XRD, another is based on the XRF analysis.

3.14.2 Specimen Preparation Procedure for Bath Samples

Using bath tongs, obtain a bath sample in the form of a solid cone, cylinder, or ball of approximately 50 g, and crush it before grinding. Crush with a hammer, or use a jaw crusher with hardened steel wear-resistant jaws, to obtain coarse material (-5 mesh). Grind the crushed material further. If samples are destined for XRD analysis, grind with a disk pulverizer, with manganese steel grinding plates. Then produce either a regular cavity slide, or prepare the specimen by briquetting the ground material at 40 t/in². for 5 s in a tapered aluminum cup. The aluminum cups are usually 40 mm in diameter. For high sample throughput, prepare briquettes using a fully automatic sample preparation equipment with a jaw crusher, grinding mill and pelletizing press, such as HBMP from Herzog [65]. If samples are destined for XRF analysis, select one of the two major methods of preparing bath electrolyte: pressed pellets or fusion. This procedure can be done in various steps using a variety of manual or semiautomatic equipment. The briquetting sample preparation technique was optimized as follows [64]: 13.0 g crushed sample and 3.0 g SX binder [59] (composed of nine parts styrene copolymer and one part wax) are ground together in a rotary mill (puck and ring) for 3 min. Subsequently, one or two 7.0 g briquettes, 40 mm in diameter each are formed in a hydraulic press from the finely ground material using a forming pressure of about 40 t/in². for 5 s.

The fusion technique is used to determine concentrations of major and minor elements. The unique feature of the fusion technique is that a calibration can be carried out on synthetic standards. Thus, an "absolute" calibration, independent of natural standards, is possible. Additional elements not covered by the natural bath standards can be determined too. The fusion specimen preparation technique developed for bath electrolyte [64] is as follows: 1.0 g of the dried sample is mixed with 6.0 g of the analytical flux composed of 0.5 g $LiNO_3$ and 5.5 g mixture of 50% $LiBO_2$ and 50% $Li_2B_4O_7$. The powder is transferred into a platinum/gold crucible and three drops of a 25% LiBr solution are added to the crucible before fusion.

3.14.3 Special Problems and Hints

In the case of the briquetting specimen preparation technique, the use of binder is mandatory. Bath electrolyte does not have a self-binding property, so the ground powder cannot be pressed in a mechanically stable briquette, even if an aluminum cup is used as a support. Also, a grinding time of approximately 3 min is necessary for a successful determination of oxygen. A so-called free alumina determination by XRF is based on the oxygen measurement. If the oxygen measurement is not required, the grinding time of 1 min is adequate. In the bath electrolyte fusion technique, the $LiNO_3$ is added to the flux to oxidize carbon present in the matrix and prevent the glass beads from cracking during solidification.

3.15 SURFACE WATERS

3.15.1 Use of Ion-exchange Resins

By far the most popular, and also the most successful, techniques for the analysis of surface waters have been those based on the use of ion exchange resins. The major advantage of most ion exchange methods is that the functional group is immobilized on a solid substrate, providing the potential to batch extract ions from solution. The sample itself can be either the actual exchanged resin or a separate sample containing the appropriate ions re-eluted from the resin. The success of the method depends to a large extent on the recovery efficiency of the resin, which in turn is determined by the affinity of the ion exchange material for the ions in question and the stability of complexes present in solution. Preconcentration factors of up to 4×10^4 can be achieved from suitable ion exchange material using around 100 mg of resin [66]. One of the most useful ion exchange resins is Chelex-100. This resin contains iminodiacetic functional groups and acts chemically very similarly to EDTA (ethylenediamine tetra-acetic acid). It shows a good recovery efficiency for a wide range of ions, and its chemistry can be predicted from prior experiments with EDTA.

On the negative side, however, it is not very successful in the separation of ions from solution that are high in iron and calcium (e.g., seawater), elements which occupy the available sites to the exclusion of the trace element ions. Many other ion exchange resins have also been used; for example, Dowex I X8 has been used for the determination of cobalt in iron rich materials [67] and Wolfatit RO resin for the separation of trace amounts of gold in cyanide liquors [68]. It is sometimes found convenient to use filter paper impregnated with ion exchange resin. This technique has been found especially useful in the extraction of low concentrations of rare earth elements [69]. The exchange resin can also be employed as a membrane through which the solution to be analyzed is passed [70]. This method is a particularly convenient means of separation for the X-ray fluorescence method, because the paper can be mounted directly in the spectrometer for analysis following the separation process. As an example of the use of Chelex-100, 200 ml of water were passed through two Chelex-100 membranes for a period of about 20 min. The enrichment factors obtained were in excess of 1000, allowing the separation of potassium, calcium, manganese, cobalt, nickel, copper, zinc, rubidium, and strontium as chlorides or nitrates at element concentrations in the range 10 ppm to 10 ppb [71]. Similar techniques have been used for the determination of uranium [72], of barium [73], and of selenium [74] in groundwater.

3.15.2 Coprecipitation Methods for Surface Waters

Another method which has been employed with some success to preconcentrate elements in natural water samples is the use of coprecipitation. This method offers a relatively simple method with the advantage of giving a fairly uniform deposit which can be easily collected. One of the earlier applications of this technique involved the use of iron hydroxide as a coprecipitant for the determination of iron, zinc, and lead in surface waters [75]. One of the more popular coprecipitants in use today is ammonium pyrrolidine dithiocarbamate (APDC). In the application of this method to the analysis of natural waters, detection limits in the range 0.4 to 1.2 ppb have been claimed for the elements vanadium, zinc, arsenic, mercury, and lead [76]. Other coprecipitants have been described, including the use of iron dibenzyl dithiocarbamate for the determination of uranium at the ppb level in natural waters, zirconium dioxide for the determination of As in river water [77], and polyvinylpyrrolidone-thionalide for the determination of iron, copper, zinc, selenium, cadmium, tellurium, mercury, and lead in waste and natural water samples. The use of this latter reagent is particularly useful where large concentrations of magnesium and calcium may be present.

3.15.3 Surface Adsorption

A third technique which is somewhat less popular but still useful is application of surface adsorption reagents. Activated carbon is generally used as the carrier after conversion of the metal ion(s) of interest to a suitable form. For example, iodide in groundwater has been determined by this method after conversion to silver iodide [78]. Neutral 8-hydroxyquinoline complexes have been employed in which again the oxine complex is collected on activated carbon [79]. Activated carbon is particularly useful where PIXE is being employed as the analysis technique because the resultant sample is ideal for direct presentation to the instrument [80].

3.15.4 Other Concentration Methods

Other preconcentration methods which have been employed in this area include electrodeposition [81], precipitation chromatography [82], liquid-liquid extraction, immobilized reagents [83], plus a variety of other techniques well known to the analytical chemist.

3.16 OIL ANALYSES

3.16.1 General Discussion of the Problem

Provided that the liquid specimen to be analyzed is a single phase and relatively involatile, it represents an ideal form for presentation to the X-ray spectrometer. Where the vapor pressure of the liquid to be analyzed is too great to allow the use of a vacuum path, a special sample cup or a helium path instrument must be used for the measurement of the longer wavelengths (e.g., >3 Å). The liquid phase is particularly convenient because it offers a very simple means for the preparation of standards. Because most matrix interferences can be successfully overcome by taking the sample into liquid solution, this technique represents the final stage of many sample handling methods. Although the majority of matrix interferences can be removed by the solution technique, the process of dealing with a liquid rather than a solid can itself present special problems which in certain instances can limit the usefulness of the technique. For example, the taking of a substance into solution inevitably means dilution and this dilution, combined with the need for a support window in the sample cell (unless special optics are employed), plus the extra background arising from scatter by the low atomic number matrix, invariably leads to a loss of sensitivity, particularly for the longer wavelengths.

3.16.2 Special Problems and Hints for Oil Analyses

Problems can also arise from variations in the thickness and/or composition of the specimen support film. The most commonly used type of film is 6 μm polyethylene terephthalate, and the attenuation effect by this film will considerably reduce the intensities of the longer wavelengths. It is possible to reduce the thickness of polyethylene terephthalate film down to about 3 μm before pinhole formation becomes a problem and use of this thinner film somewhat reduces the total attenuation. One apparently insurmountable problem with polyethylene terephthalate films is the presence of large inclusions of high absorption, consisting of, among other things, antimony salts from the copolymer catalyst used in the manufacturing process. Although the total concentration of impurity is relatively small (typical ash content 0.2 wt%) the average atomic number of the inclusions is high and anomalous results can occur when measuring wavelengths greater than about 6 Å. An alternative support film is 12 μm oriented polypropylene which is attractive because its ash content is low (usually less than 0.005 wt%) and impurities are relatively evenly distributed. Also, because its average atomic number is less than one-half that of polyethylene terephthalate, its absorption coefficient is lower by a factor of about five. Thus, even allowing for the need for a 12 μm film thickness the attenuation by polypropylene film is a factor of two less than 6 μm polyethylene terephthalate. The greater film thickness is required to overcome the tendency of the oriented polypropylene to sag under the weight of the sample.

Care should be taken to select only fully oriented polypropylene for cell windows because even partially oriented material may sag. Nonoriented polypropylene (such as is used for making thin flow counter windows) must be avoided at all costs! The process of taking a sample into solution can be rather tedious and difficulties sometimes arise where a substance tends to precipitate during analysis. This difficulty may be due to the limited solubility of the compound or to the photo-chemical action of the X-rays causing decomposition. In addition systematic variations in intensity can frequently be traced to the formation of air bubbles on the cell windows following the local heating of the sample. Despite these problems, the liquid solution technique represents a very versatile method of sample handling in that it can remove nearly all matrix effects to the extent that accuracies obtainable with the solution method approach very closely the ultimate precision of any particular X-ray spectrometer. Some care must be taken to match the mass absorption coefficient and density of reference material and sample very closely, and there is also some evidence to suggest that variations in the pH of a solution may cause significant systematic errors. Density variations can be particularly troublesome, especially in the analysis of hydrocarbon solutions. For example when determining heavy metals such as lead, differences of about 15 to 20 % are found in the relative intensities of identical concentrations of organo-metallic lead in benzene and cyclohexane. Fortunately, it is often possible to use coherently or incoherently scattered tube lines as an internal standard to correct for both density and mass attenuation coefficient variations [84]. Where liquids of more than one phase are to be analyzed, prior separation must be made before specimens are presented to the spectrometer. A typical example is the analysis of used lubricating oils where, in addition to the hydrocarbon matrix containing unused additives, decomposition product and fine metal particles may also be present. In this instance, separation can be brought about by dilution with a light solvent followed by centrifuging out the solid matter. The centrifugiate can then be treated as a heterogeneous solid, and the supernatant liquid treated as a homogeneous solution. If necessary, the light solvent can be distilled off.

3.17 COAL DERIVATIVES—PITCH AND ASPHALTS

3.17.1 General Discussion of the Problem

Pitches prepared by coal or petroleum tar distillation are used as binders in a variety of products. For example, the aluminum industry constitutes a major market for pitches as they are used in the manufacture of large scale anodes and cathodes needed in the production of aluminum [85]. A high sulfur content gives rise to environmentally objectionable SO_2, and metal oxides may affect the quality of the final metallic product. Precise determination of major and minor elements in pitch is quite difficult by conventional analytical techniques [86]. Low temperature ashing and subsequent digestion of the ash in acids or its fusion with borates, followed by spectroscopic determination, provides good results only for certain elements. However, simple calcination will tend to lead to loss of sulfur as SO_2 and of lead and zinc as the volatile elements. Neutron activation analysis offers reliable results on most trace elements but requires long cooling periods between elemental determinations and the cost of the analysis is also relatively high. Therefore, the rapid and reliable XRF analysis of pitch and other bituminous materials is of considerable interest.

3.17.2 Selection of a Representative Sample of Pitch or Asphalt

Most pitches and asphalts are stored in manufacturing plants in hot form, i.e., tanks heated above 150°C. These could be sampled directly into flat disposable aluminum dishes or pans (similar to those used for roasting chicken or baking cakes) by simply opening a valve in the tank. Solid pitches are usually supplied in the form of flakes, rods (pencil pitch), or chunks [86] packaged in bags. Asphalts are transported in liquid form and also as solids in steel barrels or buckets.

Sampling Hot Pitch or Asphalt   Place a large (approximately 8 × 11 in.) aluminum pan on a thick plywood board. Fill with a sample of hot pitch or asphalt from a tank (by simply opening a tank tap). Cover with aluminum foil or a sheet of stainless steel to protect from dust contamination and cool to room temperature. Bring the solidified pitch to the laboratory and empty the pan on a sheet of aluminum foil. The pitch or asphalt "cake" can be easily removed by gently tapping the pan surface. Wrap carefully into another larger piece of foil and break into smaller (ca. 1 in.) pieces with a plastic hammer. Store about 200 g of the crushed sample in a clean glass or steel can for further use.

WARNING: Use proper procedures and safety equipment for the handling of hot pitch or asphalt. Avoid any unstable supports, benches, or tables. Hot pitch or asphalt causes second and third degree burns! Be extremely cautious when handling molten pitch or asphalt.

Sampling of Solid Samples   Select a representative sample of the bulk by collecting flakes, rods, or small pieces from the top, middle, or bottom of the bag or container. Collect at least 200 g of sample and store in a container.

Sampling of Asphalt   If stored in a hot liquid form, use the sampling technique for hot pitch. Once the sample cools to room temperature, wrap it into a large sheet of aluminum foil, freeze to $-20°C$ and break into small pieces with a plastic hammer. Store for further use in the same way as pitch.

3.17.3 Preparation of Pitch Specimens

Grinding and Pelletizing   Pitch is submitted for XRF analysis in the form of solid chunks that have to be crushed with a hammer so that the small pieces can enter a jaw crusher [85]. A 100 g portion of the pitch is ground with the jaw crusher to approximately -5 mesh. Using a riffler, this coarse powder is divided into a 14 g subsamples. Pitch is a different material from coke. It can be finely ground and briquetted [85], just as coke can, or it can be prepared for XRF analysis by melting the coarse grains followed by casting into a cup [86, 87]. In the grinding method, grind the 14.0 g sample and 7.0 g of an organic binder together in a Herzog HSM-F 36 puck-and-ring grinder for 120 s. Then form one or two 9 g briquettes in a hydraulic press. The average thickness of a pitch briquette is 6.2 mm, and the diameter is 40 mm.

Hot Casting of Pitches or Asphalts

Apparatus   This apparatus consists of a 25 × 25 cm hot plate with temperature control, disposable aluminum dishes 60-80 mm in diameter, box of 6-μm thick precut Mylar films, 40-mm diameter plastic cups for XRF analysis of liquids, 10-mm thick

Plexiglas plate, aluminum tongs for handling crucibles, several 10 cm long rods cut from a 2-mm diameter clean aluminum or steel wire. A well ventilated laboratory fume hood.

Hot Casting (in a fume hood)

1. Weigh approximately 15–20 g of pitch or asphalt into a disposable aluminum dish (use top-loading balance only). Before you fill the dish, make a small "beak" in its rim to allow for pouring of the hot melt.

2. Place the dish on a hot plate heated at 140–160°C. The plate is covered by aluminum foil in order to protect it against spills. Six dishes can be easily placed on a 25 × 25 cm plate. Let the sample melt for 10–15 minutes.

3. While the pitch melts, prepare 40 mm XRF sample (SPEX or Chemplex) cups used for analysis of oils or liquids. Stretch 6 μm Mylar foil over the cup and secure it with the retaining ring. If there are any wrinkles on the surface, stretch film again, until the film surface is perfect. Lay the sample cups window side down on a clean Plexiglas plate.

4. Stir hot pitch occasionally while on a hot plate, using an aluminum or steel wire rod.

5. Take the hot, molten sample off the plate, swirl carefully to give the liquid its final mix. Cool the sample by further swirling to a consistency of honey or a thicker syrup.

6. Pour the hot sample into the prepared cup.
   IMPORTANT: Fill the cup three-fourths full. The sample thickness must be at least 20–25 mm. This is very important for the correct determination of heavy metals such as lead, etc. The pitch or asphalt temperature should be such that it would not burn through the Mylar window when poured into the cup. The Plexiglas plate keeps the Mylar window cool by dissipating the heat. The hot casting procedure works only with Plexiglas plates, not with metallic ones such as aluminum. When dealing with high softening point pitches (prebake) or asphalts, use 6 μm Kapton (Chemplex product) for the cup window. Kapton foil is heat resistant to 200°C and would not burn through. It is more expensive than Mylar.

7. Let the sample cool to a room temperature for at least 15 minutes.

8. After cooling, remove slowly the retaining ring which holds the Mylar window. Peel off the Mylar foil carefully from the surface of the sample. If traces of Mylar still adhere to the surface, scrape them gently with a scalpel. If necessary, examine the surface with a magnifying lens in order to spot small pieces of Mylar on the surface.

9. The surface of the pitch or asphalt must be flat and mirror shiny, without any rough defects or bubbles. Once the method is mastered, 95% of the samples have ideal surfaces for XRF determination.

10. Reattach the retaining ring on the cup and lay the sample face down on a lintless sheet of soft paper, napkin, or similar sheet of clean paper (lens cleaning tissue paper is recommended).
11. The sample is now ready for analysis in the spectrometer.

IMPORTANT: Soft pitches which flow at room temperature could be "hot cast" at lower hot plate temperatures. These casts are then easily frozen in the cup at -20 or $-40°C$. The window is stripped off just prior to analysis and while the sample is cold. The slight vapor condensation on the pitch surface is removed in a few seconds by the vacuum of the spectrometer and does not cause any interferences. The sample is analyzed while frozen and quickly removed from the spectrometer, before it starts to flow. It is recommended that only one such sample at a time is analyzed in order to avoid damage or contamination of the X-ray tube window. Start measuring counts on high atomic number elements first to avoid count fluctuations due to vacuum changes in the spectrometer. The "heavy" element count is independent of vacuum or condensed mist on the sample surface.

Thermal Stability of Pitch and Asphalt Samples (Loss of Volatiles) The thermal stability of pitch or asphalt sample is of prime interest when the hot-casting method is employed, because of possible loss of volatiles. A systematic survey of thermal behavior of pitches in air has shown that the majority of pitches and asphalts lose volatiles [86] only when heated above $200–220°C$. About 2–3% loss of volatiles was observed at $200°C$ and the oxidation of the sample started only above $250°C$. The loss of volatiles is negligible during the melting of the pitch or asphalt at $120–130°C$ and would not affect the analytical results. Overheating of the pitch is usually signaled by the white-yellow smoke cloud formed above the sample surface. When the smoke becomes yellow and thick, some sulfur may be lost.

Handling of the Samples During Sample Preparation It cannot be emphasized enough how important it is to follow all safety rules and precautions when preparing and handling the hot, molten pitch or asphalt. Severe burns are usually caused by personal distraction while manipulating a hot sample. Eye irritations are also caused by pitch or asphalt dust. All sample preparation must be carried out in a well-ventilated fume hood. Working conditions should be "extra clean." Lay aluminum foil or coated paper board on the working bench of the fume hood or tables where pitch or asphalt is handled. Dust-free conditions are desirable. Dust found in plants and laboratories contains large proportion of silicon, aluminum, iron, etc. As trace levels of metals and silicon will be determined in the sample down to ppm levels, any dust, fiber or other small particles which can contaminate the sample are undesirable. Keep the workplace clean by periodically replacing the aluminum foils or paper board.

Discussion The hot cast sample preparation method, described above, eliminates all the disadvantages associated with grinding, weighing, addition of a binder, mixing, and pressing of the pellet. The use of disposable aluminum dishes for melting of the sample completely eliminates pitch or asphalt dusts which tend to contaminate grinding equipment, laboratory benches, fume hoods, etc. The perfectly flat sample surfaces allow for excellent and reproducible results, especially for the low atomic number elements. Compared to the classical pulverization and pressing procedure, this method reduces sample preparation time by 80% and eliminates contamination of the laboratory environment.

3.17.4 Special Problems and Hints

The briquetting technique offers a significant automation potential if the Herzog automatic grinder is used. The grinder reduces substantially the analytical time required for the manual approach, and offers a reliable way for the preparation of a briquetted sample. Because sodium is one of the analyzed elements, care must be taken with respect to the organic binder used. A binder such as SX [59] contains sodium at a significant level (approximately 120 ppm) which may vary from batch to batch. A much better binder for this application is polyethylene available as a 60 μm fine powder. The usual grinding and pressing of the sample into pellets and the use of binders or additives can be completely eliminated in the casting method. However, the casting method is slower (because the operator contact time is longer) than the briquetting method employing the automatic grinder. The casting sample preparation technique gives less chance for sample contamination (binder) than the briquetting technique. On the other hand, some pitches are difficult to melt and cast, which may result in significant specimen porosity, varying density, and nonuniform thickness. Certified reference materials of coal-tar pitch (PITCH 01-07, and CAF) [4] are available from:

> Alcan International Ltd., STANDARD SAMPLES, P.O. Box 1250, Jonquiere, Québec, G7S 4K8.

DOMTAR certified reference materials coal-tar pitches (series A-D) can be obtained from:

> Domtar Inc., P.O. Box 300, Senneville, Québec, H9X 3L7 (attn. V. Kocman).

Also available commercially are pitch standards Series B-400 from:

> R&D Carbon Ltd., Le Chablé, P.O. Box 157, 3960 Sierre, Switzerland.

3.17.5 Particle Size Effect in the Analysis of Silicon in Pitch

The XRF analysis usually involves a relatively thin surface layer of the specimen. The effective specimen layer contributing to the measured analyte line radiation depends on the analyte atomic number and the matrix. The depth in the pitch matrix from which the specific radiation can emerge was calculated using Togel's formula [88]. Whenever the size of silicon containing particles exceeds the penetration depth for this element, then a particle size effect begins to play an important role and cannot be neglected. If the XRF calibration is based on synthetic standards manufactured with

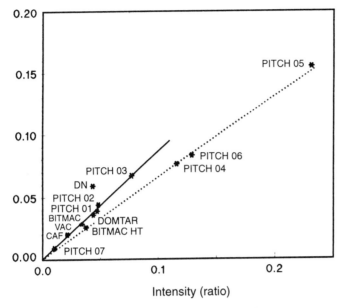

Figure 3.8. Silicon calibration graph.

the silicon organo-metallic compound as described in [86], then serious discrepancies in reporting silicon results occur that are caused by this effect. Figure 3.8 shows two calibration curves for silicon obtained using briquetted samples. In Figure 3.8 the Alcan standard DN [reanalyzed many times by atomic absorption spectroscopy (AAS)] is situated well above the solid calibration line. On the dotted line there are Pitch 04, 05, and 06 standards, which had been heavily doped with octaphenylcyclotetrasiloxane (OCTS), a Si organo-metallic compound $C_{48}H_{40}O_4Si_4$ containing 14.2% silicon. Also on this line is the Bitmac HT pitch standard, which due to manufacturing conditions (high temperature distillation) contains mostly organo-metallic silicon. The other pitches: Pitch 01, 02, 03, 07, CAF, Bitmac VAC, and Domtar are regular production samples.

In trying to explain the spread of points in Figure 3.8, four series of synthetic calibration standards were made. Standard PITCH 07 was ground first, and then doped with the organo-metallic compound of submicron particle size, with Celite (a very fine quartz) and with coarse quartz (40 μm and 250 μm) in increasing amounts. Figure 3.9 shows four calibration curves clearly. There is a separate and distinct working curve for each type of silicon source used. The distinction of the calibration curves is due to the particle size effect. The intensity level and sensitivity of the working curves increase as the silicon particle size decreases. As the size of the particles decreases, a greater number of photons have a fraction of their pathlength in more than one particle. Silicon present in quartz is very different from the rest of the matrix with respect to the X-ray absorption. It is interesting to note that, for example, a 0.15% Si addition in the form of the organo-metallic compound

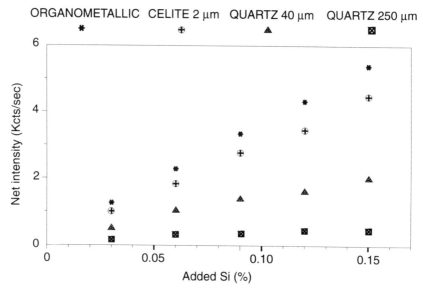

Figure 3.9. Silicon in pitch, standard calibration method (pitch 07).

produces 18 times more intensity than a similar addition of silicon as coarse quartz (150 μm). The finding also indicates that the XRF results might vary depending on the nature of naturally occurring silicon in the pitch. Consequently, it was suspected that the standard DN (Figure 3.8) is contaminated by fine sand, quartz, or other SiO_2 compounds.

It is hence clear that the pitch calibration standards cannot be prepared by a standard addition method using the organo-metallic compound (OCTS) as a source of silicon. These synthetic samples give a higher intensity for silicon than do naturally occurring pitches, and any calibration curve established with them would underestimate the amount of SiO_2 in the regular pitch shipments. If the synthetic standards are rejected and the calibration is based on natural pitch standards, then the silicon particle size effect is hardly noticed. Fortunately, most pitches happen to contain Si of a particular particle size. Exceptions may occur if a quartz-contaminated specimen or distilled pitches with high Si contents are analyzed.

3.17.6 General Discussion

Pitches and asphalts, like heavy oil bottoms, are by-products from distillation of tars and oils. Pitches are used mostly as binders for the pulverized coke in the manufacture of a variety of electrodes. The electrodes are used in electric arc furnaces for steel and alloy manufacture, or large "prebaked" electrodes are utilized in electrolytic baths (pots) for the production of aluminum. Asphalts are used in road construction, roofing, sealants, and in some special water-resistant paints. Pitches and asphalts contain between 0.1–4% sulfur [89] and trace (ppm) quantities of elements such as sodium, silicon, phosphorus, chlorine, calcium, vanadium, iron, manganese, chromium, nickel, zinc, and lead.

There are several difficulties encountered in the preparation of pitch or asphalt samples. In some cases, cryogenic grinding may be necessary to obtain a suitable sample and this grinding adds time to the sample preparation. Cleaning of the sample preparation equipment is also a difficult and time-consuming procedure and in most cases organic solvents have to be used. Most pitches and asphalts have softening points between 80–150°C and are too soft and sticky to grind or press without the use of a binder. As most industrial binders are not totally pure, unwanted trace impurities may be introduced into the sample by the binder.

One way to avoid the use of binders (and possible contamination) is to "hot cast" the pitch or asphalt directly into a plastic sample cup fitted with a thick Mylar window. After cooling, the window is carefully removed from the surface of the sample to expose a mirror shiny and perfectly flat pitch or asphalt surface ready for XRF analysis. The novel "hot casting" procedure [86] completely eliminates the use of mixing and grinding equipment and the lengthy, cumbersome cleanups. Sample preparation times are drastically reduced and the method requires only a good laboratory fume hood.

3.18   LEAVES AND VEGETATION

3.18.1   General Discussion

Traditional sample preparation of leaves and vegetation for analysis is based on dissolution (digestion) of the individual samples in acid and subsequent analysis of the elements by spectroscopic methods (atomic absorption or plasma spectroscopy). XRF spectrometry provides much better reproducibility and matches the precision of inductively coupled plasma spectrometry. Norrish and Hutton [90] were the first who described the XRF determinations in leaves and vegetation. Kocman, Peel, and Tomlinson [91] described an improved method which could be used with modern XRF spectrometers.

Specimen preparation for XRF determination of macro (calcium, magnesium, potassium, phosphorus, and sulfur) and micronutritional elements (iron, manganese, copper, and zinc) and other important elements such as sodium, aluminum, silicon, chlorine, and lead requires preparation of pressed pellet samples. Specimen preparation for XRF is simple and fast, with very little manipulation. Representative leaf samples from trees or vegetation are collected, dried, pulverized, and pressed into a pellet without any contamination. Usually 10–20 elements can be determined quantitatively in a matter of minutes. The method described here is suitable for any modern wavelength or energy dispersive XRF system.

For more details on nutritional balance of trees, vegetation, and sampling of leaves, the reader is referred to a monograph by G. H. Tomlinson and F. L. Tomlinson [92] on acid rain effects on forests of North America and Europe.

3.18.2 Selection of a Representative Sample of Leaves or Vegetation

Hardwoods Leaves of oak, poplar, and maple are harvested in North America and Europe preferably in the middle of August, when leaves are fully mature and before they start to fall off.

Softwoods (Conifers) Spruce or pine needles are harvested in October and November.

Vegetables, Grasses Harvest when necessary, preferably when leaves or blades have matured.

3.18.3 Collection of Leaves from a Tree

Imagine a tree crown divided into three horizontal and equal parts (Figure 3.10). The top layer will have fewer nutrients than the bottom. Collect separately from top, middle, and bottom by cutting small branches with scissors mounted on a long pole. If yearly analysis is required, sample only from well-marked plots and same trees. This approach allows one to study the effect of various fertilizers on the health of trees and vegetation. If the leaves are covered by dust, dirt, etc., rinse them in a large basin filled with distilled water.

Leaves and needles are most conveniently collected into 1 kg (2 lb), paper bags. These can be easily labelled, closed by twisted wire, or stapled. Bags protect collected leaves from contamination in transportation. The bags filled with leaves or vegetation are also suitable for later drying of the samples in an oven.

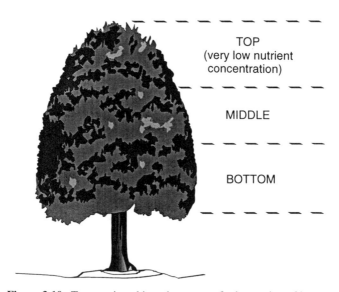

Figure 3.10. Tree sectioned into three zones for harvesting of leaves.

IMPORTANT: Handle leaves and vegetation only with surgical (thin rubber) gloves. Wash hands thoroughly in running water and rinse in de-ionized or distilled water right after you put gloves on. Most thin gloves contain talc (magnesium silicate) on their surface, which has to be washed off before it can contaminate the sample. When collecting a large number of samples, spread polyethylene (or similar film) on the ground, portable tables, benches, and racks where you store or manipulate samples in order to protect them from contamination.

3.18.4 Specimen Preparation

Apparatus   The apparatus should include a good quality laboratory-type or kitchen-type blender with hardened steel blades, a one liter glass container (such as used for making milkshakes), an aluminum foil to protect sample handling areas from dust, sharp scissors, a well-regulated electric drying oven, several approximately 50 ml acrylic or glass vials with lids for storing the pulverized samples, a 32 mm diameter aluminum backing cups, a die for pressing the pellets, a strong industrial-type vacuum cleaner and tissue paper (preferably lint free).

Drying of Leaves and Needles   Most leaves, needles and vegetation are dried in an electric drying oven at 85°C in the paper storage bags (see section 3.18.3). The 85°C temperature should be rigorously maintained when needles are dried, otherwise some essential oils may be lost from the sample. Some leaves (such as maple) may be dried at an even higher temperature of 105°C. Drying times vary from 24 to 72 hr and depend on the amount of moisture in the leaves. Vacuum drying is not recommended.

Pulverization of the Sample

1. Cut leaf stems (petioles) from leaves (this operation could be carried out when leaves are green and collected into the bags). Separate needles from twigs; these would easily fall off from twig when dried.
2. Weigh approximately 5 g of the sample and pulverize for 3 min in a laboratory blender at a high speed setting. No liquid or water should be added; this is *dry pulverization*!
3. Store the resulting −100 mesh (150 μm) powder in small hermetically tight vials for further use.
4. Clean the glass container of the blender with a strong vacuum cleaner. Move the hose around the blades to vacuum all remaining particles. The rest of the vegetation dust deposited on the inside walls of the container is wiped out by hand using a paper towel or lint-free tissue. Washing with water or solvents is not necessary between the analyses. Use water wash only at the end of the work to clean thoroughly all parts of the blender. Rinse with distilled water and air dry.

Pressing of Pellets   Weigh approximately 4 g (± 0.2 g) of pulverized leaves or vegetation into a plastic dish using top-loading balance (exact weight of the sample is not important for the XRF analysis). Transfer the powder into a supporting aluminum cup positioned in a 32 mm diameter die.

IMPORTANT: Do not add any boric acid or binders, dilutants and pressing aids to the sample.

Press at 25 tons on the ram for a duration of approximately 20 s. Slowly release the pressure over a period of 10–20 s. The resulting pellet must be at least 4 mm thick with a flat and smooth surface. Label pellet properly with a suitable sticker. Prepared pellets are presented to the XRF spectrometer for analysis, or stored for further use in hermetically closed plastic bags or vials. Longer storage in humid air will lead to deterioration of the pellet surface.

3.18.5   Discussion

The described specimen preparation method is recommended for the determination and monitoring of 15–20 major, minor, and trace nutrient elements in foliage and plant analysis. Sample specimen preparation is easy and nondestructive in the sense that the vegetative matter is reduced to a fine dry powder which could be reused later, if necessary. The use of modern microcomputer operated wavelength dispersive XRF spectrometers allow for quick analysis of a large volume of samples.

REFERENCES

[1] Ochi, H. and Okashita, H. (1987), " X-ray fluorescence analysis of ceramic materials. Comparison between the powder method and the glass bead method," *Shimadzu Hyoron*, **44**(2), 157–163, CA 107(18):160056d.

[2] Nakada, S. (1987), "X-ray fluorescence determination of trace elements in silicate rocks," *Kyushu Daigaku Rigaku Kenyu Hokoku, Chishitsugaku*, **15**(1), 37–44.

[3] Metz, J. G. H. and Davey, D. E. (1992), "Statistical comparison of analytical results obtained by pressed powder and borate fusion XRF spectrometry for process control of a lead smelter," *Adv. X-ray Anal.*, **35B**, 1189–1196.

[4] Dow, R. H. (1982), "A statistical comparison of data obtained from pressed disk and fused bead preparation techniques for geological samples," *Adv. X-ray Anal.*, **25**, 117–120.

[5] Claisse, F. and Samson, C. (1962), "Heterogeneity effects in X-ray analysis," *Adv. X-ray Anal.*, **5**, 335–354.

[6] Bonetto, R. D. and Riveros, J. A. (1985), "Intensity for cubic particles in X-ray fluorescence analysis," *X-ray Spectrom.*, **14**, 2–7.

[7] Allen, Terence (1981), *Particle Size Measurement*. Chapman and Hall: London.

[8] ASTM D 75–87. (1995), *Standard Practice for Sampling Aggregates. Annual Book of ASTM Standards*, Vol 4.02, ASTM: Philadelphia.

[9] ASTM C 702–93. (1995), *Standard Practice for Reducing Samples of Aggregate to Testing Size. Annual Book of ASTM Standards*, Vol 4.02, ASTM: Philadelphia.

[10] Ingamells, C. O. and Switzer, P. (1973), "Proposed sampling constant for use in geochemical analysis," *Talanta*, **20**(6), 547–568.

[11] Ingamells, C. O. (1974), "Control of geochemical error through sampling and subsampling diagrams," *Geochim. Cosmochim. Acta*, **38**(8), 1225–1237.

[12] Ingamells, C. O. (1976), "Derivation of the sampling constant equation," *Talanta*, **23**(3), 263–264.

[13] Ingamells, C. O. (1978), "A further note on the sampling constant equation," *Talanta*, **25**(11–12), 731–732.

[14] Ingamells, C. O. (1980), *General sampling theory*, Sampling Assaying Precious Met., Proc. Int. Semin., 2nd, 39–63. Donald A. Corrigan and Myron E. Browning, eds., Int. Precious Met. Inst., Inc.: Brooklyn, NY.

[15] Gy, Pierre M. (1986), "The analytical and economic importance of correctness in sampling," *Anal. Chim. Acta*, **190**, 13–23.

[16] Gy, Pierre, M. (1990),"Recent developments of the sampling theory," *Analysis*, **18**(5), 303–309.

[17] Kanare, H. M., Unpublished work at Construction Technology Laboratories, February 1993.

[18] Tuff, Mark A. (1986), "Contamination of silicate rock samples due to crushing and grinding," *Adv. X-ray Anal.*, **29**, 565–571.

[19] Goto, H. and Ohno, S. (1981), "X-ray fluorescence analysis of chemical components in rocks and minerals by powder method. Part I. grain size effect and contamination upon pulverization in pressed powder method." *Chishitsu Chosasho Geppo*, **32**(4), 213–226. (CA 96(6):45458d).

[20] Hickson, C. J. and Juras, S. J. (1986), "Sample contamination by grinding," *Can. Mineral.*, **24**(3), 585–589.

[21] Leoni, L. and Saitta, M. (1974), "X-ray fluorescence analysis of powder pellets utilizing a small quantity of material," *X-ray Spectrom.*, **3**, 74–77.

[22] Rose, W. I., Bornhorst, T. J., and Sivonen, S. J. (1986), "Rapid high-quality major and trace element analysis of powdered rock by X-ray fluorescence spectrometry," *X-ray Spectrom.*, **15**, 55–60.

[23] van Zyl, C. (1982), "Rapid preparation of robust pressed powder briquettes containing a styrene and wax mixture as binder," *X-ray Spectrom.*, **11**, 29–31.

[24] Baker, J. (1982), "Volatilization of sulfur in fusion techniques for preparation of disks for X-ray fluorescence analysis," *Adv. X-ray Anal.*, **25**, 91–94.

[25] Kocman, V., Foley, L., and Woodger, S. C. (1985), "The use of rapid quantitative X-ray fluorescence analysis in paper manufacturing and construction materials industry," *Adv. X-ray Anal.*, **28**, 195–202.

[26] Istone, W. K., Collier, J. M., and Kaplan, J. A. (1991), "X-ray fluorescence as a problem solving tool in the paper industry," *Adv. X-ray Anal.*, **34**, 313–318.

[27] Kocman, V. (1989), "Analysis of limestone and dolomite rocks," in *X-ray fluorescence analysis in the geological sciences*, ISBN 91–9216–38–2, Geological Assn. of Canada, S.T. Ahmedali, ed., pp. 272–276.

[28] King, Bi-Shia and Davidson, V. (1988), "Sample preparation method for major element analyses of carbonate rocks by X-ray spectrometry," *X-ray Spectrom.*, **17**, 85–87.

[29] Wheeler, B. D. (1987), "Accuracy in X-ray spectrochemical analysis as related to sample preparation," *Spectroscopy*, **3**, 24–33.

[30] Kocman, V. (1984), "Rapid multielement analysis of gypsum and gypsum products by X-ray fluorescence spectroscopy," in *The Chemistry and Technology of Gypsum*, ASTM STP 861, R. A. Kuntze, ed., American Society for Testing and Materials, pp. 72–83.

[31] Kocman, V. and Foley, L. (1987), "Certification of four North-American gypsum rock samples type: $CaSO_4 \cdot 2H_2O$, GYP-A, GYP-B, GYP-C, and GYP-D," *Geostandards Newsletter*, **11**; No 1, 87–102.

[32] West, R. R. and Sutton, W. J. (1954), "Thermography of gypsum," *J. American Ceramic Soc.*, **37**, 221–224.

[33] Kocman, V., Foley, L., and Woodger, S. C. (1985), "The use of rapid quantitative X-ray fluorescence analysis in paper manufacturing and construction materials industry," *Adv. X-ray Anal.*, **28**, 195–202.

[34] Chemical Analysis Committee of the Society of Glass Technology, *Glass Technology*, **19**, p. 93, (1978).

[35] Lithium Tetraborate Flux No. 0602, High Density, Ultra Pure, from: Claisse Scientific Corporation Inc., 2522 Ste. Foy Rd., Sainte-Foy, Quebec, Canada, G1V 1T5.

[36] Lithium Tetraborate Flux, "Spectromelt A-10," from: EM Science Corp., 480 S. Democrat Rd., Gibbstown, NJ 08027–1297.

[37] ASTM C 219–94, (1995), "Standard Terminology Relating to Hydraulic Cement," *Annual Book of ASTM Standards*, Volume 4.01, ASTM: Philadelphia.

[38] Hansen, M. (1958), *Constitution of Binary Alloys*, McGraw-Hill:New York.

[39] Elliot, R. P. (1965), *Constitution of Binary Alloys, First supp.*, McGraw-Hill: New York, 877 pp.

[40] Hatch, J. E. (1984), *Aluminum: Properties and Physical Metallurgy*, American Society for Metals, Materials Park, OH.

[41] Van der Voort, G. F. (1984), *Metallography Principles and Practice*, McGraw-Hill: New York.

[42] Avner, S. H. (1974), *Introduction to Physical Metallurgy*, McGraw-Hill:New York.

[43] Altenpohl, D. (1982), *Aluminum viewed from Within*, Aluminum-Verlag, Düsseldorf.

[44] Barrett, C. and Massalski, T. B., (1980), *Structure of Metals*, 3rd ed., Pergamon Press, Oxford.

[45] Feret, F. and Sokolowski, J. (1990), "Effect of sample surface integrity on X-ray fluorescence analysis of aluminum alloys," *Spectrosc. International*, **2**(1), 34–39.

[46] Togel, K. (1961), "Zerstrungsfreie Materialprüfung," R. Oldenburg: Munich.

[47] Samuels, L. E. (1982), *Metallographic Polishing by Mechanical Methods*, American Society for Metals, Materials Park, OH.

[48] *Metals Handbook*, 10th ed., (1990), Vol. 1–5, ASM International, Materials Park, OH.

[49] Jenkins, R. and de Vries, J. L. (1975), *Practical X-ray Spectrometry*, Springer-Verlag: New York.

[50] Black, H. R. (1953),"Analysis of bauxite exploration samples," *Anal. Chem.*, **25**, 743–748.

[51] Matocha, C. K. (1975), Proceedings from the 6th International Conference on Light Metals (ILMT) Leoben/Vienna (Austria), June 1975.

[52] Feret, F. (1987), "Characterization of aluminum bearing minerals by XRF/XRD"—XXV Colloquium Spectroscopicum Internationale, Toronto, Ontario, June 1987.

[53] Solymá, K., Sajó, I., Steiner. J., and Zöldi, J. (1992), "Characteristics and separability of red mud," *Light Metals*, 209–223.

[54] Feret, F. and Giasson, G. F. (1991), "Quantitative phase analysis of Sangaredi bauxites (Guinea) based on their chemical compositions," *Light Metals*, 187–191.

[55] Wheat, T. A. (1971), "Calcination of Gibbsite," *J. Canadian Ceramic Soc.*, **40**, 43–48.

[56] Sleppy, W. C. (1985), "Formation and properties of alumina," *J. Metals*, **40**, 22–24.

[57] Adamson, A. N. (1970), "Alumina production: principles and practice," *The Chemical Engineer*, 156–171, June.

[58] Feret, F. (1990), "Alumina chracterization by XRF," *Adv. X-ray Anal.*, **33**, 685–690.

[59] Van Zyl, C. (1982), "Rapid preparation of robust pressed powder briquettes contaning a styrene and wax mixture as a binder," *X-ray Spectrom.*, **11**, 29–31.

[60] Mannweiler, U., Schmidt-Hatting, W., and Koutny, M. (1983), "New automatic analysis of bath components in the Alumina reduction process," *Erzmetall*, **36**(6), 274–277.

[61] Baggio, S. and Foresio, C. (1980), "An X-ray method for measuring the alumina content in reduction cells for aluminum production," *Aluminum*, **56**(4), 276–278.

[62] Topor, D. and Birhal, A. (1972), "Cryolite bath ratio by X-ray diffraction," *Revue Roumaine de Chimie*, **17**(6), 961–968.

[63] Lüschow, H. M. (1972), "Röntgenspektral Analyse mit Elektronen-strahlanregung-beitrag zur Analyse van technischem Kryolith und Aluminum fluoridhaltigem Aluminiumoxid," *Aluminum*, **48**, 783–789.

[64] Feret, F. (1988), "Characterization of bath electrolyte by X-ray fluorescence," *Light Metals*, 697–702.

[65] Ouellet, J. and Page, R. (1992), "Automatic preparation and analysis of bath electrolyte samples," paper given (not published) at 40th Denver Conference on Applications of X-ray Analysis, Hawaii, August 7–16.

[66] Leyden, D. E., Patterson, T. A., and Alberts, J. J. (1975), "Preconcentration and X-ray fluorescence determination of copper, nickel and zinc in sea water," *Anal. Chem.*, **47**, 733–735.

[67] Roelands, I. (1985), "Determination of cobalt in iron-rich materials by X-ray fluorescence spectrometry after solvent extraction and anion-exchange resin," *Chem. Geol.*, **51**, 3–8.

[68] Peter, H. J., Braun, J., Dietze, U., and Volke, P. (1985), "Trace determination by X-ray fluorescence analysis after enrichment by ion exchange resins," *Z. Chem.*, **25**, 374–375.

[69] An, Q. (1985), "X-ray fluorescence spectrometric determination of trace amounts of RE in ores by ion-exchange paper," New Front. Rare Earth Sci. App. Proc. Int. Conf., Rare Earth Dev. Appl., **1**, 551–552.

[70] Campbell, W., Spano, E. F., and Green, T. E. (1966), "Micro and trace analysis by a combination of ion-exchange resin-loaded papers and X-ray spectrography," *Anal. Chem.*, **38**, 987–996.

[71] Van Grieken, R. E., Bresseleers, C. M., and Vanderborght, B. M. (1977), "Chelex-100 Ion exchange filter membranes for preconcentration in XRFA of water," *Anal. Chem.*, **49**, 1326–1331.

[72] Minkkinen, P. (1977), "Combined ion exchange paper and X-ray fluorescence method for determination of uranium in natural waters," *Finn. Chem. L.*, **4–5**, 134–137.

[73] Clechet, P. and Eschalier, G. (1981), "Trace analysis of barium in water with cation resin-loaded paper for X-ray fluorescence analysis," *Analysis*, **9**, 125–129.

[74] Robberecht, H. J. and Van Grieken, R. B. (1980), "Sub–part per–billion determination of total dissolved selenium and selenite in environmental waters by X-ray fluorescence spectrometry," *Anal. Chem.*, **52**, 449–453.

[75] Bruninx, E. and van Meijl, E. (1975), "Analysis of surface waters for Fe, Zn, and Pb by coprecipitation of iron hydroxide and X-ray fluorescence," *Analyt. Chim. Acta*, **80**, 85–95.

[76] Pradzynski, A. H., Henry, R. E., and Stewart, J. L. S. (1975), "Determination of selenium in water on the ppb level by coprecipitation and energy dispersive spectrometry," *Radiochem. Radioanal. Lett.*, **21**, 277–285.

[77] Katsuno, T. (1981), "X-ray fluorescence analysis of arsenic in river water by means of coprecipitation with zirconium hydroxide," *Nagano-ken Eisei Kogai Kenkyusho Kenkyu Hokoku*, **3**, 52–55.

[78] Howe, P. T. (1980), "Analysis for iodide in ground water by X-ray fluorescence spectrometry after collection as silver iodide on activated charcoal," At. Energy Can. Ltd., AECL–6444, 11 pp.

[79] Vanderborght, B. M. and van Greiken, R. E. (1977), "Enrichment of trace metals in water by adsorption on activated carbon," *Anal. Chem.*, **49**, 311–316.

[80] Johansson, E. M. and Akselsson, K. R. (1981), "A chelating agent-activated carbon-PIXE procedure for sub-ppb analysis of trace elements in water," *Nucl. Instrum. Methods*, **181**, 221–226.

[81] Vassos, B. H., Hirsch, R. F., and Letterman, H. (1973), "X-ray microdetermination of Cr, Co, Cu, Hg, Ni and Zn in water using electrochemical preconcentration," *Anal. Chem.*, **45**, 792–794.

[82] Zeronsa, W. P., Dabkowski, G., and Siggia, S. (1974), "Selective separation and concentration of silver via precipitation chromatography," *Anal. Chem.*, **46**, 309–311.

[83] Hercules, D. M., Cox, L. E., Onisick, S., Nichols, G. D., and Carver, J.C. (1973), "Electron spectroscopy (ESCA): use for trace analysis," *Anal. Chem.*, **45**, 1973–1975.

[84] Anderman, G. and Kemp, J. W. (1958), "Scattered X-rays as internal standards in X-ray emission spectroscopy," *Anal. Chem.*, **30**, 1306–1309.

[85] Feret, F. (1994), "The Si particle size effect in X-ray fluorescence analysis of Pitch," *X-ray Spectrom.*, **23**, 130–136.

[86] Kocman, V., Foley, L., and Landsberger, S. (1989), "Analysis of bulk coal-tar reference pitch sample by neutron and photon activation analysis, X-ray fluorescence spectrometry and atomic absorption spectroscopy," *Carbon*, **27**, 185–190.

[87] Certified Reference Samples of Coal-Tar Pitch, Alcan International Ltd., P.O. Box 1250, Jonquière, Quèbèc, G7S 4K8.

[88] Togel, K. (1961), Zerstörungsfreie Materialprüfung, Oldenburg: Munich.

[89] Kocman, V., Foley, L. M., and Woodger, S. C. (1985), "The use of rapid quantitative X-ray fluorescence analysis in paper manufacturing and construction material industry," *Adv. X-ray Anal.*, **28**, 195–202.

[90] Norrish, K. and Hutton, J. T. (1977), "Plant analysis by X-ray spectrometry," *X-ray Spectrom.*, **6**, 6–11.

[91] Kocman, V., Peel, T. E., and Tomlinson, G. H. (1991), "Rapid analysis of macro- and micronutrients in leaves and vegetation by automated X-ray fluorescence spectrometry— a case study of acid rain affected forest" *Communications in Soil Science and Plant Analysis*, **22**(19, 20), pp. 2063–2075.

[92] Tomlinson, G. H. and Tomlinson, F. L. (ed.) (1990), *Effects of acid deposition on the forests of Europe and North America*, CRC Press, Boca Raton, FL.

CHAPTER 4

SPECIMEN PREPARATION IN X-RAY DIFFRACTION

CONTRIBUTING AUTHORS: DAVIS, JENKINS, McCARTHY, SMITH, WONG-NG

4.1 COMMON PROBLEMS IN PREPARING AND PRESENTING SPECIMENS FOR X-RAY DIFFRACTION

The powder diffraction method assumes that the particles in the specimen are ideally random in orientation and that there are enough crystallites in the specimen to achieve a representative intensity distribution for these crystallites. The details of what constitutes the ideally random sample will be discussed below. However, ideally random samples are not the only manner of presentation that may be used. In fact, any orientation distribution that can be prepared reproducibly may be employed for diffraction analysis. Platy and fiber orientations are examples of conditions that may be met by specialized preparation procedures. Certain other partially oriented states with some samples may also yield reproducibility that allows even quantitative analysis when the procedure is rigorously defined and followed.

The ideally random specimen is one in which the probability of orientation for any and all Bragg planes is equally likely for any direction in space. This randomness may be achieved by having a single crystallite spinning freely in space at the intersection of the beam axis and the diffractometer axis; however, the intensity of such an experiment would be too low to be useful. It is usually achieved by having a very large number of very small crystallites whose orientations take all the possible positions, so that there are always crystallites which diffract in every position of the specimen in the diffraction apparatus. The details and limitations of this physical state will be discussed in Section 4.3.1.

The most successful departure from the ideally random specimen has been developed to perfection by the clay mineralogists. Clay minerals have layered crystal structures in which individual 7 to 10 Å thick layers are structurally coherent over several unit cells in the a and b crystallographic directions within the layers. When these layers are stacked to form the three-dimensional crystallites, the stacking may be misaligned so that the periodicity in the a and b directions is in poor registry while the c direction is regular due to the repeating thickness of the layers. The crystallites

A Practical Guide for the Preparation of Specimens for X-Ray Fluorescence and X-Ray Diffraction Analysis, Edited by Victor E. Buhrke, Ron Jenkins, and Deane K. Smith. ISBN 0-471-19458-1 © 1998 John Wiley & Sons, Inc.

are usually submicron platelets which behave like playing cards being dealt on a table when allowed to settle on a smooth flat surface. All the platelets orient with their layers parallel to the support surface. In diffractometer geometry, this type of specimen will enhance the (OOl) Bragg planes and de-emphasize the (hkO) and (hkl) planes. Because of the turbostratic stacking, the (hkO) and (hkl) planes are diffuse and of little use for most routine clay analysis, so the oriented specimen enhances the most useful component of the diffraction pattern.

To orient ideally, all the crystallites in the specimen should have their layers rigorously parallel to the specimen surface, and the a and b directions in the layers should be randomly oriented in the surface plane. The latter condition is usually achieved because the small size of the clay particles promotes the ability to achieve randomness in the absence of any shape effect that would tend to align the *a* and *b* axes, such as a bladed shape as opposed to a circular shape. Rigorous parallelism, however, is not easy to achieve because some particles will tilt when the platelets only partially overlap the lower particles. The degree of misorientation will be a function of the aspect ratio (the thickness to diameter ratio) and thickness of the specimen. The portion of the specimen in contact with the support will be the better oriented, and the orientation will deteriorate as the number of platelets in the specimen increases. Procedures for preparing oriented clay specimens will be detailed in section 5.5.

Fiber specimens may also be prepared for diffraction analysis in which the fiber directions of the particles are all aligned parallel to each other and either parallel or perpendicular to the diffractometer or camera axis. In a fiber, there is a family of Bragg planes that are all parallel to the fiber axis, and diffraction from these planes would be enhanced if the specimen had all the fibers with the long dimension parallel to the rotation axis. This family of Bragg planes is called a zone. The enhancement of the axial Bragg planes emphasizes the structure packing of fibrous molecular units in the structure, which may be regular, and deemphasizes the periodicity along the fiber axis, which would be determined with the usually less perfect perpendicular alignment. As for the clay minerals, these planes are just the ones that are enhanced and deenhanced by the imperfect crystal structure of these materials. Studies of asbestos and polymers often use this geometry for specimens. Where the fibers are long threads, specimens may be easily wound on a support to achieve this orientation. Where the particles are small (< 2 mm in length), achieving perfect orientation is more difficult.

In addition to the orientation, there are several aspects of specimen preparation that the diffractionist needs to consider when preparing the specimen for study. Probably the single most important aspect is to determine the requester question that needs to be answered. Most sample submitters need to be queried as to what they really want out of the analysis. This situation is either because they do not understand what a diffraction analysis can provide or they do not understand the limitations of such analyses. The diffractionist needs to know what elements are present in the sample; how the sample was formed (including temperature and pressure if applicable). Does the supplier have preconceived knowledge of what phases should be present? Is the run just to confirm the presence of a specific phase, to identify all phases present, or to quantify the phase composition, or is there a new material which needs to be fully characterized? Is the sample stable in ambient conditions, or do special precautions need to be taken? Is the sample expendable? If the sample is a mixture, could the material be beneficiated to concentrate the portion that is to be examined? When are the results needed? Who will be reading the report, scientists, managers or the public? The answers to these questions will provide the information which dictates the care that needs to be expended when preparing the diffraction specimen. The answers will also dictate which diffraction apparatus is the most appropriate for the study.

In a good diffraction service facility, it is usually necessary for the diffractionist to have to educate the submitter as to what information can be obtained and at what cost in money and time. It is not uncommon for submitters to say, "I know what diffraction can do. You do not need any information; it will bias your interpretation." Given such a response, the diffractionist should return the sample and say "Come back when you have the information." There is no need to limit any analysis in this manner. The diffractionist's time is too valuable.

Another problem with submitters is that the questions often get changed. After the diffractionist supplies the answers to the phase identification, the question will be how much is present of each phase. The diffractionist needs to anticipate this change and force the submitter to clarify the desires at the beginning at a time when the proper experiment is being designed.

With the appropriate information in hand, the diffractionist may proceed to prepare the specimen for presentation to the diffraction apparatus. There are several different levels of specimen quality which are dependent on the question to be answered. If the goal is only to confirm the presence and/or absence of designated phases, than almost any specimen will be suitable. However, if the specimen is to be used for quantification, considerable time and care will be required to prepare the necessary specimen. Characterization of a new phase will require a different approach than phase identification. Sometimes, as for the preparation of reference diffraction patterns, more than one specimen with different properties produces a better result than using a single specimen with nonideal properties for either measurement. It is the purpose of the sections below to elucidate many different techniques for preparing diffraction samples that will meet the needs for all the different types of samples that the diffractionist may encounter.

4.2   GEOMETRIES OF DIFFRACTION EQUIPMENT

Although there are many manufacturers of powder diffractometers, the majority of commercial systems employ the Bragg–Brentano parafocussing geometry. A given instrument may provide a horizontal or vertical $\theta/2\theta$ configuration, or a vertical θ/θ configuration. The vertical θ/θ and $\theta/2\theta$ systems are generally most advantageous for handling powder samples at room temperatures, but the horizontal system offers special advantages where a heavy sample is used or bulky accessories are placed at or beyond the receiving slit position. A two-dimensional view of the geometric arrangement is shown in Figure 4.1. A divergent beam of radiation coming from the line of focus F of the X-ray tube passes first through a divergence slit DS, then through a parallel plate set collimator (Soller slits) SS1 before striking the specimen

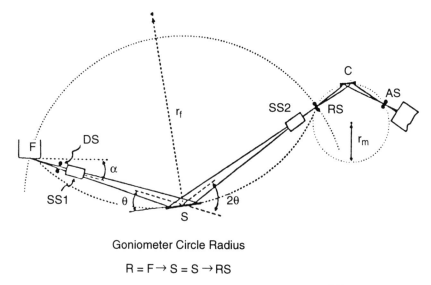

Goniometer Circle Radius

$$R = F \rightarrow S = S \rightarrow RS$$

Figure 4.1. Geometric arrangement of the Bragg–Brentano diffractometer.

S at an angle θ. The diffracted rays leave the specimen at an angle 2θ to the incident beam (and θ to the specimen surface), pass through a second parallel plate collimator SS2, through the receiving slit RS to the detector. A diffracted-beam monochromator, comprised of a crystal C and a slit AS, may be placed between the receiving slit and the detector. In order to establish the parafocusing condition, the axes of the line focus of the X-ray tube and of the receiving slit are at equal distances from the axis of the goniometer. The monochromator is generally a poorly focusing device in which the receiving slit RS, the crystal C, and the detector slit AS all lie on the circumference of the monochromator circle of radius r_m. The X-rays are collected by a suitable radiation detector, usually a scintillation counter, solid-state detector or a sealed gas proportional counter.

The receiving slit assembly and the detector are coupled together and move around a goniometer circle of radius R, centered about the specimen, in order to scan a range of 2θ (Bragg) angles. The source to specimen distance, and the specimen to receiving slit distance are both equal to the radius R of the goniometer circle. For $\theta/2\theta$ scans, the goniometer rotates the specimen about the same axis as the detector but at half the rotational speed in a $\theta/2\theta$ motion. The surface of the specimen thus remains tangential to the focusing circle r_f. In addition to being a device which accurately sets the angles θ and 2θ, the goniometer also acts as a support for all of the various slits and other components which make up the diffractometer. The purpose of the parallel-plate collimators (Soller slits) is to limit the axial divergence i.e., divergence across the specimen along the diffractometer axis and, hence, partially control the shape of the diffracted line profile. It follows that the effective scattering center of the specimen surface should be on the axis of the goniometer and this axis must also be parallel to the axis of the line focus, divergence slit, and receiving slit.

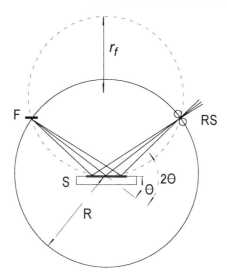

Figure 4.2. Focusing and goniometer circles in the Bragg–Brentano parafocussing diffractomter.

Two circles (actually cylinders) are generated by the Bragg–Brentano arrangement, the goniometer circle, and the focusing circle, and a cross section of these circles is shown Figure 4.2. The goniometer circle has a fixed radius R. The specimen lies at the center of this circle while the source and receiving slits lie on its circumference. Radiation from the tube is directed by the divergence slit onto the surface of the specimen, at an average angle of θ. Radiation is diffracted by the specimen at an angle 2θ to the incident beam, and the receiving slit is placed at this same angle (the "Bragg angle") to collect the diffracted X-ray photons. The source F, the specimen S and the receiving slit RS, all lie on the circumference of the focusing circle of radius r_f. It will be apparent from the figure that because the distances from source to specimen, and specimen to receiving slit are fixed, the radius r_f must vary with the diffraction angle. There is a simple relationship between the goniometer and focusing circles:

$$r_f = \frac{R}{2 \sin \theta}. \tag{4.1}$$

Some of the mechanical requirements of the parafocusing geometry are fulfilled in the design and the construction of the goniometer, whereas the others are met during the alignment procedure, for example, the take-off angle α of the beam, which is the angle between the anode surface and the center of the primary beam. Depending on the target of the X-ray tube and the voltage on the tube, the intensity increases and resolution decreases with increasing take-off angle. For normal operation the take-off angle is typically set at 3–6°. In the alignment of the diffractometer, great care must be taken in the setting of the goniometer zero and the $\theta/2\theta$ position because errors in these adjustments each may introduce errors in the observed 2θ values. Even with a properly aligned diffractometer, all measurements will be subject to certain errors; the more important of these will be discussed below, including the axial divergence error, the flat specimen error, the transparency error, and the specimen displacement error. Other inherent errors may also be present but are generally insignificant for most routine work. These aberrations include refraction and effects of focal line and receiving slit width.

While most routine diffraction work is done in the *reflection* mode described previously, it is sometimes advantageous to work in *transmission* mode. In the transmission mode the specimen is thin and planar, and the diffracted radiation passes right through the specimen. Thus, if in reflection a specimen, due to preferred orientation, showed only (*OOl*) reflections, the same specimen, in transmission, would show only (*hk*0) reflections. Special attachments to Bragg–Brentano diffractometers have been designed to allow this type of measurement [1]. While, in principle, this approach sounds like a useful idea, in practice, application of the transmission diffractometer is very difficult because of the problems in getting the specimen thin enough (because of the problems of preferred orientation and crystallite statistics) to allow a measurable portion of the X-ray beam to pass. It should be noted, however, that it is common practice to utilize the Guinier camera in transmission mode.

4.3 PROPERTIES OF MATERIALS AND SAMPLES

To analyze fully the results of a diffraction experiment, the specimen must either be ideally random or ideally oriented. What is really meant by this description, and can the ideal specimen ever truly be achieved? The following analysis of the random sample shows what needs to be considered when preparing specimens for the various diffraction devices.

4.3.1 Crystallite Statistics

Regardless of the technique employed, the conditions for accuracy, which are very specimen dependent, are the same: total randomness of the crystallite orientations; sufficient crystallites in the experimental sample to meet the statistics required by the desired sampling accuracy; and sufficient intensity measured to meet counting statistics as required by the desired level of precision. Randomness may be described by selecting an equivalent general direction in all crystallites and examining the distribution of this direction vector in space for all the crystallites in the sample. "Crystallite statistics" requires determining how many of these crystallites will diffract in the experiment and whether they are sufficient to allow intensity measurements to the desired accuracy. These conditions will be examined in more detail. This discussion will follow principles first presented by de Wolff [2] who first used the term *particle statistics*. This term should be changed to *crystallite statistics* to be more descriptively accurate.

The conditions of randomness may be analyzed by circumscribing a sphere of unit radius about the sample and plotting the intersection of the poles (direction vector or diffraction vector if the pole is perpendicular to the Bragg plane) on the surface of the sphere. Because the direction selected was general, all other directions in the crystallites should behave similarly. Randomness requires the pole density (number per unit area) to be uniform over the surface of the sphere. Randomness may be achieved with any number of poles (crystallites in the sample) as shown in Figure 4.3. When the number of poles is small, the angle between adjacent poles is large, so that only a few grains in the sample can meet the conditions for diffraction [3]. The average angle between poles is

$$\alpha_p = \sin^{-1} \sqrt{\left(\frac{8\pi}{\sqrt{3} \times N} \right)} \tag{4.2}$$

where N is the number of crystallites in the sample. The number of crystallites in a sample depends on the volume irradiated by the X-ray beam and the crystallite size. Assume the irradiated area is 1 cm^2 which is a reasonable average when using an instrument with a fixed divergence slit. The volume of the sample depends on the depth of penetration of the beam. Because of the effect of beam attenuation with path length, a good estimate of the effective depth is twice the half-depth of penetration, i.e.,

$$t_{1/2} = \frac{1}{\mu} \tag{4.3}$$

where μ is the linear attenuation coefficient for the specimen material. For SiO_2 and $Cu\,K\alpha$, $\mu = 97.6\,\mathrm{cm}^{-1}$ which is approximately $100\,\mathrm{cm}^{-1}$. Thus, the effective volume is essentially 20 mm^3.

The crystallite size in the specimen depends on how the sample was prepared. If the sample was crushed and sieved, the maximum particle size will be determined by

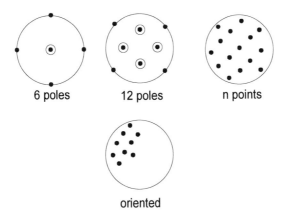

Figure 4.3. Randomness related to pole density.

TABLE 4.1. Screen Mesh and Particle Sizes

Screen Mesh	Effective Particle Size (μm)
200	74
325	45
400	38
600	25
1000	10

the screen size as shown in Table 4.1. Because the 400 mesh screen is a commonly used size, 40 μm crystallites are generally thought to be sufficiently small for accurate quantification. The effect of this size will be compared to 10 μm and 1 μm sizes in the discussion which follows. Actually, most particles are composed of many crystallites, and a 40 μm particle might be an agglomerate of 1 μm crystallites. For the discussion that follows, crystallite size will be considered equal to particle size. (See also sections 1.8 and 4.4.1.)

Table 4.2 compares the particle populations in a sample. There are fewer than a million crystallites in the 40 μm sample compared to over thirty thousand million in the 1 μm sample. The low population of the 40 μm crystallites yields an average angle between the poles of 10 minutes of angle compared to 2.5 seconds in the 1 μm sample. This difference has a significant effect under the diffraction conditions of a specimen.

Figure 4.4 shows a specimen irradiated with a beam of radial divergence angle γ. The divergence angle is much wider than any of the crystallites in the specimen, so it is not this divergence angle that defines the diffraction conditions of any individual crystallite. As shown in the figure, it is the geometric size at a given take-off angle, of the X-ray source focal spot, that limits the angular range over which a single grain may diffract. Although different crystallites within the specimen may diffract within the divergence angle γ, each crystallite is limited by the range α_s where

$$\alpha_s = \sin^{-1} \left[\frac{F + d_s}{R} \right] \tag{4.4}$$

F is the apparent width of the X-ray source, d_s is the diameter of the crystallite, R is the radius of the diffractometer, and α_s is the rocking angle of a crystallite. The rocking angle of a crystallite such as quartz is of the order of 15 s (0.0042°). Thus, α_s for a typical fine-focus diffraction source and 40 μm crystallites is around 0.044°,

TABLE 4.2. Particle Size Comparisons

Diameter	40 μm	10 μm	1 μm
Volume per crystallite (mm^3)	3.35×10^{-5}	5.24×10^{-7}	5.24×10^{-10}
Crystallites per 20 mm^3	5.97×10^5	3.82×10^7	3.82×10^{10}

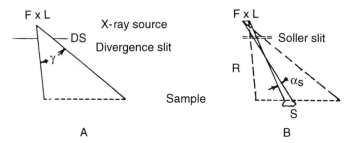

Figure 4.4. (*a*) Diffractometer optics for incident rays. (*b*) Effective ray paths for a single crystallite.

about $1/4$ of the average angle between the direction vectors for 40 μm crystallites. It is quite apparent that a 40 μm crystallite size will have too few crystallites to meet the diffraction conditions necessary for statistical significance of the intensity measurements.

To analyze fully the number of crystallites in diffraction in a particular specimen, axial divergence effects also need to be considered. The length of the X-ray source also limits the angular range in the axial direction. This length depends on the Soller slit. A medium resolution Soller slit (5°) will expose a length, L, of about 0.5 mm to a given crystallite. Thus, the number of crystallites in diffraction N_d is simply given by the ratio of the area on unit sphere (A_d) for diffraction range to the area on unit sphere per particle (A_p).

$$N_d = \frac{A_d}{A_p}$$

$$A_d = \frac{FL}{R} = 2.5 \times 10^{-4} \quad \text{(steradians)}$$

Calculating the ratio A_d to A_p in Table 4.3 gives the number of diffracting crystallites in a given specimen. The number for a 40 μm sample is a surprising 12; hardly enough for good statistics.

To achieve a specific statistical accuracy in the measured intensity, counting statistics of discrete events is a good guide to the number of counts that must be accu-

TABLE 4.3. Particle Distribution Comparisons

Diameter	40 μm	10 μm	1 μm
Area per pole (A_p, steradians)	2.11×10^{-5}	3.29×10^{-7}	6.58×10^{-9}
Angle between poles (α_p, degrees)	0.167	0.0209	0.0007
Crystallites in diffraction (number of crystallites)	12	760	38000

mulated. The same description may be applied to the number of crystallites in the sample.

$$\sigma = \frac{\sqrt{n}}{n} \tag{4.5}$$

For a standard error, 2.3σ, to be less than 1%, n should be greater than 52900 crystallites. Using this number as a guide, even the 1 μm sample fails to meet the desired condition.

4.3.2 Specimen Configurations

This analysis is for a fixed specimen with a single diffraction vector per crystallite. Actually, many aspects of the experiment modify the number of effective crystallites in the specimen. One of the factors is the diffraction multiplicity due to the crystal symmetry. Because $d(hkl)$ always equals $d(\bar{h}\,\bar{k}\,\bar{l})$ for all crystal symmetries, there are always two equivalent directions per crystallite. For crystals other than triclinic, the multiplicity may be considerably higher—up to 48 for some cubic reflections. If a variable divergence slit is used, the irradiated area is more than 2 cm². For low-absorbing materials, the depth of penetration is increased. A broader range of crystallite divergence may be obtained by using the coarser Soller slits and a broad-focus diffraction source.

Spinning the specimen in the specimen plane improves the crystallite statistics somewhat. In Figure 4.5, crystallites in position 1 may not be positioned to diffract, but as the sample is spun to position 2, the crystallite finds a position where it does diffract. Bragg planes whose tilt is within the divergence angle may find some position in which diffraction will occur. The axial divergence will also affect the range of crystallite tilt which will allow diffraction. Analyzing the effect of spinning is not simply counting all the grains within a specific orientational range as shown in Figure 4.5. Crystallite B in position 1 may also diffract in position 2, whereas crystallite A will not diffract in either position. A crystallite on the spin axis requires

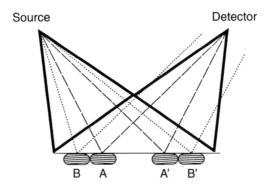

Figure 4.5. Effect of sample spinning on crystallite diffraction.

the Bragg planes to be parallel to the specimen surface, and spinning keeps this crystallite in diffraction position at all times. Spinning usually does bring more crystallites into the irradiated area because the rectangular shape of the irradiated area covers a circular area with spinning. It is evident from this analysis that sample spinning has less effect than is usually assumed. A device for rocking the sample about the diffractometer axis even a few degrees during spinning would bring many more crystallites into diffraction orientation.

The analysis of the ideally oriented specimen is more straightforward. In the case of the platy materials, the goal is to get the perpendicular to the plates to align with the perpendicular to the specimen surface. Actually, one large single crystal meets this requirement. In practice, even the best-made oriented powder specimen is far from this ideal because the mosaic spread (angle of misorientation) of the individual crystallites is usually greater than $\pm$ 5° from the ideal orientation. The effect of the misorientation is minimal if it is not too severe, and the only real problem is to determine which Lorentz correction (single crystal or powder) should be used when correcting the intensity measurement. The best oriented specimen of layered materials is a very thin preparation where most of the crystallites are oriented by the close proximity to the support surface. There are so many crystallites in the desired orientation that the diffracted intensity is more than adequate for good measurements. Rocking curves (rotation on the omega axis only) give a measure of the mosaic spread for evaluating the degree of the orientation

The fiber specimen is intermediate to the random and layered oriented specimens in that the orienting direction (the fiber axis) lies in the plane of the specimen, and there must be randomness of all the orientations around this axis. Ideal randomness around the fiber axis may be achieved by rotation of the specimen on that axis during the measurements or by having sufficient crystallites in the specimen to achieve all possible rotation positions. The number of crystallites needed to meet this requirement is far less than for the fully random specimen analyzed above. If the crystallites are less than 1 μm in cross-section, which is common in fiber studies, there is little doubt that sufficient randomness may be obtained.

The problems in specimen preparation usually appear when the particles do not meet the size criteria developed above. When the particle shape (and the related crystallite orientation) is neither nearly equant in shape and less than 10 μm in diameter or extremely shaped with aspect ratios in excess of 15, the orientation in the specimen is usually nonideal, leading to degradation of the intensity values obtainable from the specimen. Methods to overcome the problems of preferred orientation will be presented in several sections to follow.

The nature of the diffraction specimen will depend on the geometry of the diffraction apparatus employed. For the diffractometer, the specimen is prepared as a thin flat aggregate of particles. When the specimen is sufficiently thick, the nature of the support is usually unimportant. The presence of binders and any protective coatings, however, can affect the diffraction pattern. Specimens for the Guinier camera and specimens for transmission diffractometry must be thin because the X-ray beam must pass through the specimen. The support film and any binders will show in the pattern. Debye–Scherrer specimens also must be transparent to the X-ray beam or the front-reflection region is strongly absorbed. Cements and the support can make up a significant volume of the specimen and cause strong effects in the recorded pattern. In the procedures that follow, each of the different specimen conditions will be considered in appropriate sections.

When deciding on the most appropriate method for preparing the specimen, the question to be answered must be considered. If the specimen is for quantitative analysis, then extreme care must be taken to have the specimen homogeneous, random, and composed of sufficiently fine crystallites to meet the statistical requirements for producing accurate intensities. If the question is to confirm the presence of a phase, a coarse powder slurried on a microscope slide may be adequate. When the problem requires accurate d-spacings, a thin specimen of -200 mesh grains would be satisfactory and preferred orientation is not a problem in general, but when the problem requires accurate intensities, -200 mesh grains would not be adequate because there would not be sufficient crystallites and orientation would be a problem.

4.4   EFFECTS OF CRYSTAL PROPERTIES ON SPECIMENS

The nature of the elements, the crystal structures, and the crystal perfection affect the diffraction pattern in many ways. It is important for the diffractionist to understand these effects and how to combat undesirable effects while enhancing the desired ones when preparing specimens for study.

4.4.1   Crystallite Size and Particle Size

Two terms which are much confused in the diffraction literature are "crystallite size" and "particle size." Many authors use the terms interchangeably; however, they have distinct definitions. In a powder, particle size refers to the free fragment dimensions regardless of whether the fragment is composed of a single-crystal unit or several crystal units intimately attached. Crystallite size refers to the dimension of a single-crystal unit regardless of whether it is free or attached to other crystal units. Usually, when the particle size is small (<0.1 μm), the particles are made up of single-crystal units, and the dimensions are identical—hence the confusion. However, even when the dimensions are small, the single-crystal unit may be subdivided by walls of defects and domains of perfect (coherent) crystals. In this situation, the crystallite size refers to the domain dimensions, not the unit dimensions. It is the coherent domain dimension that is related to peak broadening, whereas it is the particle size that affects sieving and orientation.

4.4.2   Specimen and Sample Homogeneity

The term "homogeneity" applies to both the original sample supplied to the diffractionist and the specimen made for examination. For the diffraction results to be meaningful, the specimen must represent the sample, which in turn must represent the original unit of material being studied. Homogeneity usually implies uniformity in all physical and chemical properties that may affect the diffraction pattern. Lack of homogeneity in the specimen as seen by the X-ray beam may cause several types of variations in the diffraction trace. The powder may become stratified during preparation so that one phase may be concentrated at the surface and other phases concentrated at depth, thus altering the relative intensity response. Some phases may form the larger particles and other phases may form the small particles leading to packing effects and microabsorption. Nonspherical particle shapes may lead to preferred orientation of the crystallites. The inhomogeneities may also be chemical. All the particles of a given phase may not be chemically identical due to incomplete reaction leading to d-spacing shifts and peak distortion. Usually, inhomogeneity effects appear in the diffraction trace as peak shifts, peak broadening, and asymmetry.

There are several conditions that must be met with any specimen mount. It must be inert to the specimen and have long-time stability to be reusable or very low cost if used only once. The material used must not interfere with the diffraction from the specimen. It must be machinable or otherwise modifiable to allow shaping and have dimensional stability. The surface must be smooth and flat to allow accurate positioning in the instrument. Most diffractionists seem to forget that the accuracy of measurements always starts with the specimen, so if the specimen is not perfect, neither will be any measurements derived from the sample. In particular, the single most important dimension for the specimen mount is the surface planarity which affects the positioning of the specimen in the instrument. This flatness should be ± 0.0005 inches. This tolerance applies to top-referenced specimen holders; bottom-referenced holders will be discussed below. Obviously, the other dimensions are also important to allow the specimen location to be positioned correctly in the beam, but the dimensional tolerances are not as great. To meet this surface tolerance, all specimen supports should be lapped. Machined surfaces require some further processing. All supports prepared by stamping should probably be discarded completely.

4.4.3 Crystallite Perfection and Peak Shapes

Crystals are thought of as perfect periodic arrangements of the atoms in the crystal structure, and much of the diffraction theory is based on this state of perfection. In real crystals, there are many types of defects that contribute to departures from the ideal diffraction effects. These defects may be mistakes in the way the crystal is assembled or substitutional atoms which disrupt or distort the periodicity of the crystal and/or modify the intensity of diffraction. The peak shapes are influenced by crystallite dimensions that are less than 0.2 μm. The effect is to increase the full width at half maximum (FWHM) of the profile over that already inherent in the shape due to the source and instrument influence. Pure size effects broaden the profiles symmetrically. The smaller the crystallite, the larger the FWHM. If the d-spacings in adjacent domains formed by walls of dislocations are not identical due to interdomain strain energy, the effect is also to broaden the peak shape. The profile shape in this situation is not necessarily symmetric because it depends on the volume weighted distribution of strains among the domains. If the specimen is chemically inhomogeneous, the effect is similar to the strain effect because different crystallites will have different d-spacings due to the compositional differences. Modified intensities as well as peak shifting will result. The final diffraction profile is the sum of all the contributions, and the result could be quite bizarre.

Advanced diffraction techniques use the diffraction profiles of the full pattern to analyze this imperfect state. Profile fitting and Rietveld techniques allow the profiles to be expressed analytically, which allows them to be related to the physical source of the distortion.

4.4.4 Peak Resolution in Complex Patterns

Even when the crystallites are perfect, phases with low crystallographic symmetry develop complex profiles when the peaks overlap. If the crystallites are perfect, all the peak profiles are similar, i.e., the differences are due to optical aberrations—slit effects and broadening with increasing diffraction angle. Overlapped peaks are additive in forming the full experimental trace, so that it is possible to separate the component peaks by profile decomposition techniques. A single, well-resolved peak allows the program to define the active profile for the experiment, and this peak shape can be used to control the shapes of the profiles used in the decomposition procedure.

Even when decomposition techniques can be applied, they are most successful when the specimen yields optimum resolution of the instrument. If the specimen is prepared as thin as possible (a one-grain-thick layer on a mount surface), there will be no contribution to broadening due to specimen transparency. Thick specimens, especially for low absorption materials, lead to beam penetration, profile broadening, and distortion.

Resolution of peaks is related to the FWHM. Distinct valleys between two adjacent peaks cannot occur if the peak positions are separated less than the FWHM. However, the increased width due to the multiple peaks is evident in the trace, and decomposition procedures can fit multiple peaks to resolve mathematically the component profiles.

Increased resolution requires the reduction of the FWHM of the profiles. This reduction can be affected by using an instrument with different optics such as the Guinier camera or synchrotron diffractometers. Typical FWHM of well-aligned diffractometers in the laboratory is around $0.10°2\theta$ with the $K\alpha_2$ component removed with primary beam monochromators. Guinier cameras and synchrotron instruments have FWHM around $0.02°2\theta$. The better the instrument resolution, the more successful the profile-fitting procedures will be in separating overlapped peaks.

4.4.5 Particle Shapes and Preferred Orientation

Particle shapes affect the manner in which the particles will pack when a specimen is prepared. The larger the particle size, the more the influence. There are four categories of shapes that need to be considered and different degrees of shape influence. Shapes are usually described by considering the difference between the maximum and minimum dimensions of individual particles and the intermediate direction (mutually perpendicular to the maximum and minimum). The usual descriptive terms describing shape are *equant, tabular, bladed, and elongate. Equant* is used where all the dimensions are nearly equal as in a sphere or cubic shape. The *tabular* shape has one dimension much smaller than the other two. *Bladed* has one long dimension and one short dimension with the intermediate dimension neither short nor long. *Elongate* particles have two short dimensions and one long one. The other descriptive term is the aspect ratio, which is the ratio of the long dimension to the short one. Different terms may be used for different aspect ratios; for example, *acicular* is used for elongate particles with aspect ratios greater than 15.

The most efficient packing of particles with long dimensions is for the long dimensions to align with the surface of the support and with each other. Toothpicks and playing cards dropped on the table illustrate this effect. Toothpicks pack even better if they are aligned in parallel orientation. As the particles get smaller, the number of particles in the volume of a specimen increases. When the number gets sufficiently large, even shaped particles will show deviations from the highly oriented state as the specimen thickness increases. Thus, decreasing the particle size to less than 1 μm can have a strong influence on reducing the degree of orientation in the bulk specimen. Unfortunately, it is the surface of a specimen which has the greatest influence on diffraction intensities. If the surface formed next to the support is the surface used for diffraction, it will still contain the most highly oriented particles. This situation would occur if the powder is pressed with a glass slide to flatten the surface.

4.4.6 Absorption Effects and Depth of Penetration

All materials absorb X-rays to some degree, and most materials show significant absorption of Cu K radiation. The absorption of a specimen depends on the elements that comprise the phase and the density of packing of the particles. Absorption is usually employed as the linear absorption coefficient. This coefficient may easily be calculated for a pure phase from the tables of elemental mass attenuation coefficients found in almost every textbook on diffraction, and in the *International Tables for X-ray Crystallography*, Volumes III and IV, combined with a knowledge of the chemical composition of the phase.

The most important parameter to know in the preparation of specimens is the half-depth of penetration. This value indicates the horizon below the surface of the specimen above which half the intensity of diffraction originates, equation (4.3). The other half arises from below this horizon. In effect, this parameter indicates the transparency of the specimen. For SiO_2, a common material, the linear attenuation coefficient is 97 cm^{-1}. The half-depth is the reciprocal of the linear absorption coefficient which is about 100 μm for SiO_2. A 200 mesh sieve passes 74 μm particles, so a two-particle layer would be sufficient for a pattern with 50% of the possible diffracted intensity. Ten particle layers would diffract around 99% of the total possible intensity. Thus, it is not necessary to prepare thick specimens to achieve the maximum intensity.

As the absorption increases, the thickness of the specimen needed decreases, and usually only a few particle layers are necessary to achieve the optimum specimen.

This behavior is the principle reason that the smear mount on a microscope slide is so successful as a diffractometer specimen. For the heavy elements, the half depth is around 10 μm, and the X-ray beam rarely penetrates more than the surface layer of particles. On the other hand, for organic phases with only light elements, the penetration is so large that significant profile broadening occurs due to the penetration to the full thickness of the specimen. For organic phases, a cavity mount (4.5.3) or low background, single-crystal holder (4.7.4) is essential. In an extreme example of low absorption, LiH has a half-depth of 1 cm for Cu $K\alpha$, which really causes problems. The absence of absorption can be as much of a problem as high absorption in the preparation of diffractometer specimens.

High absorption also affects the preparation of mounts for Debye–Scherrer and Guinier cameras because they are essentially transmission techniques. In the Debye–Scherrer camera (and the Gandolfi), only the edge of a highly absorbing specimen contributes to the diffraction pattern. In the Guinier camera, only the edges of the particles would allow the beam to diffract. Thus, very thin specimens are necessary for optimum behavior in these devices. The optimum thickness for transmission specimens is usually considered to be the reciprocal of the linear attenuation coefficient.

4.4.7 Microabsorption

Microabsorption is a phenomena of masking where a particle of high absorption shields a particle of low absorption from the X-ray beam. When the particle sizes of the phases in the specimen get above 10 μm, this problem becomes critical, especially when the difference in attenuation coefficients of the different phases in the sample gets large. Brindley [4] has developed the correction curves for the effect, but it is usually more successful to reduce the particle size than attempt the correction. Microabsorption must be considered when preparing specimens for quantitative analysis or structure refinement or where the detection of trace quantities of low absorbing phase is desired.

Considering the role of absorption and transparency on the diffraction pattern helps the diffractionist to optimize the specimen preparation. The goal of the experiment must also be considered. If the specimen is to be used for accurate intensity measurements, the specimen thickness has to be sufficient to diffract fully the incident X-ray beam. This condition usually requires some type of cavity mount. On the other hand, if d-spacing accuracy is the goal, very sharp diffraction peaks are wanted, and a very thin sample will yield better profiles than a thick sample. Sometimes the quantity of material provided will override the experimental requirements. However, a good specimen for quantitative analysis may be made with little material if all the particles are concentrated in the volume of the mount where the X-ray beam is active. Thus, a slide mount may be employed with the particles concentrated in a rectangular area where the beam would hit the slide. This location and size may be determined by substituting a fluorescent screen in the specimen position and observing the bright region.

Many cavity mounts have cavities that are much too deep for the penetration conditions. Typical large cavities may require 0.5 g of material to fill. Such mounts may be partially filled with inert (amorphous) fillers before being covered with the actual material. Holders with very thin cavities may be constructed in the machine shop. Unfortunately, thin cavities may prove very difficult to use when the requirements are for a random specimen.

4.4.8 Background

Background in a diffraction pattern is considered to be all the intensity not directly attributable to the crystalline diffraction from the specimen being studied. These contributions are many and include scatter from the air molecules in the beam path, diffracted and scattered radiation whose wavelength is other than the desired wavelength, thermal diffuse scattering from the specimen, diffraction and scattering from the specimen support, electronic noise, scattering and diffraction from an amorphous phase or binder in the specimen, and fluorescence from the specimen. The goal in specimen preparation is to understand all the sources of unwanted radiation and minimize their effect.

Some sources such as the air scatter can be eliminated at a cost, i.e., by adding a helium beam path to the diffractometer. Fluorescence and secondary radiation can be eliminated by using wavelength discrimination in the detection system such as a secondary beam monochromator. Fluorescence by the same element as the X-ray source can be eliminated only by using a primary beam monochromator to cut out the exciting radiation. Specimen support interference can be eliminated by using nonscattering materials such as specially designed single-crystal mounts. Electronic noise can be minimized by fine tuning of the instrument, but it cannot be eliminated altogether.

Unwanted radiation originating from the specimen is more difficult to eliminate. Scattering from an amorphous component which is part of the sample is usually not avoidable unless the components can be physically separated prior to preparing the specimen. Scattering from binders is minimized by choosing the best binder material and using no more than is absolutely necessary to retain the particles in the specimen. Thermal diffuse scattering is minimized by fine-tuning of the detector window to eliminate the wavelength modified component.

Even with great care, some background will persist. Where necessary, the remainder of the background must be eliminated from the diffraction trace by data processing. Defining what is background at this point in the analysis is not easy. The contribution of the instrument and specimen support can be determined by running blank specimens under exactly the same conditions as will be used for the experiment. Even when working with ideal instrumental conditions, materials such as polymers and clay minerals will still show considerable low-angle scatter and other medium range noncrystalline diffraction which cannot be eliminated. When doing quantification, the user must be especially careful to retain the true diffraction component. It is possible to construct a background line that follows the base of the diffraction trace with or without any amorphous contribution or one that represents only the instrumental and support contributions. The choice depends on the goal of the experiment.

4.4.9 Surface Roughness

In diffractometry, it is the surface of the specimen which has the strongest influence on the nature of the diffraction trace. With very low absorbing phases, the surface influence is small; but with most phases encountered in the inorganic laboratory, the dominant contribution to the trace is from the top 100 μm of specimen. As a general rule, 25 μm vertical displacement of the diffraction point causes a shift of the diffraction peak of $0.01°2\theta$. Thus, the specimen surface must be smooth within 25 μm with no ridges or valleys to provide the optimum diffraction profile. Such smoothness may only be obtained when the particle size is below 25 μm. When the absorption is very high ($\mu > 1000$ cm^{-1}), this problem becomes very severe as a ridge of particles can completely mask the valley particles at low diffraction angles. This problem must be considered when preparing specimens containing high atomic number elements.

4.4.10 Crystal Perfection and Extinction

A little-appreciated effect in powder diffraction is extinction. This effect is the diminishing of strong intensities from crystals due to multiple diffraction from the same set of Bragg planes. There are two kinds of extinction, primary and secondary. Primary extinction occurs in perfect crystals; secondary extinction occurs in ideally imperfect crystals.

In primary extinction, the incident beam is diffracted at the Bragg angle; before the diffracted ray leaves the crystal, it is diffracted again because it is still at the exact Bragg angle with respect to the crystal. This multiple diffraction causes some of the diffracted energy to follow the original beam path, and thus, the diffracted beam is reduced in intensity. The magnitude of this extinction is not trivial. For example, the $(10\bar{1}1)$ peak ($d = 3.33$ Å) of SiO_2, quartz, has a normalized intensity of 100. On this same scale the $(10\bar{1}0)$ peak ($d = 4.26$ Å) has a theoretical intensity of 17. Experimental values are 100 and 22 (PDF #33-1161). What has happened is that the strong peak has been reduced by primary extinction, so that the normalization brings up all the other peaks. Primary extinction can be reduced only by reducing the crystallite dimensions to submicrometer values. In a perfect material such as quartz, this reduction is essentially impossible.

Even in imperfect crystals where domains are already small, extinction effects may still occur. Secondary extinction occurs when a diffracted ray from one domain is diffracted a second time by a different domain because the angle is still correct. The physics of this type of multiple diffraction is different from primary extinction, and the magnitude is not as large, but it still occurs in most samples.

Regardless of whether the extinction is primary or secondary, the diffractionist must be aware of the effect. One obvious situation where extinction can cause experimental problems is in the selection of reference phases. A highly perfect reference sample may not be the best choice for a standard diffraction trace when the materials to be studied will be highly imperfect. This situation could arise, for example, in phase quantification by full-trace matching.

4.5 PREPARING SPECIMENS FOR SIMPLE ROUTINE ANALYSIS

4.5.1 Goal of Specimen Preparation

For simple routine analysis of most inorganic specimens (corrosion products, zeolitic materials, clays, ceramics, glasses, minerals, ashes, etc.), it is important to prepare the specimen to meet the following criteria:

- The specimen should be representative of the bulk sample.
- The particle size should be 45 μm or less to promote random orientation of the crystallites and produce a sufficient number of crystallites (see section 4.3.1) to meet the powder diffraction criteria.
- The method used to prepare a powder should not cause distortion (or even destruction) of the lattice.

Samples already in powder form may be lightly ground in an agate, mullite, or other type of mortar and then sieved to a minimum -200 mesh. The specimen will feel smooth and will plate easily on the mortar surface. For most specimens, particularly those with low symmetry, sieving to -325 mesh to achieve the recommended < 45 μm mean particle size will produce an adequate diffraction pattern. Samples showing obvious variations in texture or color should be separated initially by color or be intimately ground, using the entire amount to ensure a uniform, representative specimen. Coarser specimens, such as metal corrosion products or rocks, may require the use of an impact mortar (tungsten carbide or sintered corundum) to reduce the sample to a powder form sufficiently for it to be finely ground in an appropriate mortar.

Samples such as metals and alloys will not be amenable to standard grinding techniques. One method for treating these samples is to abrade the sample using a metal file over several areas of the metal sample to obtain a representative sample. These "fines" should not be ground further because serious distortion of the lattice may result. Annealing may also be required to eliminate induced effects of filing.

Certain organic samples such as polymers, plastics and elastomers may be prepared as powders by cryogenically cooling the materials with liquid nitrogen and grinding in a SPEX Freezer/Mill. The resulting powder will have a sufficiently small particle size for routine XRD analysis.

4.5.2 Flat Specimen Supports

Two types of specimen supports are commonly used for routine XRD analysis: amorphous and crystalline.

- *Amorphous Support.* A plain glass microscope slide provides an adequate support medium for many specimens, particularly those for which the amount of sample may be limited. Other amorphous materials include a variety of rigid plastics such as Plexiglas, Bakelite, etc. The materials may be used as a flat surface or with a milled cavity.

- *Crystalline Supports.* These supports come in a variety of materials; aluminum is the most common. They usually contain a cavity. Certain single crystal specimen holders, including quartz or silicon cut off-axis, to yield no interfering X-ray pattern (often referred to as a "zero background" mount) are also available. These supports can be mounted into sample holders appropriate for the type of diffractometer being used. These supports can have depressions of varying depths, including having one surface flush, or nearly flush, with the sample holder face to accommodate very small amounts (micrograms) of material.

The following sections describe several methods for preparing specimens for the diffractometer. The procedures will assume the availability of the sample in powdered form and that it is stable under ambient conditions. The two most common sample preparation procedures are the cavity mount and the slurry mount on a glass slide. The methods are quick and the specimen can be recovered intact. The slurry mount requires 2–10 mg of powder; the cavity mount may require 0.5 g.

4.5.3   The Cavity Mount

In the cavity mount, the top surface of grains dominates the intensity in the diffraction equipment. Unfortunately, it is this same volume that is most susceptible to orientation in the cavity mount; see section 4.4.5. Orientation can be reduced considerably if a frosted surface, matte paper (Wattman filter paper) or other rough surface is used to press the specimen into the cavity. The surface will be less smooth, but the gain in randomness offsets the surface effects.

There are many types of cavity mounts supplied by the manufacturers for most diffractometers available today, and other types may be prepared in the local machine shop. The mounts are usually made of metal or plastic with rectangular or circular cavities 0.5 to 1.5 mm deep. The opening must be larger than the area of the X-ray beam at the lowest diffraction angle used to assure proper intensity measurements without corrections. Most mounts, however, do not subtend the entire beam at low angles. Plastic mounts may contribute both diffraction and scattered intensity to the background at these angles but crystalline metal holders usually have no peaks below 30° to cause diffraction.

Supplies   These include a mortar and pestle, microscope slide, spatula, cavity mount for diffractometer, and single-edged razor blades.

If the specimen is prepared for a horizontal diffractometer, the packing density must be as high as possible to prevent the particles from falling out when inserted in the diffractometer. A vertical diffractometer does not require such a dense packing, but the whole cavity should be filled.

Front Loading in Bottomed Cavity   The front loading cavity holder is the most common type used for diffraction analysis. Basically, a cavity is carved out from the surface of a holder made of metal or glass. A brief procedure for loading the specimen is as follows:

- First a spatula is used to transfer an appropriate amount of powdered sample into the holder cavity.
- A glass slide with a smooth edge or a razor blade could be used to assist the packing by dicing the sample in various directions.
- Press the surface, at the same time moving the edge of the slide across the sample. This movement will ensure that the surface is flat and at the same time remove the excess material.
- Repetition may be needed if a tight, smooth, and flat surface is not achieved the first time.
- The above procedure is appropriate for the d-spacing measurement, where correct relative intensities are not the objective.

If a sample spinner is available, one should prepare the intensity sample using the rotating specimen holder. The sample should be loosely placed onto the holder cavity without pressing or taping. Dice the sample in different directions using a razor blade or glass slide, then remove the surplus.

With cavity mounts it is possible to mix the specimen with an amorphous filler material, which can aid in producing random particle orientation. Powdered glass, silica gel, gum arabic, tragacanth, powdered cork, amorphous boron, gelatin, and corn starch have been used. The low absorption of these additives allows increased X-ray beam penetration into the specimen. The peaks may be broadened somewhat, and the intensities will be reduced due to dilution and some absorption by the additive. The starch and silica gel are composed of spherical particles which attract the specimen particles and create spherical agglomerates. When packing these aggregates, it is important to use only light pressure so as not to break up the clumps. A mixture of powder and gum arabic can be fused by low heat and reground after solidification to give a better packing specimen. Methods for preparing true spherical agglomerates for eliminating preferred orientation are described in section 4.5.9. Whenever additives are used, blank runs on the additive should always be made in order to interpret any diffraction interference effects.

In addition to using different binders and assisting additives, there are other ways to prepare the specimen directly with the powder to obtain more random orientations of the particles.

4.5.4 Slurry Mounts

It is first necessary to determine where to place the sample on the microscope slide. Some diffractometers need the specimen in the center of the slide, whereas other diffractometers need the specimen on one end, depending on how the slide is held in the instrument. After disaggregating the sample in the mortar, use the spatula to place a small pile of powder on the slide at the proper location. Add a few drops of the wetting reagent to the pile of powder, and spread the powder to cover the slide surface evenly with the tip of the spatula. The goal is to cover only that part of the slide "seen" by the X-ray beam with a dense, thin, uniform layer of grains whose surface is smooth and flat. If the specimen is not densely packed, the diffracted intensity will be lower than necessary. If the specimen is too thick, the diffraction peaks will be displaced to higher angles.

A variation on this method is to grind the wetting agent and powder in the mortar, let most of the liquid evaporate so that a thick slurry is produced, and scoop up and spread this slurry on the glass slide. This procedure is referred to as "grinding the sample under a particular liquid."

Supplies   These supplies include wetting reagents: acetone, alcohol, toluene, cyclohexane, water, mineral oil, etc; cements: collodion in ethyl or amyl acetate, Ambroid in nitrobenzene, rubber cement in toluene, mucilage in Karo, Duco in acetone or amyl acetate; petrographic microscope slides; mortar and pestle; and spatula.

Although alcohol or acetone will usually wet all the particles uniformly, allowing an even spread of grains on the microscope slide, some materials do cause difficulty. In such situations, a different wetting reagent needs to be found. Ideally, the reagent should wet all the particles, be unreactive with the particles, and evaporate quickly to allow rapid use of the mount. Water usually is not a good choice, but it may be used sometimes. Higher molecular weight alcohols or ketones may also work. Many reagents that were once common in the laboratory are no longer allowed because of toxicity problems.

Advantages of the slurry mount include the ease of preparation and the small amount of sample required. When there is a limited amount of powder, the whole sample should be concentrated in a narrow band which would lie on the diffractometer axis when the slide is inserted into the diffractometer. The use of a wetting reagent usually creates a cohesive bond to the substrate that allows the grains to adhere well to the glass. This type of mount is usually successful for horizontal diffractometers as well as for vertical diffractometers (see section 4.6.3 for further discussion of this type of mount).

- Deposit a small amount of finely ground powder on the glass slide (a few micrograms is adequate).
- Add 1–3 drops of acetone, methanol, or other solvent to the surface adjoining the powder.
- Using the tip of a stainless steel spatula, briskly whisk the solvent and powder to disperse the slurry uniformly, taking care not to have excessive "hills and valleys" on the surface of the specimen (ideally the surface area of the specimen should be larger than that irradiated by the X-ray beam).

If the particles have any elongate shape, the particles in the thin slurry will tend to settle with the long dimension aligned parallel to the slide surface, see section 4.4.5. Another problem is the positioning of the powder on the substrate. Because the powder is above the surface of the slide, all the particles are above the diffractometer axis, and the peaks will be displaced to higher diffraction angles. The thicker the specimen, the larger the displacement. This displacement effect may be compensated for by using a spacer in the diffractometer to shift the specimen surface. The spacer may be an appropriate thickness of tape adhered to the surface of the microscope slide where it contacts the diffractometer reference surface. If the surface is not flat, the hills will shadow the valleys at low diffraction angles and affect the intensity measurements.

Instead of using a low-viscosity medium for dispersing the powder, a high-viscosity mixture of collodion in amyl acetate, petroleum jelly, or Apiezon grease will not allow the particles to settle against the polished microscope slide. It will also act as a cementing binder to create a permanent mount. This type of preparation tends to yield less smooth surfaces, but the gain in particle randomness is worth the degradation in pattern quality, especially for phase identification. It is also possible to use a frosted surface microscope slide to decrease the orientation induced by the polished surface.

When analyzing small, flat pieces of metal, ceramic, or polymer film, cut the sample to the size of the microscope slide and provide a bond with a small amount of a sticky substance. Build up the edge of the slide with a spacer so that the top of the specimen is as close as possible to the reference focus of the diffractometer. In some cases, a rigid sample itself can be cut to the dimensions of a sample holder and inserted directly into the diffractometer. It is, of course, not possible to control either the particle dimensions or orientation, so relative intensities may be far from the correct values.

4.5.5 Alternative Cavity Mounting Techniques, Back-loading Cavity Mounts

Although many specimens may be prepared successfully with the cavity or slurry technique, alternative techniques have been employed to improve diffraction results. The primary goal of these alternate methods is to allow better reproducibly of specimen preparation and to minimize orientation effects.

Supplies These supplies include a bottomless cavity mount, microscope slides cut in half, transparent tape, mortar and pestle, spatula, filter paper, and a clip.

Prepare the cavity mount by covering the top of the mount with the filter paper backed by a microscope slide and held with a clip positioned so that it does not impede access to the open side of the cavity. Disaggregate the powder and carefully transfer it to the cavity with the spatula. Distribute the powder evenly in the cavity, being sure to fill the corners. Overfill the cavity slightly and then press the powder lightly with the small glass slide to compact it. Then permanently tape the small glass slide over the opening. Turn the assembly over and carefully remove the clip, the upper glass slide and the filter paper. The exposed surface should be flat and smooth without vacant pockets.

This mount should have less orientation than the front-loaded mount which has been pressed. The filter paper and the back-loading tend to diminish the orientation at the front surface. Although there may be some orientation at the back surface, that part of the specimen will not be active in the diffractometer.

The standard back-loading technique using aluminum cavity mounts is as follows:

- Grind the sample to a fine powder.
- Place the sample holder onto a frosted side up, or plain, glass slide and secure in place with sticky tape or a clip.
- Sift a sufficient amount of powder to completely fill the cavity.
- Using the edge of a stainless steel spatula, carefully tamp the surface of the powder, being sure to work the powder into the corners of the cavity.
- Using a clean plain glass slide, gently press down on the open surface of the powder.
- Remove any excess powder from around the cavity with the edge of a spatula or a single-edge razor blade.
- Sift on an additional layer of the powder and press again with the glass slide, being sure that the surface of the powder is completely flush with the aluminum mount.
- Clean the area surrounding the cavity with the edge of the spatula or a tissue and carefully remove the tape without lifting the mount from the frosted glass slide.
- Attach the backing plate and press down to secure in place.
- Invert the sample holder and lift the frosted glass slide straight up so as not to disturb the sample surface (if a plain glass slide was used, the surface may be "roughed" using a piece of clean cardboard or other similar material).

Several problems may arise with this mounting method. A large amount of sample is required to fill adequately a standard mount, typically 0.5 g or more. If sufficient sample is not available to produce a tightly packed mount, plastic shims or spacers of varying thicknesses may be cut to fit the cavity and secure the powder in place. If using a back-loading method, the shim must not extend above the surface of the back portion of the aluminum mount because the backing plate may not remain in place. Using shims frequently causes the powder to swell above the surface of the illuminated side, resulting in possible loss of sample and a displacement error in the data. If a front-loading method is used, such as described for the zero background cavity mounting method, pulling a glass slide across the surface will smooth out irregularities.

A very finely divided sample may "slump" upon sitting. If this occurs, simply reinvert the sample onto the glass slide, remove the backing plate, and add additional sample using the steps outlined above. Some light element samples, such as carbon black, may be transparent to the X-ray beam—this transparency will result in severely displaced peaks from the sample holder in the specimen pattern.

For diffractometers without an automatic sample changer, a variety of mounts may be constructed to hold unusually sized specimens. For the Siemens D500, a special circular aluminum cavity mount, with a diameter of 2.5 cm by 0.6 cm deep, can accommodate both large and small amounts of powdered sample. The cavity mount is placed on top of a polished aluminum face plate, secured by two pins. The powder is loaded into the cavity and tamped down with a spatula, being sure to uniformly cover the base surface. A Teflon insert is pressed into the cavity, and this entire unit is then mounted onto a specially designed press. The press secures the Teflon insert firmly in place. The mount is then inverted and the aluminum faceplate carefully removed. The powder surface may be "roughened" if necessary.

The back-loading technique can effectively reduce preferred orientation. The technique in which one fills the cell from the back and produces a plane surface is recommended by McGreery [5]. The following is an excerpt from *X-ray Diffraction Procedures* by Klug and Alexander [6].

> The sample holder is a rectangle sheet of aluminum $2 \times 2 \times 0.2$ cm in size with a 15-mm hole drilled through the center. Other items or equipment used in setting up the powder include two microscope slides cut in half, a razor blade, a paper clip, small scissors, stiff spatula, a roll of 2 cm Scotch tape, and a special small sieve made from a DeVilbiss hose-coupling sleeve, which acts on the principle of a flour sifter and is used for filling the cell. The cell-filling procedure is then as follows (see Figure 4.6A to I) (after McCreery [5]).

Figure 4.6. The cell-filling procedure for the back-packing technique is shown as in (A) to (I).

- A pencil notation on the face side of the aluminum cell serves for identification. The face is then covered with a clean glass slide and bound firmly at each end with tape. Doubling the tip of the tape on the back side of the cell will aid in removing it later.
- Place the cell, face down, on a flat surface and sift an excess of powder into the cavity (*A*) and (*B*).
- Tamp the surplus gently but thoroughly with the edge of the spatula (*C*). This step is important because it causes the cavity to fill evenly with a minimum amount of flow at the glass face plate.
- Slice off the surplus powder with the razor blade (*D*).
- Sift on a loose layer of powder about 2 mm thick (*E*). Press gently but firmly with the flat blade of the spatula to compress the powder (*F*). slice off the surplus powder again and repeat with a fresh layer (*D, E, F*).
- Loosen the tape at each end, wipe off the surplus powder, and cover with a clean slide (*G*).
- Turn the assembly over, remove the tape from one end, and slip the razor blade under the glass faceplate, thus holding the cell down while the glass is lifted off (*H*).
- Two diagonally opposite corners of the cell are then bound with Scotch tape so that only two small patches of aluminum about 3-mm^2 are covered. This size keeps the tape out of the path of the X-ray beam. Excess tape is trimmed off with scissors (*I*).

4.5.6 Alternative Cavity Mounting Techniques, Side-Drifted Specimens

Supplies   These supplies include a slotted cavity mount, microscope slides cut in half, transparent tape, mortar and pestle, spatula, filter paper, clip, and weighing paper.

The slotted mount may be a bottomless cavity mount with one edge of the cavity (the high edge when mounted in the diffractometer) cut out. The assembly is prepared by first taping the small microscope slide to the back side of the mount over the open cavity. The front opening is covered with a piece of filter paper and a slide which is clipped in place leaving the edge opening accessible. The powder is carefully added into the slot and guided into the cavity opening. Sufficient powder is added to fill the cavity to its original dimensions before the slot was cut. Be sure to fill the cavity evenly. Now the clip may be removed and the slide cover and filter paper carefully lifted off. If the loading was successful, the front surface will be smooth and flat with no open pits. Users of this method claim that it is the best way to minimize orientation when done properly.

The related side-drifted technique is designed mainly for intensity measurement. This technique was used for many years by the JCPDS-ICDD Associateship at what is now the National Institute for Science and Technology (NIST) for the preparation of standard reference patterns. To prepare the specimen, a glass slide is clamped over the top face to form a temporary cavity wall, and the powdered sample is allowed to drift into the end opening while the holder is held in a vertical position (Figure 4.7 *a*, *b*). Do not tap the holder during the sample drifting process.

4.5.7 Alternative Flat Surface Techniques, Top Dusted Mounts

Supplies These supplies include a petrographic microscope slide, Apiezon or Vaseline grease, mortar and pestle, and a sieve set.

Prepare the microscope slide by putting a thin grease film over the region where the powder must be located for the specific diffractometer system (either in the center or at one end). Place the slide in the pan of the sieve set. Disaggregate the powder and grind it to the required particle size. Using the 100 mesh sieve, spread the powder directly over the grease-covered region by gently tapping the sieve. Cover the greased region completely and uniformly. Remove the slide from the pan and wipe the excess powder outside the active area away. If the powder forms a thin uniform layer, the slide may be used directly. If there is an excess of powder, turn the slide on edge and gently tap it to induce the excess powder to fall away, leaving only those particles that are adhering to the grease film. This method is effective in minimizing orientation, but it is difficult to get a dense single particle layer to adhere fully to the surface, and so the technique results in lowered intensities. A step-wise procedure is presented.

- Grind a small amount of sample to a fine powder.
- Place the sample holder with the greased face upright on a flat surface covered by a 10×10 cm piece of weighing paper to catch any excess powder.
- Using a -100 mesh sieve held about 10 cm above the surface of the holder, lightly dust the powder onto the greased holder, taking care to uniformly cover the surface (if sufficient powder is not available to completely cover the surface, try to concentrate the powder in the center of the greased plate).
- Tilt the sample holder on edge and lightly tap it to remove any excess powder.
- Remove any powder adhering to the metal portion of the sample holder by carefully wiping around the edge with a tissue.

The dusted layer method should be used when preferred orientation effects may result from other mounting techniques or when dealing with an extremely small amount of sample. Only pure petroleum jelly should be used. Several commercial Vaseline-type products contain boric acid or other ingredients that can produce a diffraction pattern which will interfere with the sample pattern. Even pure petroleum jelly has a characteristic amorphous pattern. When analyzing a sample that may contain one or more amorphous phases, a pattern of the coating on the surface of the zero background plate should be obtained using the same experimental data collection parameters so that a comparison can be made with the sample pattern.

This step could also be done with any single or double-sided sticky tape used to mount a specimen. Most transparent mending tapes are usable but blank runs need to be made to ascertain any potential interfering lines.

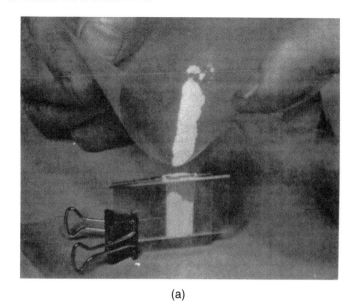

(a)

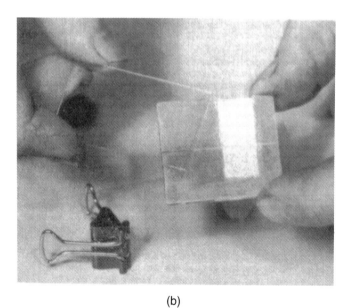

(b)

Figure 4.7. (*a*) The side-drifting technique for specimen preparation. (*b*) The surface of the specimen after the glass slide is removed.

4.5.8   Creating Spheroidal Particle Aggregates

Although the above procedures produce usable specimens for diffractometry, most mounts are still subject to considerable preferred orientation if the particles have distinctly anisotropic shapes. Modifications of these procedures have been developed to minimize the orientation partially or completely. These modifications will be described.

If the particle shapes are spherical, there is no tendency to create an orientation on packing. If a collection of shaped particles can be aggregated into spherical agglomerates, the agglomerates can then be packed into a cavity with no orientation. This technique was first suggested by Devers [7] and perfected by Snyder and Cline [8]. Commercial spray-driers may be employed for spherodizing aggregates, but their cost is a limiting factor. Also, a considerable quantity of sample is required to get a usable amount of spherical particles. A low-cost alternative method will be described.

Supplies   These supplies include a bottomed cavity mount, mortar and pestle, small diameter sieves 60, 80, and 100 mesh, clear acrylic spray lacquer, large glass plate, > 24 in., single-edge razor blades, and a glass cleaning agent.

Place the glass plate on an appropriate surface that may be cleaned after the specimens are completed. Thoroughly clean the surface of the plate with an appropriate cleaning agent. Disaggregate 2 to 5 g of sample to particle sizes < 45 μm and place the powder in the 100 mesh sieve. Dust powder over the surface of the glass plate by continuous tapping and shaking of the sieve. Try to achieve a uniform coverage. Hold the acrylic can about 12 in. above the plate and lightly spray to partially wet the particles. Allow the acrylic to dry completely. Then use the razor blade to scrape the aggregates free from the glass. Stack the sieves and sieve the aggregates completely. The uncemented particles should pass through the three sieves. Complex aggregates will be retained in the 60 mesh sieve. The aggregates in the 80 and 100 mesh sieve should be small uniform sized spheres. Depending on the material, the $+100$ mesh particles might be better spheres, but usually the $+80$ mesh collection is better. Use the best spheres to fill a cavity mount to a uniform surface level. This mount should be free of preferred orientation. Obviously, it is not advisable to compact this type of mount by pressing, so the technique works best for a vertical diffractometer. To prepare a similar mount for a horizontal diffractometer, use 6 μm Mylar film to cover the front of an open-back or open-sided mount, and load the spheres from the back or side. A small microscope slide taped on the back completes the mount.

4.5.9   Bonded Specimens

It is alternatively possible to prepare the powder into a block bound by a strong cement like an epoxy or a weak cement like collodion which is cut after setting to fit a specimen mount. The powder should be ground to -400 mesh and then thoroughly mixed with a small amount of permanent cement. About 40% of the mixture will have to be cement to get a solid product. Ideally, isostatic pressing during setting will lead to a mount with near random orientation, but setting without pressure will usually suffice. Uniaxial pressing is to be avoided. After the block is hard, it is sawed or lapped to give a flat surface and trimmed to fit in the diffractometer mount. This type of mount will show a strong diffraction band from the binder which must be subtracted from the diffraction pattern to yield the necessary information.

A slurry mount may be prepared directly in a cavity mount. A slurry with a weak cement is prepared and packed into a bottomed cavity mount. After hardening, the surface is cut with a straight-edge razor blade or lapped on a diamond wheel to give a flat surface coplanar with the mount. If the cement is just right, the surface will cut easily to give a smooth, flat finish. The cavity should be overfilled significantly, so that the differential setting of the cement does not induce any orientation of the grains during the hardening. Weak cements will usually allow an easy cut, but care has to be taken not to loosen the block from the cavity. Finding the right consistency of cement may require some experimentation.

Block mounts may be prepared by lapping a flat surface then trimming the block to fit the cavity. The support and specimen are placed face down on a smooth surface, and a supporting wax is pressed into the back to fix the specimen in position. Tackiwax is a good support. The specimen is turned over for use.

4.6 PREPARING SPECIMENS FOR REFERENCE PATTERNS

The main purpose of reference powder patterns is to serve as standards for crystalline phase identification. The patterns can be used in house, or added to the Powder Diffraction File (PDF), an internationally available database for analytical XRD maintained by the International Centre for Diffraction Data (ICDD). The goal is to obtain the best possible powder diffraction data for a particular phase. To achieve this goal, different types of specimens for the different measurements are necessary. The emphasis here is on which specimen preparation method to use for each step of data determination.

4.6.1 Preliminary Pattern

The sample from which the specimens will be prepared must be a single, well-characterized phase. Assurance of single phase material may be accomplished by petrographic microscopy and confirmed by indexing the powder pattern. Determination of the exact composition, free from solid solution variation, is, however, more difficult. If the sample was prepared in the laboratory, some assurance can be obtained from the chemicals used as precursors. In other cases, one must depend on chemical analysis, microprobe, or spectroscopic analysis.

Thus, the first step in reference pattern determination is to prepare a preliminary sample to determine its phase purity, to assure that crystalline order is adequate (i.e., that peaks are sharp), and to determine which internal standards for angle calibration are to be used. If the sample is a mineral, the analyst can do little to control impurities (especially intergrowths of associated minerals) and poor crystalline order, other than searching for a purer, more crystalline mineral specimen. For other types of materials, steps should be taken to remove or minimize impurity phases and improve crystalline order (by annealing, recrystallization, etc.). The ability to index every reflection on a list of d-spacings and indices generated from a refined unit cell (see section 4.6.2) is one of the best tests of phase purity.

The specimen for this pattern can be of the routine type described in section 4.5, such as a slurry mount on a glass slide or a pressed powder. It is not necessary to grind the sample to the final level of fineness for this exploratory diffraction pattern.

When the phase purity and crystalline order are found to be satisfactory, the sample should be reduced to particle sizes $<10\ \mu m$ in order to ensure good crystallite statistics in the specimens used through the data collection. The analyst should make another diffraction pattern to check that the grinding step has not induced phase transitions, changes in hydration state, or significantly broadened peaks. As discussed elsewhere, specimens of metals and alloys may have to be annealed to return the peaks to their original sharpness.

4.6.2   Preparing Specimens for Accurate d-spacing Measurements: Background Information

In the preparation of good diffraction patterns, the acquisition of the most accurate diffraction angles (and thus d-spacings) and the most accurate intensities may not necessarily be obtained on the same specimen. For accurate d-spacings, the diffraction profiles should be as narrow as possible from the sample available. For accurate intensities, it is more important that the specimen have a totally random orientation of crystallites and enough crystallites to satisfy the statistics of randomness within the conditions of the diffraction experiment. These two conditions are not mutually exclusive in the ideal case, but with real samples, two different specimens for the two different measurements may be advisable. The condition for sharp peaks requires that the sample be examined in a state where the crystallites are free of defects that broaden the profile, and the thickness is lower to minimize specimen transparency error. The condition that the crystallites statistics must meet usually require crystallite sizes in the 1 to 10 μm range. Using a thin smear on a glass or zero background slide will minimize peak broadening due to the specimen transparency effect.

For the measurement of diffraction angles, the goal is to achieve the sharpest diffraction profiles. Sharp profiles lead to ease of accurate measurement and to the best resolution of adjacent peaks. The width of the profile is a convolution of the X-ray source, the instrument, and the specimen. The spectral distribution of the X-ray source from a sealed tube is fixed by the physics of X-ray generation. Old tubes may have cavitated the target and distorted the source image, which would affect the focal image and not the distribution. Except for changing the tube, there is little the operator can do with the source except go to the synchrotron where the profiles are very narrow.

The instrument can affect the profile in many ways, and the user must make every effort to minimize these systematic effects. The single most important requirement is to align the instrument so that the optics are correct. Once alignment is achieved, there are still systematic aberrations that must be recognized and corrected for in the final experimental data. The specimen effects are usually greater than the aberrations from a well-aligned diffractometer, and they depend on the specimen and its manner of presentation to the beam.

Specimen effects include displacement from the diffractometer axis, beam penetration into the specimen, surface roughness, departure from focusing conditions, and fluorescence. The effect of each of these aberrations needs to be appreciated by the diffractionist when preparing specimens for study. Each aberration has a different importance when measuring accurate angles or accurate intensities. Beam penetration (specimen transparency), surface roughness, and departures from optimum focal conditions can broaden the profile, whereas specimen displacement and transparency shifts the profile, and fluorescence creates a background that affects peak-to-background ratios. All of these conditions degrade the ability to make the most accurate measurements. Specimen preparation techniques need to consider these effects in order to minimize their influence on the measurements desired.

Two very common methods are used for presenting the specimen to the diffractometer. The first approach is to fill a cavity mount designed for the diffractometer and to smooth the surface to make it coplanar with the holder (see section 4.5.3). The surface is placed in the diffractometer so that it is coincident with the axis of the diffractometer. This geometry means that the entire specimen is on one side of the diffractometer axis. Penetration of the X-ray beam into the specimen then results in a displacement of the centroid of the profile to lower angles and a broadening and asymmetry to the low-angle side which depends on the depth of penetration. The second approach is to spread the entire specimen as a thin film on the surface of a slide of glass or other material (see section 4.5.4). In this case, the entire specimen is above the diffractometer axis and the peaks are shifted to higher angles. The displacement of the specimen and the fact that the specimens are planar in both of these situations also affect the focusing conditions. Both methods of presentation need to be examined in detail. Before proceeding with the following sections, the reader should review the effect of absorption on the behavior of the X-ray beam in the specimen (see section 1.7.4).

4.6.3   Thin Specimen on Flat Support

Perhaps the most common method of specimen presentation in diffractometry is to use the thin slurry on a glass slide. The general procedure was discussed in section 4.5.4. For many situations, this technique can be very effective. It is quick and the specimen may be recovered completely for other tests. To prepare the powder, see section 1.6 on the comminution and sampling of materials. The powder is placed as a small pile on that portion of the slide that conforms to the diffracting area in the diffractometer. A drop or two of a dispersing fluid is added to the pile, and the slurry is formed and spread evenly on the slide surface with the tip of a metal spatula so that it covers an area larger than the active region and is uniform in thickness. The final specimen should be only a few grains thick, and the surface must be planar. Void regions detract from the diffracted intensity and expose the substrate so that it may contribute to the background. Hills and valleys in the surface cause uneven diffraction and affect the profile shape and intensity. For d-spacing determination, some preferred orientation is acceptable.

Several dispersing fluids are useful for preparing slurry mounts. The basic requirements include that the fluid must be unreactive with the sample yet wet the grains to create a flowing slurry. Rapid evaporation after application is convenient. Organic reagents such as several of the alcohols, acetone, cyclohexane, and ethyl acetate are often used. Water is usable, but lack of wetability is often a problem. Many other reagents used in the past have been deemed carcinogenic and are no longer recommended. Additives, such as a wetting agent in water, are not advisable because they may leave a residue on evaporation of the solvent which will cause additions to the diffraction pattern.

A more permanent mount may be prepared by using a binder in the slurry. The most common binder is some form of cellulose. Collodion, liquid cellulose, is available from most supply houses. Solvents thin the fluid and extract the carrier on evaporation leaving a rigid cellulose film binding the particles in the specimen to each other and to the substrate. The viscosity of the solvent controls the spreading and wetting properties of the mixture. The most common solvents are ethyl acetate or amyl acetate. A 5% to 10% solution of collodion in ethyl acetate will wet most aggregates of grains and allow easy spreading of the slurry prior to evaporation and hardening. It works well in cavity mounts, but too much fluid may lead to shrinkage of the final compact, requiring another treatment. Because the viscosity is very low, it may allow some grain settling during the evaporation stage and hence some preferred orientation. Amyl acetate is more viscous so a solution tends to have difficulty wetting a compact in a cavity. It does make a good slurry when spread on the surface of a glass slide. A solution of 50% collodion in anhydrous ether used to be a very good choice for a well-behaved cement, but such use of ether is no longer advisable. This cement remained thin and tacky up to the final drying and did not have a tendency to pull the specimen off the support when multiple applications were used.

To achieve the specimen which creates the smallest effect on the width of the diffraction profile, it is necessary to use thin samples unless the sample absorption is large, eliminating significant beam penetration. Under these conditions, the surface of the specimen must be very smooth and flat. Patterns from these samples allow the recognition of all of the resolvable peaks, but they do not allow correction for the angular aberrations.

4.6.4 Internal Standard Pattern

Another type of specimen has as its objective accurate interplanar spacings. The ground sample is mixed with an appropriate internal standard, usually a Standard Reference Material (SRM) from the U.S. National Institute of Standards and Technology (NIST), Gaithersburg, Maryland. Details on choosing an internal standard and applying calibrations to obtain adequate d-spacings have been provided by Wong-Ng and Hubbard [9]. The amount of standard is selected such that the intensity of the strongest peak of the standard is approximately equal to the intensities of the stronger peaks of the sample. Typically, 10–30% of the mass of the sample produces this result. The two materials should be mixed thoroughly, but intense grinding should be avoided because it may induce strain in the standard, which will alter the certified SRM peak positions.

Once the mixture of sample and internal standard is made, the specimen may be prepared by procedures already described. Some specimen displacement, such as would be produced by using a smear on a glass slide, can be tolerated because it will be corrected by the internal standard calibration procedure. Similarly, some preferred orientation can be tolerated because the intensity data are read from separate diffraction patterns.

The goal of this diffraction pattern is to read precisely the positions of as many uniquely indexed peaks as possible of the peaks identified on the specimen without the internal standard. The measured corrected peak positions are then used for the least squares unit cell parameter refinement. Peaks from the sample that are affected by overlaps with the internal standard or are too weak in this diluted specimen are read from the preliminary pattern and corrected using the positions of the nearby peaks that have been determined from the internal standard pattern.

4.6.5   Intensity Patterns

A further type of specimen is used for determining relative intensities. The objective here is obtaining correct relative intensities, i.e., without preferred orientation due to texture effects in the sample or poor crystallite statistics (see sections 4.3 and 4.4). A minimum of three replicate specimens prepared from the same aliquot of ground sample is recommended to ensure that reproducible intensities have been obtained.

The principle specimen preparation problem is to minimize orientation of the particles. This goal is, of course, of particular importance with phases with pronounced cleavage, or those with platy or acicular (needle-like) habit. Two specimen preparation methods minimize orientation. The first is to add about 50% by volume of amorphous silica gel to the sample, and to place the mixture by side drifting in a cell without packing. The side drifting method is described in section 4.5.6. An alternative method is to prepare a "dusted" slide, as described in section 4.5.7.

The use of three or more specimens gives a measure of precision of the intensity measurements. McCarthy [10] has described a computer spreadsheet method for performing the necessary statistical analysis. An indication that the specimen preparation is yielding accurate intensities can be obtained by calculating a powder pattern from published structure data. A code that produces a diffraction pattern simulation that models the instrument being used and any specimen broadening is ideal for this purpose. The POWD code of Smith [11] will produce such a pattern simulation. McCarthy et al. [12] discuss the uses of calculated powder patterns in reference pattern determination.

4.6.6   Long Duration Pattern

The preliminary, internal standard, and intensity patterns may accomplish their purposes with short to moderate count times (e.g., 1–3 seconds per step on an automated diffractometer), but both the internal standard pattern and intensity patterns using the side drifting method involve dilution of the sample, which may lessen the chance of observing weak peaks. An additional diffraction pattern designed to detect the weakest observable peaks is also recommended. With this pattern, intensity statistics are enhanced by counting at each step for five or more seconds, and collecting the pattern overnight or over a weekend. Weak peaks that are not statistically significant in the diluted specimens may now be observed with intensities well above background noise.

The choice of specimen preparation for this pattern is the same as for the specimen without the internal standard (section 4.6.3). If preferred orientation effects are severe, then one of the specimens used for the intensity patterns should be used.

The ICDD requests that reference data submitted for publication in the PDF be accompanied by a computer-readable pattern. The long duration pattern can serve this function.

4.6.7 Reference-Intensity-Ratio Pattern

In order to estimate the proportions of various components in a mixture, the reference intensity ratio (RIR) of each component phase is needed. The reference intensity ratio, also known as I/I_c (c = corundum) or RIR, has been defined as the ratio of the intensity of the strongest reflection of a sample, to the intensity of the hexagonal (113) reflection of corundum (α-Al_2O_3) in a 1:1 mixture by mass. This RIR allows the relative intensity scale usually used for reporting powder intensities to be rescaled to a relative-absolute scale.

The minimum procedure involves thoroughly mixing equal masses of the ground sample and corundum, which can now be obtained as SRM1976 from NIST. Hubbard and Smith [13] and McMurdie et al. [14] recommend that three or more mixtures in different mass ratios be used in order to improve statistical accuracy. The reader is referred to these publications for additional details.

Specimen preparation for this measurement is the same as that used for intensity determination. It is also vital to avoid preferred orientation in RIR measurements.

4.7 PROCESSING SMALL SAMPLES FOR DIFFRACTOMETRY

Standard specimen preparation methods presume the availability of sufficient material to survive any prepreparation and fill the cavity of the supporting mount. Unfortunately, this situation is not always met with the supplied samples, and some laboratories never have the luxury of sufficient sample. Fortunately, there are many methods which allow small amounts of material to be prepared for analysis that provide adequate diffraction intensity for good studies. This section will discuss the collection of small amounts of material from a limited supply of material and the manipulation of that material into a suitable specimen for study.

One of the first decisions to make when a limited sample is supplied is which diffraction technique is most suitable to get the information requested. With small samples, the Debye–Scherrer–Gandolfi or the Guinier techniques may actually be the better choice, although the diffractometer with the proper specimen can produce quite usable data from infinitesimal quantities when they are prepared properly. The proper specimen will minimize the influence of the support to the diffraction pattern while maximizing the contribution from the material, i.e., minimize the background and maximize the peak intensities.

4.7.1   Collecting or Concentrating Material for Analysis

Recommended Supplies   These supplies include a good binocular microscope, a boron carbide mortar and pestle, a pin vise with needles, a honing stone, a glass plate 6 to 12 in. square, liquid reagents, petri dishes, sieve set, dental drill or engraver, steel Plattner mortar, and a hammer.

If the material is supplied to the laboratory as a powder in a small vial, the only problem is transferring the powder to the specimen mount. If the sample is a large quantity of which a portion is to be examined, then the material needs to be concentrated. If the sample is a block, the desired material needs to be separated from the matrix and collected. The first step, regardless of the situation, is to examine the sample under a binocular microscope. The diffractionist can see whether the phase to be studied occurs as discrete grains or whether considerable effort will be needed to concentrate and clean the phase before preparing the specimen. The strategy to collect and concentrate the phase now needs to be selected.

If the phase occurs as independent grains in the sample, handpicking under the microscope is usually the most effective method for cleaning and concentrating the phase. It is usually more economical of operator time to concentrate the material prior to analysis rather than use time later to unravel the data from the diffraction pattern of the bulk mixture. The most direct method for concentrating the grains is to pick them using a needle to transfer the particles to a collecting dish. The needle may be held in a pin vise (available at any hardware store) or stuck into a wooden handle. Usually, the grain will adhere to the needle, and it can be carefully lifted and transferred. If it will not adhere, then wipe the tip of the needle on your skin to get a thin film of oil on the surface. Stubborn particles may be induced to stick if the needle is wiped with Tacki wax. Be sure to wipe excess wax off the needle before going after a particle. Usually, tapping the side of the needle on the edge of the collecting dish is sufficient to get the particle to fall into the well. A small boron carbide mortar makes an excellent collecting dish and avoids the second transfer for grinding.

In handpicking under a microscope, nothing beats steady hands. Micromanipulators are usually cumbersome and susceptible to being bumped when the movements are being followed through the microscope. Obviously the operator needs to avoid strong drink, smoking, strenuous activity, or personnel conflicts prior to picking grains.

Before further work is done on the collected particles, they should be examined for grains of unwanted materials that may have been transferred with the selected ones. Use the needle technique to remove the impurities and any phase particles that have adhering impurities. Some particles may have to be broken to get the impurity loose from the desired part.

The concentrated particles are now ready to be ground to the size needed for the specimen. Hard particles in a mortar tend to jump out when crushed, so a catcher is needed to prevent loss. This catcher is a metal washer cover with a center hole just large enough for the pestle. When the grinding is done, the powder will probably be smeared on the inside of the mortar. The tip of a spatula is used to scrape the surface and bring all the material to the center. Also scrape off the material adhering to the pestle. The powder can then be transferred to the specimen mount. If a highly polished mortar is used, the recovery can be nearly 100%. Grinding of grains of phases susceptible to damage should be avoided. Such materials may be crushed to fine sizes between two microscope slides or between steel plates if the grains are too hard. Highly polished surfaces of the platens facilitate the preconcentration of the material.

Where the sample is a block or a rock sample, the grains may have to be dislodged from the matrix. If the phase is soft, the needle may be used to separate the phase from the substrate. Care needs to be used to prevent the freed material from flying off into space. A large clean glass plate should be placed under the sample to catch particles. A drop of water or other fluid will help to retain the freed grain. The collecting dish should be placed in a position to minimize the transfer distance. Before removing the sample, it should be tapped to drop loose grains onto the glass plate. Then the glass plate should be examined for grains to be transferred to the collecting dish. Scraping the plate with a single-edged razor blade will usually bring all the material on the plate to a central point where it may be examined for useful particles. Finally, the material in the collecting dish should be checked for unwanted impurities.

Heavy probing with a needle quickly destroys the point and renders the needle useless for fine work. The points may be maintained sharp if a honing stone is handy. An excellent honing stone is a slab of alumina ceramic with a smooth surface. The needle should be repointed periodically by honing. If the need arises for a chisel shaped tip, that shape can be easily accomplished using the honing stone also.

Where the particles are hard and well bound in the matrix, more drastic measures are needed to dislodge the grains. An engraving tool with a carbide tip or a dental drill with diamond points may be used. These techniques usually cause the fragments to fly off, so a catch device is advisable. The drill may be used to undermine the desired particle, or it may be used to pulverize the phase and disperse the powdered material. The work should be done on a large glass plate, so the scattered particles may be recollected. A platform with a specimen mount backed by a plastic backstop is often successful if placed in the right position "down wind." This catcher is especially effective with the drilling technique.

It may be desirable to crush the bulk sample into smaller particle sizes to release the desired phase as single-phase fragments. Only a crushing action should be used. A steel Plattner mortar is very effective for this step. Most Plattner mortars have a 1 in. ID sleeve which fits into the anvil, and the pestle is hammered into the sleeve to break up the material. The sample should only be crushed to the size necessary to release the phase. A set of 3 in. sieves, 10, 35, 60, 80, 100 mesh, are then useful to select the most useful size range for cleaning. Sometimes it is useful to put the particles into a petri dish and wash the particles with acetone or other fluid. This washing cleans the surface of fine adhering dust. Decant the wash fluid a few times. While the particles are wet, they can be examined under the microscope for purity. Dry grains are usually easier to pick because the particles may be transferred to the concentrate without having to break the surface tension of the liquid. However, it may be easier to move the grains into small piles in the liquid and then use a suction device to collect and transfer the concentrate.

4.7.2   Preparing the Specimen

When preparing specimens with limited material, it is important to remember that the half-depth of penetration for most nonorganic materials is of the same magnitude as the particle size, so a thin layer is more effective than a pile of powder in one spot. Layers too thin to be seen with the unaided eye can still give quite strong diffraction patterns. It is important also to know just where on the specimen mount the area seen by the X-ray beam is located. It may be necessary to put a fluorescent screen into the sample position on the diffractometer to determine this active area. Diffractometers with variable divergence slits have a fixed large active area at all values of 2θ. Diffractometers with fixed slits have a smaller area at the higher angles, and it is this area that should be covered with the specimen.

The best specimen support is one which contributes no diffraction intensity to the experimental data even when it is seen by the X-ray beam. Amorphous materials such as glass or plastic generate considerable intensity which goes into the background and reduces the peak-to-background ratio for the desired data. Even polycrystalline metal holders may contribute to the diffraction experiment if they are in the X-ray beam. Single crystal supports are available, made of quartz and silicon, which are cut so that no diffraction can be detected in the standard diffractometer experiment. These supports are the best for tiny amounts of material. One source for single crystal mounts is The Gem Dugout. If the single-crystal plates are not available in the laboratory, then glass is the second best choice.

For small samples, the powder is transferred using the tip of a spatula to the active area on the support, and then the powder is spread out into a monograin layer as densely packed as possible. A drop of an appropriate fluid (alcohol, acetone, cyclohexane, etc.) will assist in the spreading of the grains. The tip of a spatula should be used to control the spreading of the grains. Usually, the wetting of the particles is sufficient to get the particles to adhere to the support for the diffraction run even when the diffractometer is of the horizontal type. The addition of a permanent binder may contribute to the background. If a binder is necessary, especially for the horizontal diffractometer configuration, then a blank mount with only binder should be examined to evaluate the magnitude of its contribution. Thin layers of grease or collodion in ethyl acetate are sometimes used for binders.

If the decision is made to use the Debye–Scherrer–Gandolfi or Guinier techniques, the reader is referred to those sections of this book for the preparation of the specimen (e.g., Chapter 7).

4.7.3 Concentrating Phases in Large Bulk Samples

Many times large samples are supplied for analysis where the information desired only concerns one phase in the mixture. Samples where the goal is phase quantification should never be separated into component parts. In those situations involving the identification of only one phase, the question must be considered whether it would be more efficient to separate out the phase and concentrate it or simply run the bulk material and attempt to isolate the diffraction information from the composite diffraction pattern. There are many methods for the concentration of certain fractions of a sample which may be applicable and require a minimum of effort, so preconcentration should always be considered. Some of these methods are discussed in other sections, e.g., handling minerals, but the more general techniques will be described here.

Separation techniques depend on the nature of the sample. No sample may be effectively beneficiated if the individual particles are not monophasic. The next consideration is the particle size. If the particles are large (> 2 mm), handpicking under a microscope will be efficient. Even particle sizes down to 0.5 mm are amenable by handpicking. Sizes down to 0.1 mm are suitable for separations based on physical properties. Density, magnetic moment, electrical conductivity, and surface affinity may be used for concentration of phases.

4.7.4 Preparation of Mounts for Diffractometry

Surface Perfection   All diffractometer manufacturers supply specimen mounts for their diffractometers. Alternatively, it is very common for laboratories to prepare their own mounts for various reasons. The following remarks are meant to guide the user in the considerations that must be included in any specimen mount design.

The choice of suitable materials is limited to glass and various plastics, metals, or certain single crystals. Lucite and Bakelite are common plastics which have been used. Steel, aluminum, and brass are common metals. The metals and most plastics can be machined easily to desirable shapes. Glass has to be ground with diamond tools, sand blasted or etched chemically to create cavities. Single crystals require special techniques which will be discussed below.

In general, specimen holders are rectangular or circular to fit the instrument. A hole or cavity is excavated at the position for the specimen powder. The opening should be slightly larger than the area seen by the X-ray beam. The opening may be bottomed or open. A bottomed holder should have a depth of at least ten times the half-depth of penetration of the material with which it is to be used. An open holder will have to be backed when in use, but it allows filling from the back side.

This basic design may be modified to meet specific needs. The opening may be made smaller or the cavity bottom more shallow to limit the quantity of sample required. One edge of the holder (the higher side for either the horizontal or vertical diffractometer) can be slotted to allow side loading. See the discussion of side loading in Section 4.5.6. Whenever the opening is made smaller than the beam area, the operator has to remember that the holder may contribute to the diffraction pattern or vignetting may occur, affecting intensity values. Single-crystal holders with small circular cavities are subject to vignetting.

Single-crystal holders may be made from several materials, but silicon and quartz have been the most successful. Silicon cut parallel to the (510) planes seems to be the ideal choice. There are no possible diffractions that can occur within several degrees, so even rotation of the plates will not give rise to spurious peaks. The silicon (511) cut is generally useful for most purposes because the (511) peak for Cu $K\alpha$ is at $96°2\theta$. Other orientations have been used [such as (100) and (111)] so that the characteristic peak can be used for calibration purposes. Quartz is cut tilted $6°$ from the c-axis in the X plane. The (00l) reflections are at $6°$ and are not encountered unless a wide divergence slit is used. Occasionally the forbidden (001) and (002) peaks will be detected if the plate is spun in its plane. These peaks appear very sharp due to multiple diffraction which may occur in highly perfect materials. Sources for these plates are given in Chapter 8. For small amounts of sample, the single-crystal plate is covered with a thin layer of grains at the active area. Quantities as small as micrograms can be successfully used for good diffraction patterns.

Cavities may be ground in single-crystal holders so that they may be used in the same manner as described above. Correctly positioned small cavities are useful for locating the proper spot to load the specimen. However, where there is a small quantity, it is more efficient to spread out the powder over the whole active area. In this way none of the grains is masked by absorption. Larger cavities allow reproducible amounts to be loaded for quantitative methods.

Single-crystal holders do have the drawback that they do not like to be heated. Quartz is very heat sensitive because it is a poor conductor and because of the phase change at $573°C$. Silicon is a better conductor, but uneven heating creates strains that activate some of the cleavages. Quartz is more chemically inert than silicon, and its surface can be easily cleaned. Silicon may react with some materials at ambient conditions. In spite of these drawbacks, these materials make excellent specimen supports.

4.8 PROCEDURES FOR SPECIMEN PREPARATION BY AEROSOL SUSPENSIONS

4.8.1 Introduction

The principle behind filter aerosol suspension and collection is based on three factors:

1. Particles in a turbulent air stream tumble as they move within the zone of flow.
2. Particles are impacted and held within an irregular fiber matrix, further enhancing randomness of orientation.
3. The void dimensions of the filter are similar to the diameters of the particles being trapped. As the layer of particles builds up, the randomness of particle orientation propagates beyond the top of the filter matrix so that the matrix interference is reduced (but not entirely eliminated).

Because the layers prepared by the method are thinner than the "infinite" thickness required to maintain constant flux of irradiation over all scattering angles (fixed slit arrangement), equations based on filter and sample absorption parameters and load densities are available to correct the intensities used in quantitative analysis applications [15].

The following sections describe several devices based on these principles which are useful for preparation of random powder layers on filters for either qualitative or quantitative XRD analysis. In all applications described below the sample particle size should be reduced to an average of about 5 μm diameter or less.

4.8.2   Simple Aspirator Bottles

Large Capacity Aspirator Bottles   Figure 4.8 demonstrates the 4- or 8-l aspirator bottle for aerosol suspension. All that is needed is a rubber gasket placed over the top inlet and a Buchner funnel or filter cassette connected to an air circulation pump. The best pump for any size bottle is a 1/6 hp, 1725 rpm GAST or equivalent. It operates on 115 v and draw 4–5 amps at 30 l/min. maximum capacity. They are readily available in any general laboratory supply warehouse (such as Sargeant-Welch or VWR Scientific). The cassette or Buchner funnel should be 37 mm, 42.5 mm, or 47 mm diameter to fit the opening of the aspirator bottle, but a smaller size could be used if the inlet can be reduced.

The typical procedure for loading and mounting the sample is:

*   Place 50–100 mg of powdered sample on the juncture between aspirator floor and wall, opposite the hose inlet. Leave hose inlet unconnected and unblocked.

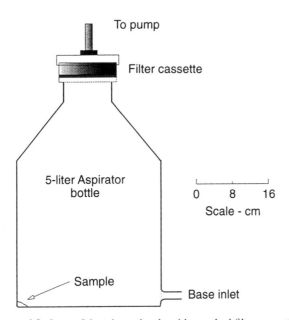

Figure 4.8. Large 5-l aspirator bottle with attached filter cassettes.

The exact amount of sample depends on several things, but one should plan on about 50–75% loss to aspirator bottle walls for some materials.

- Place the preweighed filter in the cassette or funnel and lock in place; where a Buchner funnel is used the pump must be started to hold the filter in place.
- Create the aerosol "cloud" with a lab air pressure line (with a small metal tube reducer connected through tubing to the air valve, lowered to within one inch of the sample inside the bottle) applying a short spurt of air. In place of the lab air line the pressure side of another air circulating pump may be used.
- Quickly remove the air line, turn on the cassette circulating pump, and place the loaded cassette over the aspirator bottle (top!) inlet, rotating the bottle under the filter every 5–10 s, as the pump pulls the aerosol into the filter. About 30 seconds collection time is usually sufficient but this time is also dependent upon the density of the aerosol cloud and the total load desired.
- The loaded filter may be then reweighed and cut with a circular cutter that has been machined to the size needed, or simply cut with scissors to fit the diffractometer holder. A small amount of Vaseline is useful to hold the filter to the diffractometer mounting surface, especially for horizontal goniometers. If prepared and cut properly, and the load is not over 2000 μg cm^{-2}, the filter load should not spall off the filter during slow spinning in a horizontal goniometer (where the sample plane is vertical). A touch of highly diluted rubber cement around the edges only of the cut loaded filter will suppress spalling because it invariably begins around the disturbed layer and not in the load center.

Small-Capacity Aspirator Bottle   Figure 4.9 illustrates a 250 ml bottle which works on a slightly different principle than the large bottles described previously and requires much less sample (a few milligrams). Only 25 mm filters and cassettes can feasibly be used with this size bottle. No air pressure line or auxiliary pump is required for this system. The procedure for loading is the following:

- A plug or cork is fitted with a 1 mm capillary with a 15–25° bend at the interior end and inserted into the floor inlet.
- Place the sample into the bottle from the top and shake it into the floor inlet so that it rests against the plug.
- Tape the loaded cassette securely to the neck of the aspirator bottle (Some buildup of the neck width with masking tape may be necessary to do this task).
- Unscrew the vacuum side bleed in about halfway and then turn on the circulating pump connected to the cassette.
- The cloud is formed by the drawing action of the circulating pump on the cassette while the operator is withdrawing and rotating the capillary outward to the plug wall where the sample is resting. Momentarily pulling the capillary completely out may be necessary to get all of the sample suspended.

It will take more practice to achieve a uniform sample layer with this smaller bottle than with the other systems, in part because no rotation during loading is possible.

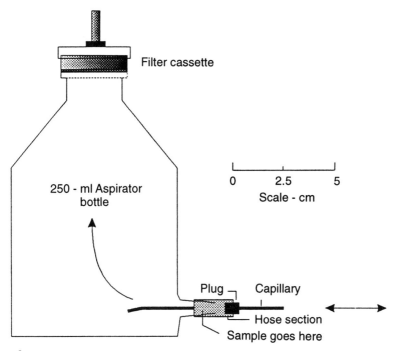

Figure 4.9. Small 250-ml aspirator bottle showing plug and capillary inserted into base inlet and with cassette attached.

The system works well with VM-1 filters where sample quantity is limited (Note: VM-1 filters are flatter and more featureless and hence may not be as effective as are fiber filters in suppressing preferred orientation).

4.8.3 Tubular Aerosol Suspension Chamber (TASC)

Laboratory experience shows the tubular aerosol suspension chamber (Figure 4.10) has proven to be the most effective apparatus for obtaining uniform layers of random particles on filter substrates[1]. We believe that this effectiveness stems from the long vertical path of the aerosol stream and geyser-like convection cell of spherical 500 μm glass beads that breaks up clumps and generates a more dilute aerosol. Construction of this apparatus requires more effort and is more expensive, but the benefits are well worth this effort. A full description of the TASC appears in reference [16]; however, the description and procedures given here are intended to supersede those of the 1986 paper because considerably more experience in application of this device has been gained.

---

[1]More recent designs, but based on the same fluidized bed suspension principle, are now commercially available through the Davis Consulting Group, 4022 Helen Ct., Rapid City, SD 57702

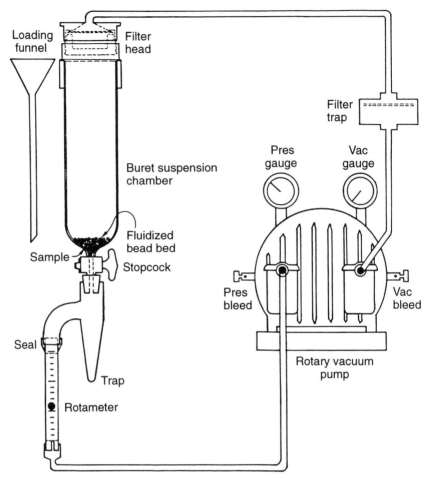

Figure 4.10. Tubular aerosol suspension chamber (TASC) and associated vacuum and circulating pump, filter trap, cassette, loading funnel, distillation receiver trap, and rotameter.

It is important to load the beads to a level of about 1–2 cm above the base inlet of the burette, sandwiching the sample in between. This step is accomplished by loading the beads through a long stem funnel exiting about 2 cm above the burette base inlet. Control of the flow is important: the bleed valves of the pump should be adjusted so that the line air pressure that passes through the distillation tube trap below the burette inlet is equal to or slightly less than the vacuum flow applied to the cassette attached to the burette. If such is not the case a partial vacuum could develop within the burette, which creates nonuniform flow, especially if the seal about the cassette and other connections are not good. Positive over-pressure in the burette is not good either, creating more of a "push" to the aerosol motion to the filter rather than pull which more securely imbeds the particles within the filter fibers. It is extremely important that the burette and stopcock be clean and dry for each application.

In regions where humidity is high (above 60%, but depending also on the material being loaded) it is good to include a drierite desiccator column in the line between the pressure exit of the pump and the distillation tube trap. The filter trap between vacuum side of the pump and cassette is an absolute must for pump protection; it should be checked frequently and the filter replaced whenever clogging significantly reduces vacuum levels.

Relative to building the device, the gas burette purchased through chemical suppliers will have a stopcock at each end. The burette may be cut exactly in half with a diamond saw and the length extended with cardboard tubing (obtainable at hobby shops) to about the 15-cm height recommended. This procedure provides a second burette column for use as a spare. Plastic tubing has not been tried as an extender; whatever is used should not allow static charge buildup. Even the spherical glass beads within the glass burette sometimes transfer static charges to some samples, which results in layer buildup on the lower walls and slows the loading process. A properly prepared sample should only require 15–30 seconds for loading; however, some samples, because of the charging effect mentioned previously, may require up to 15 minutes for loading. Loading larger sample quantities for such materials helps to overcome this problem.

The glass spheres obtained from Potters Industries[2] are very uniform in size and have never been observed breaking up or reaching the filter surface under proper flow settings. The 0.5 mm size is critical in that sufficient convection cell action may take place without individual spheres reaching the filter surface. The procedure for TASC operation consists of the following steps:

- Unscrew the bleed valves of the pump so that they are completely backed off, but not removed.
- Load the beads to just above the burette inlet using the funnel.
- Load the sample and then another batch of beads so that the resulting sandwich extends about 1–2 cm above the burette inlet. Remove the funnel.
- Load the cassette with a preweighed filter (Whatman GF/C is recommended) and lock the cassette cover down firmly. Seal the cassette to the burette (or burette extension) with masking tape.
- Turn on the circulating pump and reduce the vacuum bleed so that the gauge shows about 1–2 divisions reading above the "peg."
- Open the stopcock and adjust the vacuum bleed to give 3–4 divisions (few cm Hg) vacuum reading. Quickly adjust the pressure bleed inward until the rotameter ball just begins to respond upward.
- If the beads of the geyser convection cell action rise more than three-fourths of the way up the TASC column, the bleed valves should be backed off slightly. This adjustment will ensure that the beads are not reaching the filter surface.
- After loading is complete, back off the bleed valves except leave the vacuum bleed in enough to maintain a slight convection cell action within the beads.

[2]Potters Industries Inc., Brownwood Division, Highway 279 North, HG30, Box 20, Brownwood, TX 76801.

Leave the pump running! Then, untape the cassette and place it with filter upward in a convenient holder, such as a ring clamp or hose clamp attached to a ring stand. Now, close the stopcock, remove the cassette cover, and shut off the pump, in that order.

• Remove and weigh the filter. A strong light (i.e., microscope lamp) placed beneath the loaded filter provides a good indication of the uniformity of loading.

Common sense (and care!) should be used in cleaning the apparatus. Only the burette and its stopcock need be washed between each run. The extension (if any), cassette, and loading funnel can be blown clean with an air jet under a hood for most applications.

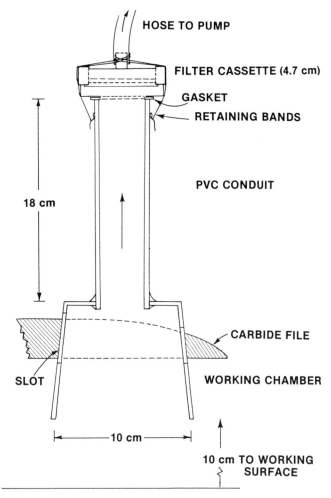

Figure 4.11. Microhood for diminution of particles with carbide files and suspension through a PVC column-to-filter cassette.

4.8.4 Microhood for Special Materials

One cannot easily grind micaceous or asbestiform materials, except under liquid nitrogen. An alternative to this approach is to cut the material with a good carbide file under a small hood and column, such as shown in Figure 4.11. The entire apparatus should be placed under a laboratory fume hood with strong draft, and the operator should also wear a protective dust mask. The device is easily constructed from plastic dishware parts, cardboard tubing, and a rubber gasket for cassette mounting. The same 30 l/min. circulating air pump used for the other suspension devices may be used here.

The cassette can either be taped to the tubing or held down on its gasket with rubber bands to hooks mounted partway down the tube. In the TASC apparatus one file may be mounted in slots cut within the microhood when mica sheets are rubbed over the file. Adjusting the flow (vacuum side only to cassette) allows rejection of the larger particles. Multicomponent samples containing asbestos or mica are loaded in a small pile on one file with a second file providing the cutting action. The extent to which selective cutting and/or loss of components takes place in multicomponent samples is not known from the limited experience we have with this device. It has been used mainly for pure phase mounting of asbestiform or micaceous materials for reference intensity ratio measurement. The length of the tube above the hood is important—the longer the tube the more uniform the load collected.

The only operation steps are:

- Loading and mounting of the cassette.
- Turning on of the pump (bleed valves entirely in).
- Diminution action with the files beneath the hood.

REFERENCES

[1] Alexander, L. E. (1969), *X-ray Diffraction Methods in Polymer Science*. Wiley, New York, 582 pp.

[2] de Wolff, P. M. (1958), "Particle statistics in X-ray diffractometry," *Appl. Sci. Res.*, **7**, 102–122.

[3] Smith, D. K. (1992), "Particle statistics and whole-pattern methods in quantitative X-ray powder diffraction analysis," *Adv. X-ray Anal.*, **35A**, 1–16.

[4] Brindley, G. W. B. (1945), "The effect of grain or particle size on X-ray reflections from mixed powders and alloys, considered in relation to the quantitative determination of crystalline substances by X-ray methods," *Phil. Mag.*, **36**, 347–369.

[5] McCreery, G. L. (1949), "Improved mount for powdered specimens used on the Geiger-counter X-ray spectrometer," *J. Am. Ceram. Soc.*, **32**, 141–146.

[6] Klug, H. P. and Alexander, L. E. (1974), "*X-ray Diffraction Procedures for Polycrystalline and Amorphous Materials*," Wiley: New York, 372–374.

[7] Bloss, F. D., Frenzel, G., and Robinson, P. D. (1967), "Reducing orientation in diffractometer samples," *Am. Miner.*, **52**, 1243–1247.

[8] Smith, S. T., Snyder, R. L., and Brownell, W. E. (1979), "Minimization of preferred orientation in powders by spray drying," *Adv. X-ray Anal.*, **22**, 77–87.

[9] Wong-Ng, W. and Hubbard, C. R. (1987), "Standard reference materials for X-ray diffraction, Part II. Calibration using d-spacing standards," *Powd. Diff.*, **2**, 242–248.

[10] McCarthy, G. J. (1988), Laboratory Note: "A LOTUS 1-2-3 spreadsheet to aid in data reduction for publication of X-ray diffraction data," *Powd. Diff.*, **3**, 39–40.

[11] Smith, D. K., Nichols, M. C., and Zolensky, M. E. (1983), "POWD10, A Fortran IV program for calculating X-ray powder diffraction patterns," Dept. of Geosciences, Pennsylvania State University, University Park, PA.

[12] McCarthy, G. J., Martin, K. J., Holzer, J. M., Grier, D. G., Syvinski, W. M., and Nodland, D. W (1992), "Calculated patterns in X-ray powder diffraction analysis," *Adv. X-ray Anal.*, **35**, 17–23.

[13] Hubbard, C. R. and Smith, D. K. (1976), "Experimental and calculated standards for qualitative analysis by powder diffraction," *Adv. X-ray Anal.*, **20**, 27–39.

[14] McMurdie, H. F., Morris, M. C., Evans, E. H., Paretzkin, B., Wong-Ng, W., and Hubbard, C. R. (1986), "Methods of producing standard X-ray diffraction patterns," *Powd. Diff.*, **1**, 40–43.

[15] Davis, B. L. (1992), "Quantitative phase analysis with reference intensity ratios." NIST Special Pub. 846. *Proc. Conf. Accuracy in Powder Diffraction*, Gaithersburg, MD, May 26–29.

[16] Davis, B. L. (1986), "A tubular aerosol suspension chamber for the preparation of powder samples for X-ray diffraction analysis," *Powd. Diff.*, **1**, 240–243.

CHAPTER 5

SPECIFIC AREAS OF SPECIMEN PREPARATION IN X-RAY POWDER DIFFRACTION

CONTRIBUTING AUTHORS: BLANTON, HAMILTON, IYENGER, JENKINS, McMURDIE, NELSON, RASMUSSEN, RENDLE, ROACH, RYBA, SMITH, WONG-NG

5.1 CERAMICS

In the broad sense, ceramics can be considered as solid materials which are largely composed of, and which have as their essential component, inorganic, nonmetallic phases. This definition includes not only materials such as pottery, porcelain, refractories, structural clay products, abrasives, porcelain enamels, cements, and glass but also nonmetallic magnetic materials, ferroelectrics, semiconductors, superconductors, manufactured single crystals, glass ceramics, and a variety of other products which were not in existence until a few years ago and many which do not exist today [1].

In order to describe specimen preparation and mounting of ceramic materials for XRD study, it is, however, more convenient to classify ceramics according to their physical form and chemical properties. These ceramics will be described in the categories of powders, ceramic compacts, thin films, or fibers. Mention will also be made of whether or not the material is a stable common material or is toxic or environmentally unstable, with solvent of crystallization, extremely hard (refractory material), or soft.

5.1.1 Powders of Stable and Common Ceramics

Preparation of Powders of Appropriate Crystalline Size   When dealing with powdered materials, the size of the crystallites must fall within an acceptable range for the information desired (see section 4.3). For the d-spacing measurements, normally the particle size is such that the material will pass through a 200 mesh sieve. For intensity measurement, in order to achieve reproducibility of intensity, and to reduce the preferred orientation problems, prolonged grinding for obtaining

A Practical Guide for the Preparation of Specimens for X-Ray Fluorescence and X-Ray Diffraction Analysis, Edited by Victor E. Buhrke, Ron Jenkins, and Deane K. Smith. ISBN 0-471-19458-1 © 1998 John Wiley & Sons, Inc.

crystallites small enough to pass through at least a 400 mesh sieve (or about 40 μm) is important. If samples are not obtained in a fine state, one needs to reduce the size of the particles by crushing and grinding. This reduction in size of coarser materials reduces the preferred orientation effect for samples which have pronounced cleavage, such as mica, or other tendencies to develop a nonequant shape. After the material has been ground, a sieve should be used to separate that portion of material which has an undesirable particle size. For quantitative analysis, such sieving is undesirable because phase separation may occur which will bias the results. If necessary, a brush may be used to gently guide the powder through the sieve.

Because most ceramics are brittle, (e.g., oxide crystals, chemical precipitates, smoke, and dusts), the common way to grind a sample is to use a mortar and pestle. To reduce the particle size of a large quantity of material, ball milling may be performed, and hard materials, such as borides, silicides, and other refractory materials, can be ground with a mortar and pestle made of tungsten carbide. For very soft materials, materials that may change form or chemical composition during the grinding process, materials with layer-type structure, and materials which have water of crystallization, care should be taken not to deform the structure if excessive force is applied. A crushing procedure is usually less damaging than one involving grinding. The pattern of a damaged specimen will consist of broad and diffuse lines. One common way to handle soft material is to chill it to low temperature (i.e., liquid air or dry ice temperature), then grind or crush in the usual way, which will produce powders with reduced strain.

5.1.2 Specimen Mounting of Ceramic Materials

Various techniques have been used for specimen mounting. These techniques include front loading, back loading, side drifting, dusting on a slide, and mixing the specimen with inert solvent in order to achieve a slurry (section 4.5). For toxic or unstable materials, special procedures which will be discussed below have also been developed by various diffractionists.

5.1.3 Atmospheric-sensitive or Toxic Ceramics

For special specimens such as the environmentally unstable or toxic materials, different care has to be applied during every specimen preparation step. Several techniques have been used:

1. *Flood the specimen chamber with inert gas.* One common practice being used is grinding and mounting the specimen inside a dry box or glove bag, then transferring the assemblage to the X-ray specimen chamber. While inserting the specimen, one needs to flood the X-ray specimen chamber with inert gas such as helium or nitrogen. The chamber lid must be closed tightly, and the inert gas must be passed continuously through the chamber during the X-raying process. Note that argon should never be used as the inert gas because of its high beam attenuation. The drawback of this method is that any momentary

exposure to atmosphere after removal of the specimen from its inert atmosphere storage may cause deterioration of some materials for subsequent tests.

2. *Protect the specimen with thin films.* Another way of avoiding reaction of specimen with atmosphere is to cover the regular holder with thin films on the top and bottom sides. For example, the bottom of the holder could be covered with the Kel-F film and fixed with warm wax. After the wax is cooled, the holder can be transferred into a dry box and packed with well-ground specimens. Before taking out the holder and specimen from the dry box, the assemblage should be covered with a plastic wrap. Warm wax is then applied around the edges of the holder to secure the wrap. Before running the specimens, the empty, modified specimen holder should be scanned to obtain the background correction that will be applied to the specimen run.

3. *Use an inert atmospheric cell.* The most convenient way to mount the specimen is to use a portable, self-contained inert atmosphere cell inside a dry box, then seal up the cell before transferring to the X-ray chamber. J. Ritter of NIST [2] has designed such an inert atmosphere cell, which consists of the following features:

- Compact, lightweight aluminum construction.
- Positive specimen placement on the diffractometer.
- Integral hermetic "O" ring seal.
- Simple press-together assembly for ease of manipulation in a dry box or glove box.
- Total specimen recovery without exposure to atmosphere.
- X-ray transparent window material such as polyimide.

The drawing of the inert atmosphere cell components, which include the cell body, specimen holder, end closure, and X-ray transparent window is shown in Figure 5.1. The procedure for using this holder is as follows:

- Place the cell and specimen in a dry box or glove bag.
- Load the specimen in the machined cavity of the cell (Figure 5.1*b*), using a spatula, and smooth out the specimen surface with the use of a glass slide.
- Keep AD assembled with BC until the O-ring seal is fully engaged.
- Remove the assembled loaded cell from the inert atmosphere.
- Insert into the diffractometer chamber and adjust for centering position.
- Recover the specimen in an inert atmosphere after the X-ray pattern has been run.

4. *Diffractometer inside a glove box.* Another way to perform X-ray analysis on air-sensitive materials is to have a desktop type of diffractometer set up inside

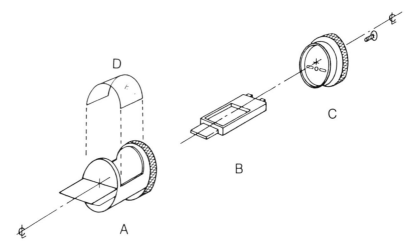

Figure 5.1. The assembly drawing of the inert atmosphere cell components (after Ritter, 1988).

a glove box, which can be filled with the desired atmosphere, (i.e., oxygen, nitrogen), or specially mixed gas. Helium is not a good choice because its high dielectric constant allows potential high voltage problems in the detector and X-ray tube. Argon should not be used because of its high attenuation of the X-ray beam. In this case, both the process of specimen mounting and the X-ray characterization will be conducted inside the dry box. One does not need to use a hermetic cell if a positive pressure is maintained. An automatic specimen changer will be convenient for insertion of specimen into the X-ray specimen chamber.

5.1.4 Ceramic Compacts or Ceramic Thin Films

For ceramic compacts or ceramic thin films, the most convenient way of mounting the specimens is to use a holder with a special support as shown in Figure 5.2*a* to *d*. The procedure to mount the specimen is:

- Apply sufficient clay, Tackiwax or other material that does not show any X-ray diffraction peaks to fill the space in the cavity (*a*).
- Set the specimen face down on a flat and clean surface (e.g., a wax paper on the surface of a flat table) (*b*).
- Secure the clay to the specimen, making sure the surface of the specimen is flat and level with the surface of the holder (*c*).
- Remove excess clay around the surface area of the holder which may interfere with the diffraction experiment (*d*).

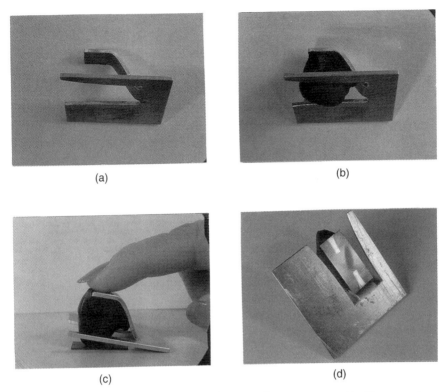

Figure 5.2. (*a*)–(*d*) Mounting procedure for ceramic compact or thin-film ceramics—specimen holder designed by J. Cline.

5.2 PREPARATION OF BULK METALS AND ALLOYS

5.2.1 General Discussion of the Problem

Metals and alloy specimens pose a considerable problem for the diffractionist for several reasons. They are typically supplied in block form, and to reduce them to a powder for study is not an easy task. Most metals and alloys are easily deformed during preparation, which degrades the diffraction patterns and masks the information desired. They are also very susceptible to surface reactions with the atmosphere, forming unwanted compounds which hide the unreacted material due to the high absorption of the elements involved.

5.2.2 Specimen Preparation Procedures for Metals

The nature of the desired information usually dictates whether diffraction patterns may be obtained from the specimen as supplied, or if it must be reduced to a powder by some means. For example, grain size, texture, and residual stress determinations must be performed on the specimen as received. On the other hand, if the question posed is one of phase confirmation, and the specimen block is of a suitable size to fit directly in the diffractometer, then the preparation is minimal and consists of generating a flat surface and cleaning it of any reacted materials. However, as received block specimens are usually unsuitable for phase identification studies because the grain size may be too large, intergrain strains may be present which distort the diffraction pattern, or the grains may be too highly oriented to yield satisfactory diffraction patterns.

Direct crushing with a mortar and pestle rarely works with metallic materials unless they contain significant amounts of intermediate phases, which are usually very brittle. Filing with a clean fine-toothed metal, carbide or diamond file is the most common method for obtaining powders of metallic materials. The file should be clean to prevent contamination, although powders prepared in this way usually contain minute amounts of the file material as an impurity. Filings are collected on a sheet of weighing paper or other suitable surface. Usually, the best particles are those that are retained in the teeth of the file, so these particles should be collected by periodically cleaning the file into a collecting dish with a brush or sharp needle. If enough material can be collected in this manner, then the specimen can be annealed and packed in a mount by one of the methods mentioned in section 4.5. If not, the other cuttings need to be employed. Sieve the cuttings to get only the smallest fraction that produces sufficient material for analysis. The particles should then be annealed by putting them into a furnace and heating to at least one-half the melting temperature in an inert atmosphere. In the absence of an atmosphere-controlled furnace, the particles may be sealed in an evacuated quartz tube for heating, although silicon as an impurity may thereby be introduced into the specimen. This annealing is necessary to reduce the deformation in the filings to a minimum and restore the diffraction pattern quality. The diffractionist must remember that many alloys are purposefully brought to a metastable condition through heat treatment, and annealing such materials will probably change the phases present and/or their amounts. For example, annealing promotes the transformation of metastable martensite to ferrite and carbides and the decomposition of any retained austenite present in hardened steels. In addition, the diffractionist is well advised to consult any available phase diagram information to see what phase changes may take place upon annealing in equilibrium alloys.

Other methods for reducing a block specimen to a form usable in diffractometer mounts includes machining the material on a mill or lathe and collecting the cuttings. These cuttings are usually long ribbons that need to be cut into short lengths to be used. The harder and more brittle metals will usually break into short pieces; softer metals may be cut with a razor blade. Some materials may be drawn or extruded as wires which may be packed in parallel in a specimen mount. These specimen materials should be annealed prior to examination, if possible.

The contamination of metal surfaces by atmospheric reaction can be retarded or prevented by retaining the specimen in a special enclosure after preparation and transferring it to the diffractometer in a sealed condition. Very few diffraction facilities are equipped with such special facilities, however, so alternative means are necessary. A standard cavity specimen mount may be modified by sealing a thin film of Mylar over the top opening using epoxy cement or polymer glue. Stretch the Mylar to maintain a smooth flat structure until the cement sets. Then load the cavity from the back and seal the back with a cemented glass slide or more Mylar. Mylar tapes with a cement already attached are available. A window in the center of the tape for the X-ray beam can be prepared by using a solvent such as benzene and a cotton swab. Alternatively, the specimen may be enclosed in a cement (collodion in ethyl acetate) or coated with an oil, such as moisture-free kerosene or mineral oil, as it is packed in the specimen mount.

Many metal specimens are supplied as metallographic mounts in plastic that may be used directly in a specially designed diffractometer specimen spinner. This device spins the specimen, which helps with the grain statistics but not with preferred orientation. These mounts are usually given a light polish and/or etch just prior to use to eliminate any surface reactants. If the reactions with the atmosphere are a problem because they are too rapid, the surface of the polished mount may be sprayed lightly with an acrylic lacquer to seal the metal from the atmosphere. Alternatively, a Mylar window may be cemented over the surface as described. Regardless of the cover used, blanks should always be run to determine the diffraction contribution of the covering material. Even though the materials are essentially amorphous, they still contribute to the pattern, especially in the Bragg–Brentano condition.

For multiphase alloys, chemical extraction of a particular phase for diffraction analysis may be possible. How this step is accomplished is largely a matter of trial and error (experience), but generally involves the use of a weak acid or base which slowly and selectively dissolves one of the phases, leaving very fine particles of the remaining phase or phases.

5.2.3 Special Problems and Hints

Most metals and alloys can be readily identified by their X-ray powder diffraction patterns, although information about trace phases will not be obtained. The diffractionist may be called upon to analyze specimens of metals and alloys ranging in size from traces to large ingots. However, if analysis of a large block of metal is required, then it may be more convenient to remove a small sliver by means of a sharp scalpel blade or a few pieces by careful filing with a needle file.

One alloy, brass, is very common, and it occurs occasionally as a forensic specimen (see section 7.10). XRD can be used to distinguish the various phases of brass and can provide a reasonably accurate (1%) estimate of the copper–zinc ratio within a phase [3]. However, certain precautions have to be taken before accuracies of this order, and greater, can be obtained. Filings and slivers of brass require annealing before analysis. Strains within the metal as a result of machining and casting operations, and indeed filing to provide the specimen for XRD, manifest themselves as broadened diffraction lines on powder diffraction photographs, particularly in the back reflection region, which will hinder accurate d-spacing measurement.

Annealing the specimen will relieve the strain, resulting in sharp, well-defined diffraction lines on the powder photograph, enabling the measurement of accurate d-spacings. The slivers or filings are first sealed inside a normal (thick-walled) glass capillary and heated in a muffle furnace to 500°C for 2–3 hours, then allowed to cool slowly back to room temperature inside the furnace over 12–15 hours. The capillary tubes are then broken open and the specimen, if filings, is placed inside a thin-walled glass capillary tube as though it were a powder or, if a sliver, is mounted on the end of a glass fiber with rubber cement, as for a paint flake (section 7.10).

It is essential that the annealing process for most materials be carried out in an atmosphere devoid of oxygen to prevent oxidation of the metal. The procedure outlined above does not meet this criterion precisely, but sealing the filings or slivers in a glass tube limits considerably the flow of air around the specimen during the heating and cooling cycle. Not all metallic specimens need to be annealed; for instance, the softer metals such as aluminum and lead do not need annealing because any strains in them are quickly relieved at room temperature.

When sampling a metal or alloy, care should be taken to specimen as uncontaminated an area as possible. The main contaminants will be oxides, carbonates, sulfates and the like, depending on the metal itself and its environment. This special care will not apply in the case of forensic specimens (see section 5.10) where it is important to maintain the integrity of the original specimen.

5.2.4 Step-by-step Procedure for Metals

1. Decide whether to file the metallic object or to remove a tiny sliver using a scalpel.

2. Take a piece of plain paper, roughly 4 in. × 4 in. fold it in half, then open it and press it flat on the work bench. If filing, use a fine needle file (available from any good quality tool shop or hardware store) and file the object gently such that the filings fall on the piece of paper. If removing a sliver, carefully slice off a piece of metal, using a sharp scalpel. Keep the dimensions of the sliver under 1 mm.

3. Take one of the 100 mm glass capillary tubes and cut it, using a glass cutter, into four 25 mm lengths. Take one of these 25 mm long tubes and seal one end by heating the end in a roaring bunsen flame until it melts and forms a solid globule of glass at that end. Allow the tube to cool and then pour the filings into the tube through its open end, using the folded paper. Place any slivers of metal in the tube.

4. Holding the sealed end of the capillary tube between finger and thumb, seal the other end in the bunsen flame as before. Prepare one capillary tube in this manner for each specimen.

5. Place the sealed capillary tubes inside a muffle furnace at room temperature and then switch on the furnace such that it heats up to roughly 500°C in 2–3 hours. Switch off the furnace and allow it to cool to room temperature over a period of 12–15 hours.

6. Remove the capillary tubes from the furnace, wrap them in individual layers of tissue paper, and break open (with care!) the tubes using fingers and thumbs. Pour the annealed metallic specimen out onto another folded, opened piece of paper.

7. Scoop up the annealed filings in a funnel of a thin-walled glass capillary tube for XRD analysis (one end is already sealed and the other end has a funnel-shaped opening to facilitate entry of the specimen). Ensure that the length of capillary tube occupied by the filings is approximately 5 mm, then cut off the capillary with a match flame or hot wire such that the piece left containing the filings is roughly 10 mm in overall length. The specimen can now be placed in the Debye–Scherrer camera.

5.3 MINERALS

Preparing specimens from most mineral samples usually has the same problems as preparing specimens from laboratory preparations and ceramics; however, there are some differences, primarily due to the nature of the materials. This discussion will exclude the study of clay minerals and zeolites which are covered in other sections. A rock is an aggregate of minerals assembled by one or more geological processes. These processes may be chemical, physical, or both. Chemical processes tend to produce assemblages which are in chemical equilibrium at the time of formation. This situation would mean that the individual mineral phases should be unique and homogeneous; however, as crystals form, the interior of the crystal may be masked from the reacting surface and zoning may occur. Thus, such minerals could show a range of physical and chemical properties. Subsequent changes in the geological environment will superimpose further changes on the minerals that may alter their chemical and/or physical state. They may partially or wholly react to form new minerals or they may modify their structure or chemistry slightly in response to the changing pressure and temperature. The solid-state changes do not occur instantaneously, so portions of the minerals may be trapped in their intermediate forms. The resulting rock is a hybrid of the multiple processes, and the minerals may also be a range of states from the initial form to the form stable in the last environment.

With few exceptions, minerals are solid solutions based on an ideal composition and structure. Deviations from the ideal mineral are the rule rather than the exception. It is, thus, probable that no two minerals in nature are exactly identical nor are they identical to the synthetic counterparts formed in the laboratory. Sapphire, Al_2O_3, is an example. In the laboratory, the pure material is colorless and can be grown into very large, essentially perfect single crystals. In nature it is almost always colored, including browns, reds, blues, and greens, due to chemical variations and is usually included with various other mineral impurities. Usually the differences between powder diffraction patterns of the synthetic and natural materials are very small, but there are many examples where they are not. One example is the many feldspars.

When studying minerals in the laboratory, one does not usually have to take special precautions such as isolating the sample from the atmosphere. The specimen has usually been in a bag since collection, and any changes that could have taken place have probably already occurred. Geological reactions usually require considerable time to complete. However, when interpreting results, one must always remember that alterations could have taken place. When the lunar rocks were returned to Earth, considerable precaution was maintained to prevent the Earth's atmosphere access to the specimens until they could be studied. The same may be true of specimens brought from mines or wells. Usually though, mineral specimens may be processed in the open under the microscope when preparing the specimens.

5.3.1 Analysis of Minerals

Fortunately, with mineral analysis, one usually has an idea which minerals are involved in a particular study. Where new minerals are being characterized, one needs to isolate the minerals and work with single-phase patterns. Where the goal is to identify a small mineral grain, it is also wise to isolate that grain and work with a single-phase pattern. Where the goal is to understand the rock assemblage of minerals, one may isolate the minerals or work with the whole rock. As always, once the question is defined, the best method of analysis is also defined.

In mineral analysis, it is possible to generate subsets of reference patterns to use with most analyses. Just knowing the specimen is natural allows a subset of the Powder Diffraction File to be used. This subfile is known as the Mineral Subfile. In 1998, this file contained about 4,300 mineral patterns representing over 3200 mineral species. Except when the mineral is known to be new and undescribed, the probability is very high that appropriate reference patterns are in the Mineral Subfile. Further subdivision may be made based on geological processes, on chemical content, or on mineral structural groups. A useful subdivision could be rock-forming minerals, which would contain all minerals that might occur in common igneous, sedimentary, and metamorphic rocks. Such a file would contain about 500 of the patterns in the PDF. Further subdivision to only sedimentary minerals would bring the size down even further. The smaller the size, the higher the probability of matching the pattern of an unknown with a reference pattern.

The other way to improve the probability of locating a pattern match is to expend considerable effort in preparing the specimen for study. One of the most common mistakes of inexperienced diffractionists is to expect to get all the answers on a specimen prepared from the bulk specimen. If mineral quantification is the goal, there is no choice but to use the bulk specimen. However, if identification and characterization is the goal, time should be taken in the initial steps to isolate the mineral and prepare a single-phase specimen. This time will be recovered in the analysis time.

5.3.2 Beneficiation of Minerals

Mineral beneficiation may be carried out in several ways. The most direct (and sometimes the most efficient) is to handpick the particles under a good binocular microscope. The specimen may be bulk or crushed to coarse fragments. Fragments should be small enough to be monomineralic. In fine-grained rocks such as volcanics, the crystallites sizes are below 0.5 mm and are hard to separate directly. Coarser rocks, however, can be processed to yield a collection of specific mineral(s) in a reasonable time. The best picking tools are a sharp needle held in a pin vise or a vacuum tube with an air trap for the particles. If the fragments are large, sharp tweezers could be used, but be careful because it is easy to launch a fragment into outer space by squeezing the tweezers too hard. The needle is usually safer. A small mortar is a good collecting bowl. Loose fragments should be retained in a small diameter (60 mm) petri dish. Sieves of 10, 35, and 60 mesh are useful for concentrating particle sizes that are the most efficient for picking. Place the open petri dish under the microscope and, using the needle, touch the tip to the desired fragment. If you are lucky (or good), the fragment will adhere to the tip of the needle. You may now transfer the fragment to the collecting dish. Tap the needle lightly on the rim of the collecting dish, and the fragment will drop into the dish. If the fragment will not adhere to the tip, wipe the needle with your fingers or on your skin to generate a thin grease film. This trick will usually get the fragment to behave. At several times during the collection, the contents of the dish should be examined and undesirable fragments removed. It takes considerable time to collect sufficient fragments for a standard cavity mount, so most studies in mineralogy are done with smear mounts or microcavity mounts. The single-crystal low-background support is especially useful here.

If the mineral is to be isolated directly from the bulk specimen, the needle can be used to loosen the desired mineral growths. The directed pressure needs to be applied carefully so that the particle will not fly off. If the specimen is placed over a dish when the fragments are picked, the fragments can be directed into the dish. Once a large number of fragments have been loosened, the material in the dish should be examined. Before the rock specimen is removed, it should be turned over, letting any loose fragments fall into the dish. The lifetime of needles that are used for digging out specific grains is short. Needles can be quickly resharpened if a good honing stone is part of the specimen preparation kit.

Mineral beneficiation may also be carried out in other ways. Density liquids may be used if the specific gravities are in the right ranges (typically 2.2 to 3.3). Prepare a column of the density liquid (usually methylene iodide or bromoform used in a fume hood) and add the powder. Then dilute the liquid with a solvent (toluene) until some fragments start to sink. Carefully adjust the specific gravity with the solvent until you get the desired separation. You may have to use two steps: separating the heavier particles first then lowering the specific gravity just enough to get the desired phase to sink. This technique can be very effective if the fragments are monomineralic.

Sometimes one can play prospector with a sample. Place the fragments into a petri dish and add a fluid (water, alcohol, or acetone). Gently swirl the dish to get the fragments moving slowly around the perimeter. Fragments with different densities will move at different rates, and the lighter or heavier particles may concentrate sufficiently to be collected with a spatula.

Magnetic separators are another way to separate samples into their component minerals. While one might have to play with the settings a while to get the most effective angle, tilt and field strength, the separations can be surprisingly clean. By changing the field strength several times, you may be able to separate all the component minerals. Sometimes the sensitivity can be surprising, yielding a phase present in very small quantity which was unsuspected prior to separation. Magnetic separators work best on fragments of uniform size around 0.5 mm.

Processing the most common mineral in the Earth's crust can be frustrating. In igneous rocks, it is common that two feldspars coexist at the time of formation (a K-feldspar and a plagioclase). On cooling, both these feldspars change composition and structural state in response to the lower temperature and pressure. Each feldspar may evolve into two distinct species creating four different minerals in the same rock. Due to the small physical dimensions of the evolution, it is usually impossible to separate all these minerals. In metamorphic rocks, the feldspars may be more distinct, as they form in response to rising temperature and pressure, but there still will usually be two species present. In sedimentary rocks, the situation is worse because these rocks form by a physical process and the collection of feldspar fragments depends on the source rocks. Hence, there may be many different compositions and structural states represented.

To get the best results in feldspar studies, it is usually wise to work with as few fragments as possible for each specimen. Single fragments will result in the minimum amount of confusion caused by the presence of multiple states of the minerals present. The usual measurements desired are accurate d-spacings and peak resolution, so a small pile of fine powder spread as a one-particle layer on a single-crystal low-background substrate will provide an ideal specimen for analysis. This situation is also the time to consider using the Guinier camera rather than the diffractometer because of the increased resolution and the need for less specimen.

Except for quartz, the other major rock-forming mineral groups all cause problems in determining the exact species that is present, and success is usually dependent on preparation techniques to get the specimen as monomineralic as possible. Moderate success in the speciation of pyroxenes is possible, particularly if the cell parameters can be refined from the measured data. Amphiboles are more difficult because of the very wide range of compositions, so independent chemical information is extremely valuable in amphibole speciation. Micas present different types of problems. The muscovite family may be directly identifiable. The basal spacing distinguishes the Na form from the K forms, and the polytype is distinguished by the diffraction pattern. The biotite micas are more of a problem because of the chemical variability and the many polytypes that occur even within a single-crystal fragment. Independent chemical information is essential for speciation. The olivines are relatively straightforward, but the presence of trace elemental substitutions in the normal Mg–Fe series complicates the interpretation somewhat. Cell parameters along with some chemical information are usually sufficient for speciation.

For concentrating the pyroxenes and amphiboles, the magnetic separator is very effective. The presence of transition elements in the minerals usually leads to sensitive magnetic susceptibilities which allow quite good separations. Density separation usually works better with the micas, more because of the shape than the composition. Fortunately, most rock samples are coarse enough in grain size that sufficient material can be hand separated. It is usually the laziness of the operator that causes the other physical separations methods to be employed when the picking method would be more satisfactory.

Metamorphic minerals, in addition to the ones mentioned, also have similar problems of chemical and structural variability. The garnets, scapolites, staurolite, and aluminosilicates all require good measurements to obtain speciation. Independent chemical information is invaluable.

5.3.3   Identification of Sedimentary Minerals

Identification of sedimentary minerals gives problems similar to those with other minerals. Detrital mineral fragments are treated the same as their igneous–metamorphic counterparts, but the authigenetic minerals also have problems. The carbonate family has several chemical series, ordered and disordered states, and some compositions have polymorphs. Evaporite minerals are usually chemically pure, and grain isolation leads to easy identification. The clay minerals will be discussed in another section (see section 5.5). The biggest problem in identification is probably the cementing phases that hold the grains of sedimentary rocks together. They are coatings on grains and interstitial material which are usually of low crystallinity and impossible to isolate. When preparing the specimen for this study, try to minimize the volume of the large grains by separating them before the specimen is assembled.

Isolation of a coating phase is not easy. One trick to get the diffraction pattern of a coating phase on a grainy substrate is to use a single grain as the specimen. The substrate grain, being single crystal, will only diffract if the grain orientation is just right, but the coating which is polycrystalline will yield a normal powder pattern. This trick can be used with the Debye–Scherrer camera or with the diffractometer if several grains are aligned on the diffractometer axis.

To characterize fully an inhomogeneous sample, it is better to try to isolate the minerals prior to study and record individual patterns than to try to interpret the data in a pattern obtained from the mixture. Because a rock is really very inhomogeneous, it is usually more useful to know exactly which grains are which mineral, so local intergrain reactions can be interpreted rather than just knowing a particular mineral is present in the rock somewhere. Although it does require time, the true picture may only be obtained by examining each type of grain independently rather than running one bulk diffraction pattern.

If mineral identification is the main goal and if the individual mineral grains are small, the Debye–Scherrer–Gandolfi camera is still very useful. A single fragment may be isolated from the sample with the needle and mounted on the tip of a glass rod for this camera (see section 5.10.2). Fragments as small as 25 μm can be processed this way. Such a fragment has a good chance to be monomineralic, and the identification of the pattern will usually not be difficult. Actually, a specimen of several grains (> 4) usually provides a better film image than a single-crystal fragment because of a blind region in the Gandolfi motion. Many diffractionists look down their noses at this old camera, but it certainly can save time.

5.3.4   Determination of Crystalline Silica

The measurement of the abundance of the crystalline phases of SiO_2 has become an important role of many diffraction laboratories in industry because of the health hazard of these materials. There are several published procedures for the determination of respirable quartz and cristobalite using air sampler filters (HSE MDHS 51/2, NIOSH-7500, NIOSH-7501), which should be consulted for the details. The specimens are transferred from the original filters to smaller diameter membranes to collect the material and concentrate it to an area observed in the diffractometer. These procedures may also be applied to bulk sample analysis, but they are designed for air-quality monitoring. The results from these methods are recognized for legal proceedings. Even when using one of these methods, it is imperative to plan the sampling so that the specimen is representative of the test to be performed. Air samplers do survey considerable volume of air, so the sampling is usually representative of the atmosphere. It is the solid materials that are difficult to sample, especially where the source material is natural rocks which have considerable variability. Section 1.4 should be consulted for a detailed discussion of sampling procedures and analysis.

A general method proposed by Elton, Salt, and Adams [4] for the determination of quartz in kaolins and carbonates may be used for detection and quantification to the 0.1% level.

- Grind 5 g samples with 10 ml water for 15 minutes in a McCrone micronizing mill. Use corundum elements in the mill. Filter the sample using a paper filter in a Buchner funnel, applying vacuum to the filter flask. Place in an oven at 100°C until dry.
- Transfer the dried powder from the filter paper to a Janke and Kunkel A10 grinding mill and mill for 30 s.
- Weigh out a predetermined amount of specimen. Backfill this specimen into the appropriate specimen holder for the available instrument. Apply no pressure. Place the backing support in place and turn the specimen over for use. The amount of specimen should be just sufficient to underfill slightly when pressed. This amount needs to be determined for each type of specimen holder.

When using the standard Philips circular specimen holder, the amount of kaolin is 0.500 ± 0.005 g. The tool illustrated in Figure 4.3 allows specimens to be packed in a very reproducible manner. Specimens which are not weighed and which are packed by hand show significantly poorer intensity reproducibility than specimens prepared in the manner described. The tool exerts very little pressure on the powder, and there is no evidence to suggest that it introduces preferred orientation more than traditional handpacking. Use of the tool depresses the back of the powder by about 0.5 mm. To fill the gap between powder and the specimen holder backing plate, a card disk of the appropriate diameter and thickness is inserted. This disk supports the powder when the specimen is turned right way up.

Where there is insufficient specimen to meet the needs of the above preparation, an alternative method is used.

- Mill available sample in a McCrone micronizing mill as described above.
- Evaporation of the slurry may be better than filtering to ensure complete recovery of the powder. Alternatively, use a small membrane filter to collect the powder and scrape the filter while still wet to a glazed tile for oven drying.
- Crush the dried powder in a polished mortar and pestle.

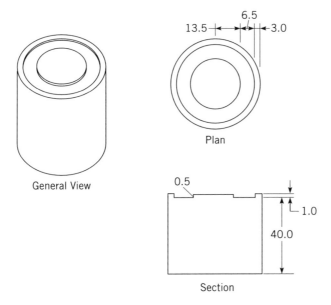

13.5 |← 6.5 →|←3.0

Plan

General View

0.5

1.0

40.0

Section

Figure 5.3. Tool for packing specimens in standard Philips circular holders.

- If there is still insufficient specimen to fill the specimen holder shown in Figure 5.3, the dimensions may be modified to provide for thinner specimens.

Quartz may be detected in specimens that do not have interfering diffraction peaks down to the 0.03% level. Carbonate rocks are an example where interference is usually at a minimum. However, if mica or mica clay is present, the detectability is severely limited by the (003) peak interference at 26.6°2θ. A good review of the whole silica determination problem is given by Smith [5].

5.4 ALUMINUM ORES

The analysis of bauxite is beset with difficulties including the large number of minerals which may be involved, the very fine-grained nature of most of the minerals, and the usual imperfect crystallinity of the crystallites. The goal of analysis is to differentiate the species and to semiquantify their abundance for plant process control. There is the need to maximize the intensity response (and hence the counting statistics) and minimize the preferred orientation of the crystallites so that the results are interpretable. Speed of analysis and turn-around time are of importance.

Bauxites are the result of tropical processes of soil formation by the weathering of the base rocks. The extreme conditions lead to the leaching of most elements in the original rock and the concentration of aluminum as oxyhydroxides along with other elements such as Fe, Ti, and Si. The resulting soil is generally soft and friable and composed of as many as a dozen or more minerals as submicrometer crystallites. XRD coupled with elemental analysis is the best approach to analysis.

5.4.1 Grinding Samples of Aluminum Ores

The first step is grinding the sample to a very fine powder to assist in improving the crystallite statistics and reducing preferred orientation. The size required to overcome these two considerations depends on the nature of the sample. Materials consisting of fine microcrystallites, e.g., goethite, are not particularly sensitive to either grinding or preferred orientation; coarse crystalline materials, e.g., some forms of gibbsite, are highly sensitive. For the former, grinding to -45 μm is perfectly adequate; for the latter, grinding to -5 μm may be required. Often mineral mixtures contain both coarse and fine fractions, hence the usual practice is to grind fine.

The problems associated with fine grinding are the time requirement, the potential of overheating the sample, and the potential of causing lattice distortion. The latter can adversely affect peak intensity; and in the most severe cases, it can cause a phase change. The overheating problem is readily overcome using a coolant, and a coolant should be used in all mechanical grinding methods. It is also good practice, though not essential, to use a coolant fluid such as acetone when grinding small amounts of sample in a mortar and pestle.

The effect of grinding time on the integrated intensities for a bauxite are shown in Table 5.1 for two different grinding methods. There is considerable variation in peak intensity with grinding time. The intensity for the platy muscovite, which is very susceptible to preferred orientation, decreased significantly; goethite and hematite, which have crystallite sizes on the order of 300 to 500 Å, were hardly affected. These data indicate that hand grinding may be very operator sensitive. Hence, mechanical methods are preferred as they give a consistent grind for a given sample. Ideally, the grind should be optimized for each material analyzed if quantitative analysis is being sought. Of the grinding equipment tested, the micronizing mill has proved the most consistent, the Tema mill (ring and puck) a close second, and the ball mill the least

TABLE 5.1. Effect of Grinding Time on XRD Integrated Intensity for Bauxite Minerals. Sizes are in angstrom units.

Grind Time	Muscovite	Kaolin	Gibbsite	Quartz	Goethite	Hematite
Tema mill						
30 Seconds	133	137	6036	335	308	181
1 minute	210	162	6930	443	325	175
3 minutes	68	110	6339	364	301	213
6 minutes	52	99	5590	411	276	158
9 minutes	27	43	3379	355	239	192
Micronizing						
5 minutes	103	123	6516	395	289	195
10 minutes	112	134	6844	419	305	208
20 minutes	73	109	5534	394	293	202

consistent. The following parameters should be kept constant for grinding: sample weight, coolant volume, grind time, and ball-rod charge if relevant. The procedure will not guarantee a perfect grind for all materials, as samples will vary in hardness and size. However, it removes operator dependence in grinding.

5.4.2 Packing of the Specimen

The specimen presented to the diffractometer should have a flat surface in the required plane and be sufficiently thick to approximate an infinitely thick specimen. The major problem is how to overcome the preferred orientation that occurs in the packing of the specimen. The three common packing methods are the "microscope slide" method, the use of a press, and side-drifting. The microscope slide method is normally a backfill method and carried out manually. The degree of preferred orientation and packing density obtained are operator sensitive. The latter consideration requires the use of an internal intensity standard for any quantitative or comparative work. Roughening the surface after mounting is claimed to further reduce preferred orientation, but again, it is operator sensitive. Mechanical pressing does cause preferred orientation, but the degree can be made operator independent by using a fixed weight of sample and a constant pressure. This method also removes the problem of variations in packing density and can enable comparative data to be obtained without the need for an internal intensity standard. Side-drifting significantly reduces preferred orientation, but it is impossible to obtain consistent packing densities, and thus an internal intensity standard is essential. Side-drifting, although reducing preferred orientation, is an operator-sensitive procedure.

One new method which is claimed to overcome the preferred orientation problem is spray drying [6] (see section 4.5.6). The sample must be ground, spray dried and then reground, but only sufficiently fine for packing. The spray drying results in the fine particles forming spherical clumps, thus potentially overcoming the preferred orientation problem. At present laboratory equipment is not readily available; also the method is currently too slow for routine analysis.

For routine work, the preferred packing method is the one which is least sensitive to the operator. It is not necessarily the best method in terms of overcoming preferred orientation. A procedure which involves the pressing of a constant weight of sample at a constant load in a press (e.g., a Hertzog press), has been found to give reproducible intensities; in cases where preferred orientation does occur. If handpacking methods are to be used for quantitative or comparative work, then it is imperative to check the sensitivity of the output to the packing procedures for that material.

5.5 CLAYS

5.5.1 Introduction

Specimen preparation is one of the most important requirements in the analysis of powder samples by XRD. This statement is especially true for soils and clays that contain finely divided colloids, which are poor reflectors of X-rays, as well as other types of materials such as iron oxide coatings and organic materials that make characterization by XRD more difficult. Specimen preparation includes not only the right sample treatments to remove undesirable substances, but also appropriate techniques to obtain desirable particle size, orientation, thickness, etc. Several excellent books are available that deal with appropriate specimen preparation techniques for clays and soils [7, 8, 9, 10].

5.5.2 Specimen Preparation of Clays

Drying   Soils and clays are dried before they are ground and separated into various particle sizes. The temperature at which they are dried depends on the material and types of clay. Soils are air dried before size and clay separation is attempted. Clays containing the smectite group of minerals are freeze dried to avoid formation of hard clumps. Clays and soils that contain other species of clay minerals are dried at 110°C in a forced air oven. Clays are usually extremely fine grained and do not require too much preparation. Soils, on the other hand, contain not only clay-sized materials but also other minerals in silt (2–50 μm) and sand size (> 50 μm) range. They do require grinding to attain appropriate particle size for X-ray analysis.

5.5.3 Pretreatment of Clays

The ground soil and clay sample may require certain pretreatments before they can be analyzed by XRD. This step is to remove undesirable coatings and cements either to improve the diffraction characteristics of the sample or to promote dispersion during size fractionation. Relatively pure clay samples may not require any of these pretreatments. Soil samples, on the other hand, will require most of these treatments.

The pretreatment can be grouped into the following:

1. *Removal of Soluble Salts and Carbonaceous Cements.* The carbonaceous cements inhibit proper dispersion of colloids and, thus, separation of fine fractions (e.g., < 5 or 2 μm). The carbonaceous cements are removed by treating the sample with either mild HCl or buffered sodium acetate (NaOAc) solution [11, 12] as follows:

    - Place a suitable amount of ground specimen in a glass beaker or centrifuge tube.
    - Add 1 N sodium acetate buffer solution (82.0 g of NaOAc and 27 ml of glacial HOAc per liter, adjusted to pH 5.0) buffered at pH 5.0 in 1 to 10 soil:buffer ratio. Note: The pH of highly calcareous specimens should be lowered to pH 5.0 using 1 N HCl.
    - Digest the suspension in a boiling water bath suspension for 30 minutes with intermittent stirring.
    - Repeat the second and third steps two to three times (if highly calcareous) or until the bubbling subsides.
    - Centrifuge and decant the clear supernatant solution.

2. *Removal of Organic Matter.* Organic matter can be removed by treating either with 30% hydrogen peroxide (H_2O_2) or sodium hypochlorite [7].

Hydrogen Peroxide

- Mix a known amount of specimen with a sufficient amount of water or sodium acetate buffer (pH 5) in a beaker to make it soupy.
- Add 30% H_2O_2 in 1:0.5 soil:peroxide ratio.
  CAUTION: Avoid contact with H_2O_2 because this reagent can cause severe skin burns.
- Stir the mixture and let it react. Maintain constant stirring to prevent frothing. If the reaction is too rapid, apply a jet of distilled water from the wash bottle
- Heat gently on a hot plate and maintain stirring.
- Continue additions of small amounts of H_2O_2 (4–5 ml) and heating until the specimen loses its effervescence.

Comment: Hydrogen peroxide treatment induces acidic environment, and the pH may approach as low as 2.0. This treatment could result in the degradation of clay minerals [12].

Sodium Hypochlorite ("Chlorox")

- This treatment is effective in removing organic matter from specimens where low pH is a problem, without affecting carbonate minerals, phosphates (apatite) or MnO_2 [7].
- Start with a sufficient amount of specimen (3–5 g) in a 250 ml beaker.
- Add sodium hypochlorite solution (adjusted to pH 9.0 to 9.5 with 1 N HCl) in 1:3 proportion under a well-ventilated hood. Adjust the pH of hypochlorite solution just before use. CAUTION: Perform the whole experiment under a hood.
- Repeat the treatment one or two more times.
- Centrifuge and decant the supernatant liquid.

3. *Removal of Sesquioxide Coatings.* Several extractants are available for sesquioxide (Fe_2O_3, Mn_2O_3) removal. The most commonly used extractant is a mixture of sodium carbonate, sodium bicarbonate and sodium dithionate [7]. The method, called the CBD method, is described below:

- Start with a suitable amount of specimen (5 g or less if the specimen is high in Fe_2O_3) in a 100 ml test tube or a 250 ml beaker if the iron content is high.
- Add 40 ml of 0.3 M Na-citrate solution and 5 ml of 1 M $NaHCO_3$ solution.
- Place the tube with its content in a water bath maintained at 75–80°C (do not exceed 80°C).

- Add 1 g of sodium dithionate ($Na_2S_2O_4$) and stir constantly for one minute, then occasionally every 5 minutes.

- If high iron oxide is suspected or brown or red color persists, add one more gram of $Na_2S_2O_4$ while constantly stirring. (Avoid exceeding 80°C because it results in the formation of FeS.)

- At the end of 15–30 minutes of digestion, add 1 N NaCl, wash it once, and centrifuge.

- Add distilled water and proceed with size separation.

The CBD treatment will remove certain crystalline iron oxide minerals such as goethite (α-FeOOH), lepidocrocite (γ-FeOOH) and hematite (Fe_2O_3).

Other extractants that are used to remove iron oxides include ammonium oxalate, sodium citrate, and nascent hydrogen sulfide.

4. *Removal of Oil Contamination.* Core samples from the petroleum reservoirs or soil samples from chemically contaminated sites are often coated with "oily" materials. They are removed either by the time-consuming Soxhlet extraction or treatment with organic reagents such as xylene, toluene, or methylene chloride. Considering the carcinogenic nature of xylene and toluene, methylene chloride is probably the preferred reagent. A known amount of specimen is washed with organic solvent (two to three times in a glass test tube or a beaker) and dried at a low temperature (about 60 to 70°C) in an oven.

5.5.4 Size Separation of Clays

Size Ranges Minerals occur in various size ranges. The USDA system of size classification is shown in Table 5.2. It is often necessary to fractionate soil and clay samples to one of these size ranges to elucidate and quantify specific mineral species present. Coarse size fractions are obtained either by wet or dry sieving through a standard set of sieves. The fine fractions, starting from very fine sand, are separated by a combination of wet sieving and gravity or centrifugal sedimentation.

TABLE 5.2. USDA System of Size Classification

Material	Size (mm)
Very coarse sand	2.00–1.00
Coarse sand	1.00–0.50
Medium sand	0.50–0.25
Fine sand	0.25–0.10
Very fine sand	0.10–0.05
Coarse silt	0.05–0.005
Fine silt	0.005–0.002
Clay	< 0.002

Optimum Size What is the optimum size range to study clay minerals? This choice depends on the type of clay mineral present. Generally, samples are separated to less than 2 or 5 μm to examine various clay minerals. Speciation of authigenic clays often requires less than 0.25 μm particles. Figure 5.4 shows the XRD patterns for four different size fractions from a sandstone rock. In this specimen, kaolinite and mica were present even in the 5–50 μm range, and the detectability of smectite increased with a decrease in size range examined.

Dispersion The pretreated sample is dispersed for size separation either by physical or chemical desegregation methods. The physical method includes using an ultrasonic probe or a tank, the former being more effective. In the chemical method of dispersion, the sample is mixed with a dilute sodium carbonate or metaphosphate ("Chlorox") solution to achieve dispersion of particles.

Separation Techniques The dispersed sample is wet sieved through 325 mesh sieve (about 50 μm) to separate very fine sand. The suspension or centrifuge method is then used to separate other fine (clay and silt) fractions either by gravity sedimentation or centrifugation. The settling times required to separate various size ranges are based on Stokes' law and are presented in extensive nomographs and tables by Jackson [7].

The times required to obtain particles of various diameters from a dispersed powder sample in distilled water (10 cm high) placed in a test tube (with a dispersing agent) by gravity or by centrifugation [7] are shown in Table 5.3. These calculated times

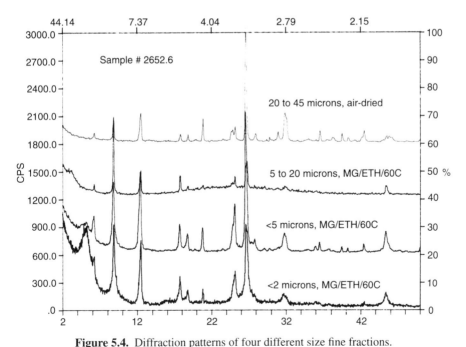

Figure 5.4. Diffraction patterns of four different size fine fractions.

TABLE 5.3. Dispersion Times

Size range (esd)	Gravity	Centrifugation
< 20 μm	4 min 38 s	—
< 10 μm	18 min 33 s	—
< 5 μm	1 hr 14 min	3 min at 300 rpm
< 2 μm	7 hr 44 min	3 min at 750 rpm

assume a particle density of 2.65 g cm^{-3} and a water temperature of 25°C. As is evident from the table, for finer particles (such as < 2 μm) centrifuge technique is better because gravity sedimentation takes a longer time. For separating particles < 5 or 10 μm, however, gravity settling is probably better since the accuracy with which lower speeds can be attained with the centrifuges is questionable.

Separation by Gravity Settling

- Take a known amount of specimen or sample suspension in a centrifuge bottle (100 or 200 ml capacity).
- Fill with distilled water (with a dispersant) to 10 cm from the bottom (top of the sediment).
- Disaggregate and mix by an ultrasonic probe 1 to 2 minutes (see Table 5.3) or by thorough stirring, and allow to settle for the appropriate time period.
- Decant the supernatant into another large container without disturbing the settled material.
- For quantitative work, the process of mixing and settling (second, third, and fourth steps) is continued until the supernatant solution is nearly clear. Note: Five to eight washings may be required to accomplish a complete separation.
- The collected fine fraction suspension is then flocculated to concentrate the sample.

Separation by Centrifuge Method

- Transfer the sample or sample suspension to a centrifuge bottle or tube.
- Add water (with a dispersant) to 10 cm high from the bottom and mix thoroughly.
- Centrifuge for 3 min at 750 rpm for separating less than 2 μm particles. Note: Centrifuging time may be increased by raising the settling depths from 10 to 15 centimeters.
- Decant supernatant carefully into a large beaker.
- Repeat the second, third, and fourth steps (process of adding water, mixing, and centrifugation) until the supernatant is clear.

Flocculation A complete clay separation will result in large volumes of clay suspension and is often necessary to concentrate the suspension for further use. The flocculation of colloids is usually accomplished by either lowering the pH or adding 0.5 g of $CaCl_2$ or $MgCl_2$ salts. The clear supernatant is then siphoned off.

5.5.5 Slide Preparation and Analysis of Clays

Two important criteria to keep in mind while preparing slides for XRD analysis are the amount of specimen and orientation of crystallites.

1. *Amount of Specimen.* The amount of specimen should be enough to produce an "infinitely" thick specimen [8, 9]. Infinitely thick specimen is defined as that thickness of the specimen that yields optimum diffracted intensity. Increasing its thickness (by adding more specimen) will result in a negligible increase in diffracted intensity. An infinitely thick specimen will ensure not only optimum intensity of diffraction peaks from each phase, but also minimize any interference from the substrate. The amount required varies with the type of material being examined. Brindley [8] tabulates Rich's work on the amounts of clay (with a mass attenuation coefficient of 50 cm^{-1}) required to obtain "infinitely" thick slides on a ceramic substrate.

2. *Orientation of Crystallites.* The types of mounts normally employed are dictated by the nature of crystallite orientation required and can be grouped into random or oriented mounts.

   (a) Random Mounts: Random mounts are preferred when the mineralogy of the whole soil or clay is required. In this type of mount, particles are packed to assume different orientations and ensure reflections from various ($hk\ell$) planes. Types of random mounts available include: spray drying, front loading, back loading, side loading or drifting, and Vaseline or Apiezon coated. These methods are described in section 4.5.

   (b) Oriented Mounts: Several methods are available to prepare oriented mounts. The Clay Minerals Society has published a workshop proceedings outlining these methods.

Methods for Preparing Oriented Mounts

Slurry Mount

- Place an aliquot of suspension on a substrate (such as a glass or quartz slide).
- Allow the specimen to air dry in a dust free environment.

Comment: This method is the easiest, but it is difficult to saturate the specimen with various ions, if required.

CAUTION: It is very important to make sure that the specimen covers the entire slide and goes under the clip when mounted on a goniometer. Otherwise, the *d*-spacing will be shifted because of specimen displacement.

Centrifuge on Tile

- Place a slide at the bottom of a wide mouth, 250 ml centrifuge tube.
- Transfer the suspension into the centrifuge tube.
- Centrifuge at 1000 rpm for 5 min.
- Pour or siphon off the suspension and carefully remove the slide.

Comment: This technique produces a very well-oriented clay mount.

Suction on Tile

- Place a ceramic tile or other porous material (porous steel), cut to appropriate size on a suction apparatus.
- Pour a known amount of clay suspension and apply vacuum.
- Saturate with desired cation, if required, by filtering through appropriate salt (1 N) solutions (see section on ion saturation).
- Wash off excess Cl^- ion by washing thoroughly with distilled water.
- Remove the slide and store.

In this method, specimens can be heated to very high temperatures. Rich and Barnihsal [13] describe this method in detail. The disadvantages of the method are the nonavailability of ceramic substrate and that more than a few milligrams is required to cover the entire substrate.

Suction on Millipore The specimen suspension is filtered onto a Millipore filter made of cellulose or silver and transferred onto a glass slide [14]. It is a good technique if ion saturation is required. If silver membrane is used, the specimen mounts can be heated to various temperatures without transferring onto a slide.

- Place a 0.45 μm pore size filter on a flat fritted glass surface of a Millipore filter holder.
- Connect the filter assembly to vacuum, and wet the filter with small amounts of distilled water.
- Pipette an aliquot of well-dispersed clay suspension and spread on the Millipore filter. Note: The suction should be on during this stage.
- Allow the free water to filter through. Note: Do not let the specimen dry out.
- Saturate with desired cation, if required, by filtering through 2 ml of one normal (1 N) appropriate salt solution (see section on ion saturation).
- Repeat last step 3–4 times.
- Remove excess Cl^- ion by washing thoroughly (2 ml, at least 4 times) with distilled water.
- Remove the filter membrane and place it, while still moist, clay side down on a clean glass slide.
- Transfer the clay on to the slide by gently rolling the back of the membrane with a cylindrical rod (for example, a test tube). Peel off the membrane slowly, leaving the clay film on the glass slide.

Molybdenum Slides   Slides made of molybdenum metal have been recommended for use as a substrate for oriented clay preparations. The Mo is microcrystalline, which produces a very low background and there are very few diffraction peaks because of the small body-centered cubic cell. The lowest diffraction peak for Cu K radiation is at 40.5 2θ, which does not interfere with most clay analyses. The molybdenum is inert to many treatments and allows the specimens to be heated.

Sources for molybdenum sheet suitable for diffraction mounts:

Alfa Products, division of Johnson Matthey, P. O. Box 8247, Ward Hill, MA 01835-0747. A 100 mm $\times$ 100 mm $\times$ 2.5 mm sheet is listed as stock number 10047.

Strem Chemicals, Inc., P. O. Box 108, Newburyport, MA 01950. A 100 mm $\times$ 100 mm $\times$ 2.5 mm sheet is listed as stock number 42-0080.

Ion Saturation   Why ion saturation? The $(00l)$ spacing and intensity from an oriented clay depends on the type of cations in the interlayer region. In nature, clays and soils are saturated with various types of cations including Na^+, Ca^{2+}, Mg^{2+}, K^+. It is easier, for proper identification, to make them monoionic. This exchange is accomplished by washing the sample with dilute salts (one normal is usually sufficient) of appropriate cations, and washing the excess salt with deionized water. The most commonly used cation is Mg^{2+} ion.

Glycolation   For identification of expandable clay minerals, several organic reagents are used for intercalation. The most commonly used solvents are ethylene glycol (EG) and glycerol. In the former treatment, oriented slides are placed in an ethylene glycol chamber maintained at about 60°C for vapors to intercalate. A drop of the solvent is placed on the edge of the oriented slide before placing it in a chamber. A glass desiccator is an excellent chamber. In the glycerol case, the solvent (5 ml of 20% glycerol solution) is washed through the oriented slide on the suction apparatus. The choice of a solvent depends on the type of information sought. For identification of interstratified clays, ethylene glycol treatment is extensively used [12].

Heat Treatment of Clays   Clay mounts on ceramic substrates, Si crystals, silver membranes or certain glass substrates can be heated to various temperatures to assist in the identification of mineral species. The temperatures normally used are 120–150°C (overnight), 300°C (4 to 5 hr) and 550°C (4 to 5 hr). Heating to 120 to 150°C drives one of the two solvent layers off the interlayer region and further confirms the presence of smectite. This treatment is especially useful when small amounts of an expandable mineral are present. Heating to 300°C is employed to elucidate 2:1 layered clays that are intercalated with hydroxy-Al polymer species. This mineral is of common occurrence in soils. The slides are heated to 550°C for 4 to 6 hr to distinguish chlorite from kaolinite. Presence of a 14 Å peak after this heat treatment confirms the presence of chlorite.

5.6 PREPARATION OF ZEOLITES FOR X-RAY DIFFRACTION

5.6.1 Introduction

Zeolites are crystalline aluminosilicates based on an infinitely extending framework of AlO_4 and SiO_4 tetrahedrally linked to each other by the sharing of oxygen ions. The framework structure contains channels or interconnected voids that are occupied by cations and water molecules. The cations are mobile and may undergo ion exchange, and the water can generally be removed reversibly leaving intact a crystalline host structure [15]. Because both the cations and the adsorbed water in zeolites affect the intensities of the diffraction lines, special precautions must be taken for their analysis.

5.6.2 Dehydrated Zeolites

Zeolites that have been calcined or dehydrated, i.e., "activated zeolites," show significant variations in the intensities of diffraction lines at low two-theta angles versus zeolites that are fully hydrated with adsorbed water. Generally the low angle lines at $< 25°2\theta$, (Cu radiation) for a dehydrated zeolite show an increase in relative intensity versus the hydrated form, and for the largest d-spacing line, the increase can be dramatic. Moreover, if it is desired to analyze a dehydrated zeolite, the analysis must be done in a controlled, anhydrous atmosphere, because the dehydrated form can rapidly adsorb water with a loss of intensity for the low angle diffraction lines. On the other hand, a fully hydrated zeolite shows a rather stable distribution of relative intensities which change little with small changes in ambient humidity.

5.6.3 Zeolite Equilibration

In order to achieve powder patterns with reproducible intensities, it is common practice to deliberately expose zeolite specimens to a high humidity atmosphere before the analysis, to fully hydrate them, and to stabilize their intensities. This process is referred to as "equilibration" of the zeolite because the specimen to be analyzed is placed in a high humidity atmosphere until it reaches equilibrium conditions of water adsorption. Choice of the relative humidity (RH) conditions for equilibration of the zeolite can be a matter of personal preference, and equilibrating atmospheres in the range of 50–80% RH are common. O'Brien [16] lists 88 different salts or salt mixtures whose saturated solutions can be used to produce a range of relative humidities at temperatures from 2–50°C and above. For an atmosphere that can be varied from 0–95% RH, Kuhnel and van der Gaast [17] constructed a device which blends dry helium with moist helium and applies the moisturized gas directly to the specimen during the analysis. Another method is to use a glass laboratory desiccator containing a saturated solution of $Mg(NO_3)_2.6H_2O$ as the equilibration chamber. This mixture gives an RH of 52.0% at 24.5°C. For total hydration the prepared zeolite specimen is exposed to this atmosphere for 16 hr. An alternative approach was employed by Meyers et al. [18] who used a controlled laboratory air atmosphere of 60% RH, 16 hr exposure, for total equilibration prior to XRD analysis. Corbin et al. [19] used a saturated solution of NH_4Cl (79% RH) to equilibrate zeolites in preparation for X-ray fluorescence analysis. Duration of the equilibration depends to a large extent on the initial hydration state of the zeolite and the amount of specimen surface exposed. In another study, Whittington and Milestone [20] reported equilibration times of several days for the analysis of calcined zeolites. Nevertheless, experience shows that if the prepared specimen slide is placed in the equilibration chamber with the diffraction surface exposed, total hydration, as evidenced by a constancy of relative peak intensities, takes place in 16 hours regardless of the initial hydration state of the material.

5.6.4   Effect of Cations in Zeolites

The other major factor affecting intensity which is especially important in quantitative analysis is the cation type and content of the zeolite. Ordinarily for qualitative purposes, the correct identification of a zeolite type can be made even though the cation type and content of the specimen may be different from the reference zeolite. This identification is possible because in the majority of zeolites, the structure and unit cell dimensions of a given zeolite type are largely determined by the configuration of the SiO_4 and AlO_4 tetrahedral in the framework. Hence, for a given zeolite type, cation differences generally result in only small changes in the unit cell dimensions, thereby simplifying the identification. (There are exceptions to this general rule whereby cation differences can cause symmetry changes which may complicate and even obscure the identity of the zeolite, but these cases are less common in routine applications.) However, cation differences also influence selected peak intensities and the effect may be relatively minor or of considerable significance. On the one hand, a given family of zeolites containing alkali and alkaline-earth cations can have very similar diffraction patterns and be mainly distinguishable by minor differences in relative peak intensities of certain reflections. On the other hand, when cation substitution results in replacing a low atomic number with a high atomic number cation, such as replacing Na with a rare-earth ion, or when guest molecules are the substituting moieties, the intensity differences may be major. For example, Fischer and Tillmanns [21] showed that when the channels of a zeolite Rho were occupied not with cations but with various methyl amines as guest molecules, the resultant patterns had quite different intensity distributions and at first inspection appeared to have unrelated structures. For quantitative analyses of zeolites, the distinctive differences in both relative and absolute peak intensities between different cation substituted zeolites of the same family can be the source of large errors in the determination of the "percent crystallinity" of the zeolite. For instance, to quantify a Ca X zeolite using a Na X zeolite as reference standard would result in a large error in the percent crystallinity of the Ca X of about 20% absolute (low). For a partially exchanged Ca,Na X zeolite, the quantification can still be significantly in error if the Ca,Na X zeolite is compared to a 100% Na X reference. The errors are even greater when the substituted cations are heavy metals, such as Ba and the rare earths, and these heavy cation zeolites are compared to lighter cation zeolites [22]. Therefore, for best quantitative results, it is important to compare the zeolite sample to a matching reference standard which has the same cation type and content.

5.6.5 Particle Size and Preferred Orientation in Zeolites

Additional important considerations in the quantification of zeolites are the particle size of the ground sample and the avoidance of preferred orientation in the prepared specimen. Controlling particle size by efficient grinding without introducing line broadening has already been discussed in Section 1.5. The recommended procedure is to wet grind 3.0 g of the zeolite using methanol and corundum grinding elements for one minute in a McCrone Micronizing Mill. This treatment should yield a specimen with a particle size in the 1–10 μm range, preferably around 5 μm or smaller. Preferred orientation can be avoided to a large extent by exercising care in the preparation of the specimen mount. However, complete removal of preferred orientation may not always be possible, especially in the analysis of natural mineral zeolites having a plate-like morphology. This condition was encountered by Taylor and Pecover [23] who diluted the natural zeolites, heulandite and stellerite, with Al powder and an epoxy resin in an attempt to reduce the preferred orientation to an acceptable level. Other techniques for reducing preferred orientation are the use of glass capillaries to contain the specimen in Guinier, Debye–Scherrer, and parallel beam geometries, as well as spheroidizing the particle aggregates by spray drying (section 4.5.7). Quantitative procedures such as the full pattern type [23], profile fitting type [24], and Rietveld refinement [25] may also be used for estimation of preferred orientation effects, and in some cases numerical corrections can be applied.

Recommended Procedure for Preparation of Zeolites

1. Prepare a humidity conditioning, i.e., "equilibration," chamber by making an aqueous saturated solution of the desired salt in a glass laboratory desiccator [16, 17, 18]. Add an excess of the salt, stir, cover, and allow the atmosphere to reach equilibrium. [$Mg(NO_3)_2.6H_2O$] is a convenient, relatively inexpensive and stable salt for this purpose, and it produces an RH of 52.0% at 24.5°C].

2. Grind the zeolite to a particle size of 1–10 μm using the wet grinding method of Section 1.5.

3. Pack the ground zeolite in a standard cavity type holder using a back-packing technique and a frosted glass slide as the support face for the diffraction surface.

4. Place the prepared slide in the equilibration chamber with the diffraction surface exposed for 16 hr.

5. Remove the slide from the equilibration chamber and examine the packing to ensure that swelling has not lifted the specimen surface above the diffraction surface. If swelling has occurred, the equilibrated specimen must be repacked and returned to the equilibration chamber for two hours prior to analysis.

6. Place the equilibrated specimen in the diffractometer and scan without undue delay.

5.7   AIR-SENSITIVE SAMPLES

Air-sensitive materials are common in many laboratories and many specimen holders have been devised to protect the specimen during analysis. Some of these special devices are covered in the bibliography. Only a few special techniques, not involving special specimen holders, will be described here.

5.7.1   Alkali and Alkaline-earth Metals

Metals such as sodium and potassium are usually stored under oil, and the following procedure has been found to give good results.

A small piece of the sample, approximately 0.5 mm on a side, is cut from the bulk and washed several times in toluene to remove the oil. The clean sample is then rapidly transferred to a drop of rubber solution and thoroughly coated with a thin layer of the adhesive. Take a Debye–Scherrer camera stub with a thin coat of plasticine or tacki wax on its large end, and poke a length of glass fiber (approximately 0.1–0.2 mm diameter) through the center of the plasticine such that it goes through the hole in the stub. Leave roughly 5 mm of glass fiber protruding from the plasticine and using a pair of tweezers, snap off any excess fiber that may be sticking out of the back end of the stub. Important: Push the fiber in from the plasticine side, because if it is pushed through from the back end, the part of the fiber that will bear the specimen, and thus be in the X-ray beam, will be contaminated with plasticine. Plasticine contains calcite and this mineral will show up strongly in the diffraction pattern.

Dip the end of the glass fiber into the adhesive coated specimen and pick it up. Try to keep the specimen right on the end of the fiber; don't let it slide down on the side of the fiber. The stub is now ready for insertion in the Debye–Scherrer camera (see also section 7.1.4).

5.7.2   Samples Stored under Water

The procedure described above is used but a water-soluble adhesive such as gum arabic is used in place of the rubber solution. Gum arabic, a white powder, is mixed into a fairly stiff, clear paste with water and then used to envelope the analytical specimen.

5.7.3   Other Samples

Either of the above methods, depending on the susceptibility of the samples to water or organic solvents, may be adopted. For samples kept under inert atmospheres, a glove box or glove bag with a flow of suitable gas may be used to protect the sample until it has been coated.

5.8 THIN FILMS

5.8.1 Introduction

Thin layers deposited onto a variety of substrates are used for many research and commercial applications. For clarity, we can categorize thin layers as films or coatings. Thin films are defined as a thin layer which has been deposited on a substrate atom by atom. Examples include chemical-vapor deposition, electron-beam deposition, and radio-frequency sputtering. Thin coatings are formed when a thin layer is deposited in bulk on a substrate. Examples include precursor solutions coated for metalloorganic decomposition, spray, brush, or roller painting, and nano-particles dispersed in a binder coated onto a support. For the discussion of specimen preparation, all samples of either type will be referred to as thin films. In a thin-film specimen, the final film thickness is typically a few micrometers (10^{-6} m) down to a few angstroms (10^{-10} m) in thickness.

For optimal performance, these thin structured materials are required to have specific mechanical, optical, electrical, and/or chemical properties. Knowledge of the microstructure of thin films is important not only for understanding what deposition conditions provide the necessary optimum properties, but it is also important for being able to reproducibly manufacture these thin-film devices. XRD techniques can provide information regarding crystallinity, phase identification, planar orientation, crystallite size, film thickness, and residual stress of thin films. XRD is common to many laboratories, and for many thin film specimens the preparation and analysis is nondestructive. Depending upon the thin film thickness, the film composition, degree of crystallinity, preferred orientation, and substrate, conventional Bragg–Brentano $\theta/2\theta$ or grazing incidence diffraction techniques can be utilized for data collection. In regard to film composition, inorganic thin films generally provide better XRD patterns than organic thin films due to the higher atomic number of inorganic elements (higher atomic scattering factors) and the tendency of inorganic materials to more easily form crystalline phases in the thin-film state. There are conditions, however, in which inorganic films can be amorphous.

It is important to understand what microstructure characteristics one is trying to measure in a thin-film specimen. For example, if a thin film is analyzed to look for an amorphous diffraction pattern, the film should not be deposited on an amorphous substrate such as glass. Collecting a diffraction pattern of a neat (uncoated) substrate is also an important part in understanding diffraction data from a thin-film specimen. The diffractionist will be able to note which diffraction peaks (if any) are due to the substrate, and can determine if the substrate diffraction pattern will interfere with obtaining useful information from the thin-film component of the specimen.

The examples discussed in this section assume that diffraction data are collected using reflection mode geometry. Transmission geometry generally will not work with thin-film analysis because the substrate is quite often thick enough to absorb much of the X-ray beam.

5.8.2 Specimen Holders

Specimen holders provided by the diffractometer manufacturer can accommodate many of the thin-film specimens that are analyzed. If a special holder is needed, you can discuss your specimen with your instrument manufacturer and have the manufacturer make the holder for you (for a cost) or design the holder yourself and have your machine shop or a local machine shop fabricate your design. For making specimen holders, aluminum and polycarbonate are very easy to machine, common to machine shops, and low in cost. If you do not have access to a machine shop, it is recommended that contact be made with a local machine shop that is able to make custom accessories for XRD applications.

5.8.3 Specimen Preparation Methods for Thin Films

Careful specimen preparation will significantly improve the ability of a diffractionist to obtain good data from thin-film specimens. However, it is important to remember that the first step in collecting good diffraction data is to start with a well aligned diffractometer!

Examples of specimen preparation used for XRD data collection on thin-film specimens are shown in Figure 5.5. The easiest specimen to prepare is a self-supporting flat specimen in which the thin film/substrate can be placed in the diffractometer specimen chamber as received (see Figure 5.5a). An example would be a film coated on a wafer or other large area substrate. Care must be taken when the specimen rests against a reference point in the specimen chamber. This contact point could result in damage to the thin film. If the as-received sample will not fit in the specimen chamber, a portion of the specimen can be removed. One way to perform this removal for glass or single crystal substrates is to scribe the specimen with a diamond scribe (VWR Scientific) and snap off the unusable portions. Note that when a specimen is reduced in size, this XRD experiment is no longer completely nondestructive. The analyzed specimen can be returned, but the total as-received sample has been altered.

If concern exists in regard to protecting the thin film from damage resulting from the method shown in Figure 5.5a, the specimen chamber does not allow for specimen preparation of this type (as is the case in some automatic specimen changer systems), or the as-received specimen area is not large enough to be presented into the diffractometer specimen chamber, one can use a holder with a clay filled well for specimen mounting (see Figure 5.5b). The selection of clay may depend upon personal preference and availability. One type of clay that has been found acceptable is a modeling clay made by Van Aken International. Tacki wax is also viable. This clay remains soft during use so that it does not place any significant stress on the specimen, yet does not show any evidence of creep (shrinkage or expansion) after the specimen has been prepared (run a diffraction pattern of the clay for reference). To prepare a specimen, place enough clay in the specimen well so that some clay protrudes above the specimen holder reference plane. Leave some room in the well to allow for lateral expansion of the clay (after pressing). Place the back of the specimen substrate onto the clay. Carefully place a flat piece of glass (i.e., microscope slide)

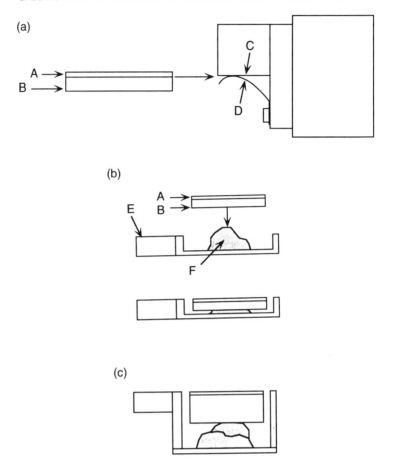

Figure 5.5. (*a*) Self-supporting flat thin film specimen which can be loaded directly into a diffractometer specimen chamber: A = thin film, B = substrate, C = diffractometer reference plane for a specimen, D = spring to hold specimen in place. (*b*) Thin-film/thin substrate specimen mounted on clay in a specimen holder: A = thin film, B = substrate, C = specimen holder reference plane which will coincide with the diffractometer reference plane, F = clay. (*c*) Thin-film/thick substrate specimen mounted on clay in a specimen holder.

on the thin-film side of the specimen and press gently until the specimen surface is level with the reference surface of the specimen holder. If the substrate is very thick, increase the depth of the well (Figure 5.5*c*) to compensate for this thickness. If the overall specimen thickness is too large for the diffractometer specimen chamber, you may need to grind away some of the substrate on the backside of the specimen. When analyzing thin films deposited on single crystal substrates, you can eliminate part or all of the strong peaks due to the substrate by tilting the specimen 1 or 2 degrees off-axis from the specimen plane. Another way to accomplish this single crystal peak removal is to offset θ 1 or 2 degrees when the diffractometer scan is set up.

Thin films which have been deposited onto a flexible support such as a polymer sheet or are on a very thin support have a tendency to curl. Because it is important for the specimen to remain flat during reflection mode data collection, we need to prepare the specimen so that it will remain flat. One approach is to mount the specimen onto a solid base. This base could be comprised of a zero-background quartz disk (The Gem Dugout), glass, silicon wafer, aluminum, etc. The choice will depend upon the thickness of the specimen and diffraction region of interest (run a diffraction pattern of your base materials for reference). If the total specimen is very thin, the base could contribute to the diffraction pattern. As an example, if the region of interest is 2° to 35° 2θ using Cu $K\alpha$ radiation, aluminum would be a useful base because the first aluminum diffraction peak occurs at about 38.5° 2θ. To prepare this type of specimen, place an adherent on the base material. This adherent could be a glue (Duco cement, glue stick, etc.) or double-sided sticky tape (run a diffraction pattern of the adherent, for reference). If glue is used, make sure it is spread uniformly on the base and is not too thick. Place the back of the specimen onto the prepared base making sure the specimen goes on uniformly without air pockets. You can then mount the back of the base onto a clay specimen mount as shown in Figure 5.5b. If an adequate solid base material is not available, one can mount the specimen onto a base with the middle removed. With the center of the base removed, the adherent is placed on the remaining outer edge. In this case the base is usually made of a metal or plastic because quartz, glass, and silicon wafer materials are not as easily machined. When removing the center of the base, keep in mind that not enough removal may allow the base to be observed by the X-ray beam. Too much removal may not provide enough support to keep the specimen flat. The adherent for attaching the specimen to the base is again a glue or double-sided tape. When mounting the base to the specimen holder, the clay is placed at two points on the backside of the base before being pressed flat with the specimen holder reference plane.

In some instances, interference from the support prevents obtaining a usable diffraction pattern for the thin film. If the specimen does not need to be returned to the originator, nor does the specimen need to be archived, removal of the film from the support can enhance evaluation of the thin film. As seen in Figure 5.6a, a coating on polyethylene terephthalate, PET, could not be adequately analyzed due to the intense diffraction pattern obtained from the highly oriented PET. The thin film was scraped off of the PET using a sharp razor blade. The scraped residue was then placed on a zero-background quartz disk and the disk was mounted as shown in Figure 5.5b. To keep the residue on the disk, a very thin film of silicone grease (Dow Corning) was placed on the quartz (remember to run a reference pattern of a thin coating of silicone grease on the quartz disk). The resultant diffraction pattern shown in Figure 5.6b allowed for complete phase identification analysis of the thin-film components.

When looking at thin films which have been coated from a dispersion, emulsion, or other liquid media, it can be useful to obtain some of the original liquid matrix and place a few drops onto a zero-background quartz disk or if the phase of interest is organic, an aluminum base is also useful. After the liquid specimen dries as a film, the base can be mounted as shown in Figure 5.5b. This technique has been found to be useful when as-coated specimens are too thin to provide a useful diffraction

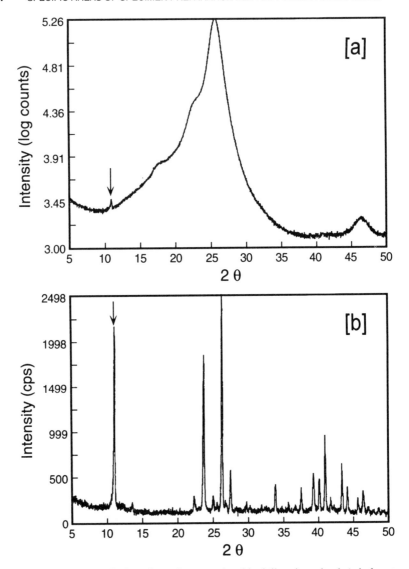

Figure 5.6. (*a*) Nano-particulate dispersion coated on biaxially oriented poly(ethylene tereph-thalate). (*b*) Nano-particulate dispersion sample as before, after scraping film off poly(ethylene terephthalate) and depositing scraped residue onto a zero-background quartz plate.

signal. The specimen prepared as described here provides a thicker specimen, which should provide a better diffraction signal. Keep in mind that the thicker specimen could have a microstructure which differs from a thinner specimen.

Although it is preferable to have a flat specimen for diffractometer analysis, in some instances it is impossible to have such a specimen (e.g., thin film coated on

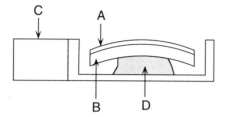

Figure 5.7. Nonflat thin-film specimen mounted on clay in a specimen holder: A = thin-film, B = substrate, C = specimen holder reference plane which will coincide with the diffractometer reference plane, D = clay.

a cylinder). Analysis of a nonflat specimen will require narrowing of the X-ray beam hitting the specimen (using smaller divergence slits or using the spot rather than the line of the X-ray tube) to reduce diffraction peak broadening. Thin films deposited onto a curved substrate will require some modification for the diffraction experiment. The region to be analyzed will have to be defined. The specimen must fit in the specimen holder and may require cutting a piece from the bulk specimen. Mounting the specimen, Figure 5.7, is similar to mounting a flat specimen as shown in Figure 5.5b. The curved specimen will have a reduced reference plane which may make it more difficult to position the region of interest in the correct position for optimum diffraction. The diffraction peaks (if observed) due to the substrate should be evaluated for broadening due to contribution of the nonideal aligned portions of the specimen.

These specimen preparation techniques should prove useful in many thin-film specimen applications for X-ray diffraction analysis. The examples presented here can and should be modified to meet the specimen preparation one may encounter in an X-ray diffraction laboratory.

5.9 MICROSAMPLES

Microsamples are considered to be those of less than 1 mg in weight, i.e., in the microgram range. In the case of powders, they are, as before, ground carefully in an agate mortar. A glass fiber (approximately 0.1 mm diameter) mounted in plasticine (modeling clay) on a brass stub is dipped lightly in rubber solution, and the end bearing the glue rolled in the finely ground powder. The powder specimen is thus presented to the X-ray beam in the form of a small-diameter cylinder. The main disadvantage here is that the beam has to pass through the glue and glass fiber as well as the specimen; see Figure 5.8.

A variation on the above is to roll a tiny (<0.5 mm diameter) ball of rubber solution in the ground-up specimen using a needle. When the ball of glue has taken up most of the powder, it is speared by a thin glass fiber mounted as described above. This arrangement of fiber and specimen means that the X-ray beam passes mainly through glue and powder, not glass.

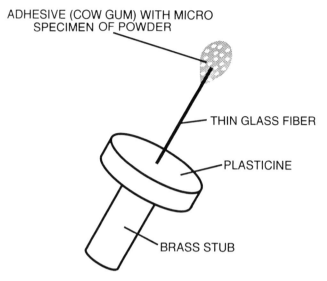

ADHESIVE (COW GUM) WITH MICRO
SPECIMEN OF POWDER

THIN GLASS FIBER

PLASTICINE

BRASS STUB

Figure 5.8. Mounting a microspecimen of powder for a Debye–Scherrer camera.

For monolithic or sliver specimens, the procedure of specimen mounting is as described for paint flakes (see section 5.10.1). An extra degree of care is required when attempting to mount such small samples on the tip of a glass fiber, and it may be unavoidable that the X-ray beam passes through part of the fiber as well as the specimen.

Another mode of specimen preparation for powdered specimens is required for the Guinier camera which, unlike the Debye–Scherrer or Gandolfi camera, is a focusing camera. A particular form of this camera, known as a Guinier–Haag camera, has as its specimen holder a thin metal washer of diameter 1.5 cm with a hole of diameter 5 mm.

1. Lay a sheet of paper towel on the bench top and place on this a sheet of 6 μm thick Mylar (polyethylene terephthalate). Place one of the thin metal washers on the Mylar film as a template, and, using a sharp scalpel, cut round the washer so that a circle of Mylar film is produced, 1.5 cm in diameter.

2. Smear one side of the Mylar film circle very lightly with Vaseline (petroleum jelly or KY jelly) and place a washer on the Vaseline side so that the Mylar film and washer stick together.

3. Take the powdered microspecimen and sprinkle it into the hole of the washer such that it adheres to the Vaseline on the top side of the Mylar film.

4. Turn the washer upside down and shake off the excess powder, leaving a thin layer of powder stuck to the Vaseline-coated Mylar film in the center of the specimen holder (Figure 5.9). 3M Magic Mending Tape is an alternative to the Mylar film.

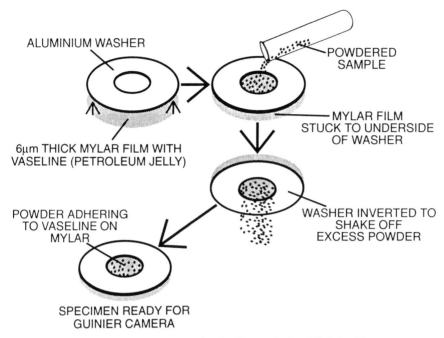

Figure 5.9. Microspecimen preparation for the standard and Guinier–Haag camera.

5.10 FORENSIC SAMPLES

Forensic science, perhaps more than most branches of science, imposes stringent demands on analytical techniques. The ability to be able to analyze very small specimens is a prime requisite, and the need to be nondestructive wherever possible in order to preserve evidence (and to enable the specimens to be passed on for complementary analysis) is also of great importance. Fortunately, powder diffraction meets both these criteria. It also has two important additional assets, namely its ability to identify compounds rather than just elements (cf. XRF, electron probe microanalysis, and emission spectroscopy), and its versatility—it can be used to analyze inorganic, organic, and metallic substances, and mixtures thereof. In theory, any polycrystalline substance, or something containing a polycrystalline substance, could be a forensic sample, and the following list gives an idea of the range of substances encountered in forensic casework: minerals, metals and alloys [3], paints, plasters, putties, ointments, cosmetics, drugs [26], sugars, papers, safe ballasts, detergents, pigments [27], dyes, plastics, abrasives, explosives, and to a lesser extent soils, gemstones, bricks, under seals, cements, mortars, and concrete.

By their very nature, forensic samples tend to be small and so the instrumentation for their analysis has to be able to cope with small samples. In XRD terms, this restriction means, first and foremost, powder cameras (Debye–Scherrer, Gandolfi or Guinier—see Chapter 8) and then diffractometers. Specimen preparation for forensic analysis follows a golden rule—do as little as possible! In truth, each sample should be treated on its merits and on the circumstances of its origin. Nevertheless, it is good practice to examine the sample first under a low power visible light microscope to observe its physical characteristics. For example, if it is a powder does it contain more than one component, is it contaminated, does its texture tell you anything? If it is a paint flake, how many layers are visible, are they colored differently, can they be separated with ease? Finally, do the nature and circumstances of the case dictate the way in which you should treat the sample? Is identification required, or merely a comparison of diffraction patterns? Is nondestructive analysis absolutely essential, is the specimen required for other analytical techniques? If so, what is the best way of mounting the specimen?

5.10.1   Preparation of Debye–Scherrer Specimens for Forensic Work

Powdered Samples (approximately 5–10 mg)   The following technique applies equally to organic and inorganic specimens with the exception of explosives, very hard abrasives, and single crystals which for one reason or another should not be ground.

Grind the coarse powdered specimen in a small agate mortar until the particle size is such that the resulting fine powder can be easily loaded into a 0.3 mm diameter thin-walled glass capillary tube to a depth of approximately 5 mm. See section 7.1.5 for details on preparing capillary specimens. Cut the tube so that a length of approximately 10 mm contains the powder. Seal the open end with a flame or hot wire. Mount this tube in plasticine (modeling clay) or Tacki wax on a brass stub which fits in the Debye–Scherrer camera (see Figure 5.10).

Samples in this category would be drugs, sugars, minerals, plasters, cosmetic powders, safe ballasts, detergent powders, pigments, dyes, soils, cements, and mortars.

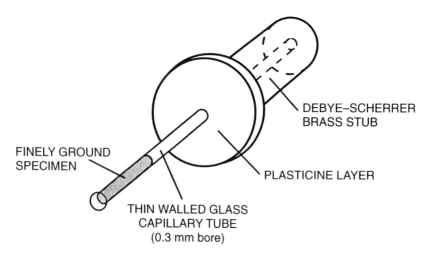

Figure 5.10. Mounting powder in a capillary tube for a Debye–Scherrer camera.

Monolithic Specimens (e.g., paint flakes)   If the paint flake is multilayered, the layers will have to be separated and analyzed individually, otherwise the components identified by XRD cannot be assigned to a specific layer. A fine scalpel, or better still a microtome, can be used to separate the layers.

1. Using a scalpel with a new, sharp blade, grip the layered paint flake with a pair of sharp nose tweezers, and carefully separate or scrape the layers so that a specimen of each layer is available for analysis. This process is best done using a low power binocular microscope (2X–4X magnification).
2. Take a Debye–Scherrer camera stub with a thin coat of plasticine on its large end and poke a length of glass fiber (approximately 0.1–0.2 mm diameter) through the center of the plasticine such that it goes through the hole in the stub. Leave roughly 5 mm of glass fiber protruding from the plasticine and using a pair of tweezers, snap off any excess fiber that may be sticking out of the back end of the stub. Important: Push the fiber in from the plasticine side, because if it pushed through from the back end, the part of the fiber that will bear the specimen, and thus be in the X-ray beam, will be contaminated with plasticine. Plasticine contains calcite and this mineral will show up strongly in the diffraction pattern.
3. Dip the end of the glass fiber in cow gum, pick up a tiny blob of the glue, and touch a paint flake (of rough dimensions 0.5 × 0.5 × 0.1 mm) with this blob. Try to keep the paint flake right on the end of the fiber; don't let it slide down on the side of the fiber. Tease the flake with the sharp end of a pair of tweezers or a needle probe until it remains in a vertical position on the end of the fiber (figure 5.11).
4. The stub is now ready for insertion in the Debye–Scherrer camera.

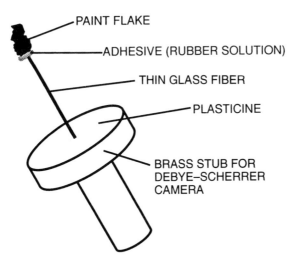

Figure 5.11. Mounting monolithic specimens (e.g., a paint flake) for a Debye–Scherrer camera.

If the paint flake is required for further analysis, thought should be given to the use of another glue for mounting. For example, if analysis of the paint resins by Pyrolysis–Mass Spectrometry is required, it would be unwise to use cow gum as a mountant. A water-soluble glue, gum arabic, should be used and the paint flake washed with water after XRD analysis and prior to analysis by PyMs.

5.10.2 Preparation of Gandolfi Specimens

These samples are usually single crystals of materials which could not, or should not, for one reason or another, be ground. The reasons could be that the crystal is explosive, too hard to grind (an abrasive material), too valuable (a gemstone), liable to change its crystalline form (polymorphism), or that its evidential value as a single crystal would be lost if it were crushed [28].

The crystal (usually 0.1–0.5 mm in any dimension) is mounted on the tip of a fine glass fiber (as for paint flakes) and placed in the Gandolfi camera. In the case of real gemstones it may be possible, if the stone has a fairly sharp culet, to mount the stone directly on plasticine such that the culet just enters the X-ray beam (see section 5.12).

Powders such as barbiturates exist in many polymorphic forms and it is known that some forms transform, upon grinding, into other crystalline modifications. If such a powder requires analysis by XRD and it can be loaded directly (without grinding) into a thin-walled glass capillary of the type described, the capillary can be centered in either a Debye–Scherrer or Gandolfi camera.

5.11 PAPER AND PLASTICS

Paper and sheet plastics may be examined directly with diffraction methods. The specimen mounting techniques are relatively straightforward.

5.11.1 Small Samples

Small pieces of paper and plastic (<1 mm^2), may be mounted and analyzed by the same method as that described for monolithic specimens such as paint flakes (section 7.10).

5.11.2 Intermediate Sized Samples

Paper and plastic samples up to approximately 1 cm^2 may be mounted on a silicon or quartz crystal substrate using a suitable adhesive (e.g., 3M Spray Mount Adhesive) which is placed directly in a powder diffractometer. Grip the paper or thin plastic specimen (up to 1 cm^2) by one corner using a pair of sharp tweezers and spray one side of it with, for example, 3M Spray Mount Adhesive. Press the specimen onto the center of the diffractometer specimen holder (the silicon crystal substrate) and place the holder in the diffractometer.

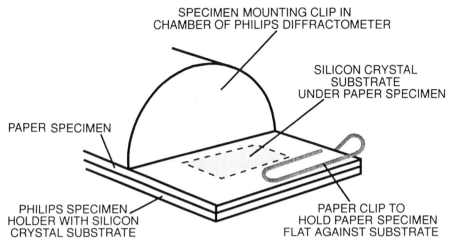

SPECIMEN MOUNTING CLIP IN
CHAMBER OF PHILIPS DIFFRACTOMETER

SILICON CRYSTAL
SUBSTRATE
UNDER PAPER SPECIMEN

PAPER SPECIMEN

PHILIPS SPECIMEN
HOLDER WITH SILICON
CRYSTAL SUBSTRATE

PAPER CLIP TO
HOLD PAPER SPECIMEN
FLAT AGAINST SUBSTRATE

Figure 5.12. Specimen mounting of a large paper specimen in a Philips diffractometer.

5.11.3 Large Samples

Specimens larger than 1 cm^2 may be held in place on a silicon or quartz crystal substrate in the diffractometer by ensuring that the "inner" edge of the specimen is gripped by the specimen mount clip in the diffractometer. The specimen is held flat by a paper clip, or any other suitable clamp, to hold the "outer" edge firmly against the substrate. Care must be taken to ensure that the specimen is held as flat as possible and to ensure that the clip used does not encroach on the area irradiated by the X-ray beam (see Figure 5.12).

5.12 GEMSTONES

Powder diffraction methods are seldom used for routine identification of gemstones for the following reasons:

- The stones tend to be single crystals.
- The stones may be set in a mounting which will limit access to the X-ray beam.
- There are many other methods of analyzing gemstones [29].

However, the occasion may arise when the authenticity of a unmounted stone (particularly a diamond) has to be verified and perhaps only a Debye–Scherrer or Gandolfi camera is available. In this instance, either camera will suffice but the Gandolfi camera will provide a clearer (less spotty) photograph, and it will be possible to distinguish between, for example, a real diamond and a diamond substitute.

Small gemstones ($<$ 5 mm from culet to table) that have well-defined culets may be mounted directly (i.e., stuck on plasticine on a brass stub) in a Gandolfi camera so that the culet is just in the X-ray beam (see Figure 5.13). With larger stones that

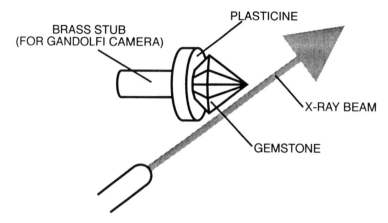

Figure 5.13. Mounting a small gemstone for the Debye–Scherrer or Gandolfi camera.

are not suitable for mounting in this way, it is frequently possible to scrape the edge of the stone (depending on the type of stone, and its value) with a scalpel blade or diamond point, and mount the resulting scrapings on the end of a glass fiber in the manner described for microsamples (Section 5.9).

5.13 ENERGETIC MATERIALS

Energetic materials, explosives, propellants, and pyrotechnics, should not be handled by individuals until they have had appropriate safety training. All operations with energetic materials involve risk and present safety hazards for both personnel and equipment. To minimize these hazards, only the smallest quantities of material consistent with the scientific objective of the experiment should be handled. The quantities used should not exceed the amounts given in Table 5.4.

Familiarization with the known hazards and sensitivity behavior of the material being handled should occur before specimen preparation. If the sensitivity of the material is unknown, assume that the material is extremely sensitive and limit the quantity handled accordingly. Quantities of energetic material in excess of that actually being examined should not be stored in the laboratory but must be stored in an approved energetic materials storage container or in a magazine. Upon completion of a diffraction scan(s), any unused material along with the specimen should be returned to the original owner for explosive ordinance disposal or chemically destroyed in an approved fashion. All personnel handling energetic materials must have previously undertaken appropriate formal training in the safe handling of energetic materials. The procedures described in this section are intended as guidelines during analyses by X-ray diffraction and are *not* a substitute for safety training and appropriate operating procedures.

TABLE 5.4. Maximum Quantities of Explosive Materials Allowed in a Laboratory without Additional Safety Precautions[1]

Hazard	Amount, gm
Unknown or unstable material	0.05
Very sensitive (UNO Class 1.1)[1]	0.05
Moderately sensitive	5
Slightly sensitive	50
Flash burning (Materials which burn so rapidly that the operator cannot react)	5
Slow burning (Materials which burn sufficiently slowly that the operator can escape after observing ignition)	50

[1]The indicated allowed quantities are for use under conditions in which the materials are known to be stable.

5.13.1 Preparation of Specimens from Energetic Materials

The preparation of specimens of energetic materials for XRD analyses pose problems due to hazards in handling them. In addition, the more sensitive energetic materials are often not pure materials but are desensitized with diluents or stored wet to decrease their sensitivity during shipping and handling, posing further difficulties during interpretation of data. Handling of energetic materials and specimen preparation should be kept to a minimum. Problems of preferred orientation and particle nonrandomness are much easier to deal with than loss of fingers or hands. Even a few milligrams of a primary explosive, such as lead styphnate, will generate a loud pop on detonation and will send glass fragments flying. Energetic materials have that nomenclature for a reason. Metal azides, styphnates, fulminates, and metal tetrazoles, as well as some organic peroxides and highly nitrated organics such as erythritol tetranitrate, are notoriously unpredictable, very sensitive to impact shock, friction (i.e., grinding!), static electricity, and electrostatic discharge and are quite brissant. All specimen mounting of energetic materials should be accomplished behind a safety shield while wearing either heavy leather gloves (sensitive materials) or heavy cotton gloves (less sensitive materials), while wearing safety glasses with side shields (or goggles).

5.13.2 Special Precautions in Dealing with Energetic Materials

Specimens known to contain (or suspected of containing) a sensitive energetic material should be handled only while wearing safety glasses or goggles (or better still, a full face shield), heavy leather gloves, and a fire retardant-impregnated long-sleeved cotton laboratory coat or overalls. If possible, a grounded metal plate behind the safety shield should be used as the specimen mounting surface to reduce electrostatic discharge. Sensitive primary explosives are often stored wet to minimize detonation from static electricity and friction; mount a minimal amount of the wet material on an off-axis-cut "zero-background" substrate such as quartz (ZBQ) by carefully selecting an aliquot of a few drops and then dripping onto a substrate. Allow the specimen on the substrate only to dry enough so that particles are not actually floating around in liquid droplets on the substrate. Thin strips of "Post-Its" can be placed on the edges of the substrate to compensate for the height (thickness) of the specimen. Alternatively, if data with a low background are not absolutely required, a single-well glass slide can be used to hold a few drops of the wet specimen. The observed relative intensities will not be stable over time because as the liquid in the specimen evaporates, the particles will continue to shift their orientation. This effect will be especially noticeable in wet specimens containing metal cations. Because the amount of specimen used for obtaining diffractometer data may be several mg (perhaps tens of milligrams), obtaining diffraction data from a thin coating of specimen on a glass fiber in a Debye-Scherrer camera or from a very small clump of powder on the tip of a glass fiber in a Gandolfi camera is much safer for particularly sensitive materials. However, camera data will have insufficient resolution to distinguish between different polymorphic or compositional forms of some energetic materials.

5.13.3 Reducing the Size of Larger Energetic Specimens

If an energetic material contains large crystallites, but is not highly sensitive to shock and friction, the large crystallites can be lightly crushed to a finer powder in a solvent in which the material is insoluble. The resulting powder can be mounted wet as described above. If the material is also not particularly electrostatically sensitive when dry, it can be mounted as a thin layer of dried powder on a thin smear of petroleum jelly on a zero-background substrate. If the specimen remains horizontal throughout the scan, the petroleum jelly is unnecessary. Again, thin strips of "Post-Its" can be placed on the edges of the substrate to compensate for the height (thickness) of the specimen. As many commonly used energetic materials are organic compounds, zero-background substrates should be used to minimize systematic errors due to specimen transparency when recording data with a powder diffractometer.

Use good laboratory practices in disposing of samples after completion of the analyses; these practices do *not* include rinsing the sample down the drain (exploding pipes are difficult to justify) or putting the sample in the trash (the custodian will not be likely to appreciate an unexpected deflagration or detonation). After analysis, the sample can be chemically decomposed to less hazardous forms and then deposed of, if appropriate, as hazardous chemical waste.

To handle energetic materials safely, the key is knowing the likely hazards ahead of time, assuming the worst case—i.e., the material could detonate, and absolutely minimizing the amount of sample being handled by using camera methods or by using a less than optimal amount for a standard powder diffractometer scan. Additional information about the properties of common explosives and energetic materials can be found in references [30, 31, 32]. However, these references will not contain information and properties on new energetic materials; talk to an energetic materials chemist or engineer actually working in the field or consult recent issues of journals specializing in energetic materials for the latest information on specific energetic compounds and their properties.

REFERENCES

[1] W. D. Kingery, Bowen, H. K. and Uhlmann, D. R. (1976), *Introduction to Ceramics*, 2nd ed., p. 3. Wiley: New York.

[2] Ritter, J. J. (1988), "A hermetically sealed inert atmosphere cell for X-ray diffraction," *Powder Diff.*, **3**, 30–31.

[3] Rendle, D. F. (1981), "Analysis of brass by X-ray powder diffraction," *J. Forens. Sci.*, **26**, 343–351 (and references therein).

[4] Elton, N. S., Salt, P. D., and Adams, J. M. (1992), "The determination of low levels of quartz in commercial kaolins," *Powd. Diff.*, **7**, 71–76.

[5] Smith, D. K. (1995), "Evaluation of the detectability and quantification of respirable crystalline silica by X-ray powder diffraction methods," in *Review Papers On Analytical Methods* sponsored by the Crystalline Silica panel, Chemical Manufacturers Association, Washington, D.C., pp. 1–77.

[6] Cline, J. P. and Snyder, R. L. (1983), "The dramatic effect of crystallite size on X-ray intensities," *Adv. X-ray Anal.*, **26**, 111–117.

[7] Jackson, M. L. (1979), *Soil Chemical Analysis—Advanced Course*, 2nd ed., Published by the author, Madison, WI.

[8] Brindley, G. W. and Brown, G. (1980), *Crystal Structures of Clay Minerals and Their Identification*, Mineralogical Society Monograph No. 5. Mineralogical Society, London, 495 pp.

[9] Moore, D. M. and Reynolds, R. C. Jr. (1989), *X-ray Diffraction and the Identification and Analysis of Clay Minerals*. Oxford University Press.

[10] Bish, D. L. and Post, J. E. (1989), *Modern Powder Diffraction, Reviews in Mineralogy*, Vol. 20. Mineralogical Society of America, Washington, D.C., 369 pp.

[11] Kunze, G. R. (1965), "Pretreatments for mineralogical analysis." In C. A. Black, ed. *Methods of Soil Analysis. Part I. Physical and Mineralogical Properties Including Statistics of Measurement and Sampling. Agronomy*, **9**: 568–577. Am. Soc. of Agronomy, Madison, WI.

[12] Douglas, L. R. and Fiessinger, F. (1971), "Degradation of clay minerals by H_2O_2 treatments to oxidize organic matter," *Clays and Clay Minerals*, **19**, 67–68.

[13] Rich, C. I. and Barnhisel, R. I. (1977), "Preparation of clay samples for X-ray diffraction analysis," In *Minerals in Soil Environment*, pp.797–808, Soil Science Society of America, Madison WI.

[14] Drever J. R. (1973), "The preparation of oriented clay specimens for X-ray diffraction analysis by a filter-membrane peel technique," *Amer. Mineral.*, **58**, 553–554.

[15] Breck, D. W. and Anderson, R. A. (1981), "Molecular Sieves," *Kirk-Othmer Ency. of Chem. Technology*, 3rd ed., Vol. 15, Wiley: New York, p. 638.

[16] O'Brien, F. E. M. (1948), "The control of humidity by saturated salt solutions," *J. Sci. Instrum.*, **25**, 73–76.

[17] Kuhnel, R. A. and van der Gaast, S. J. (1993), "Humidity controlled diffractometry and its applications," *Adv. X-ray Anal.*, **36**, 439–449.

[18] Meyers, B. L., Ely, S. R., Kutz, N. A., and Kaduk, J. A. (1985), "Determination of structural boron in borosilicate molecular sieves via X-ray diffraction," *J. Catalysis*, **91**, 352–355.

[19] Corbin, D. R., Burgess, B. F., Jr., Vega, A. J., and Farlee, R. D. (1987), "Comparison of analytical techniques for the determination of silicon and aluminum content in zeolites," *Anal. Chem.*, **59**, 2722–2728.

[20] Whittington, B. I. and Milestone, N. B. (1992), "The microwave heating of zeolites," *Zeolites*, **12**, 815–817.

[21] Fischer, R. X. and Tillmanns, E. (1991), "Ambiguities in the interpretation of powder patterns," *Proceedings of the First European Powder Diffraction Conference*, Pt. 1, Materials Science Forum Vols. 79–82, (Trans Tech Pub., Switzerland), pp. 47–52.

[22] Bolton, A. P. (1976), *Molecular Sieve Zeolites, Experimental Methods in Catalytic Research*, Vol. 2, Academic Press: New York, pp. 7–10.

[23] Taylor, J. C. and Pecover, S. R. (1988), "Quantitative analysis of phases in zeolite bearing rocks from full X-ray diffraction profiles," *Australian J. Physics*, **41**, 323–335.

[24] Smith, D. K, Johnson, G. G., Jr., Schieble, A., Wims, A. M., Johnson, J. L., and Ullmann, G. (1987), "Quantitative X-ray powder diffraction method using the full diffraction pattern," *Powder Diff.*, **2**, 73–77.

[25] Bish, D. L. and Howard, S. A. (1988), "Quantitative phase analysis using the Rietveld method," *J. Appl. Cryst.*, **21**, 86–91.

[26] Clarke, E. G. C. and Schmidt, G. (1962), in *Methods of Forensic Science*, Frank Lundquist, ed., Vol. 1, Interscience: New York.

[27] Curry, C. J., Rendle, D. F., and Rogers, A. (1982), "Pigment analysis in the forensic examination of paints. I. Pigment analysis by X-ray powder diffraction," *J. Forens. Sci. Soc.*, **22**, 173–177 (and references therein).

[28] Rendle, D. F. (1983), "A simple Gandolfi attachment for a Debye–Scherrer camera and its use in a forensic science laboratory," *J. Appl. Cryst.*, **16**, 428–429.

[29] Anderson, B. W. (1980), *Gem Testing*, 9th ed., Butterworth Scientific: New York.

[30] Meyer, Rudolf (1987), *Explosives*, 3rd ed. VCH Publishers: New York, 452 pp; and references therein (pp. 405–413).

[31] Urbanski, T. (1985), *Chemistry and Technology of Explosives*, Vol. 1–4, Pergamon Press: Oxford.

[32] *Encyclopedia of Explosives and Related Items* (1960–1983), Vol. 1–10, Picatinny Arsenal: Dover, NJ.

CHAPTER 6

SPECIAL PROBLEMS IN THE PREPARATION OF X-RAY DIFFRACTION SPECIMENS

CONTRIBUTING AUTHORS: BASSETT, CANO, COX, ELTON, LOWE-MA, PREDECKI, SALT, SMITH, WANG

6.1 HIGH PRESSURE STUDIES

Most high pressure–temperature devices fall in one of two categories:

- Diamond anvil cells (DAC)
- Large volume presses (LVP)

Essentially all of these devices have been modified for X-ray diffraction. Because pressure is defined as force divided by area, pressures can be increased by increasing force or decreasing area (specimen size). In the DAC, high pressure is produced by using a very small specimen size, typically tens of micrometers across and weighing nanograms. The modest forces that are needed can be generated by a screw. LVPs have specimens that are typically millimeters across and weigh tens of milligrams. They need forces that are much greater, usually requiring a hydraulic ram. Each type of device has its own unique advantages as well as specimen preparation requirements. No less patience is required for one than for the other.

6.1.1 Diamond Anvil Cells

In the diamond anvil cell (DAC) the specimen is placed between two flat culet faces (0.3–1.0 mm across) of faceted, gem quality diamonds ranging in size from 1/8 carat to 1/3 carat (Figure 6.1). Diamonds make excellent anvils because they are very hard, very transparent to visible light, very transparent to X-rays in the wavelength range most useful for diffraction (1.0–0.2 Å), and are single crystals which produce few interfering diffraction spots of their own. In addition, they are relatively inexpensive and readily available in a shape that with minor modification is especially suitable for use as anvils. Heaters consist of molybdenum wire wound around the diamond anvil supports. Thermocouples attached to the diamond anvils

A Practical Guide for the Preparation of Specimens for X-Ray Fluorescence and X-Ray Diffraction Analysis, Edited by Victor E. Buhrke, Ron Jenkins, and Deane K. Smith. ISBN 0-471-19458-1 © 1998 John Wiley & Sons, Inc.

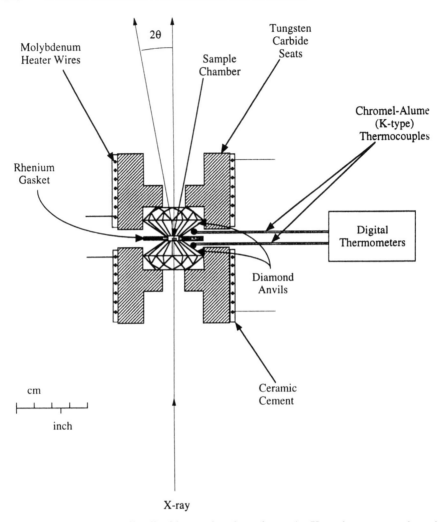

Figure 6.1. Diamond anvil cell with a gasketed specimen. An X-ray beam enters through one diamond anvil and passes through the specimen. Scattered X-rays exit through the other diamond anvil to a film or detector. The diamond anvils are normally driven together by a screw or lever.

give accurate temperature measurement at the specimen because of the excellent thermal conductivity of diamond.

A polycrystalline specimen placed between the parallel, flat anvil faces and squeezed develops a pressure gradient from low at the edges to maximum at the center and is subjected to a considerable deviatoric stress. The X-ray beam normally passes through one of the diamond anvils, through the center of the specimen, and out through the other diamond anvil to a film or detector [1]. Specimens can also be loaded in a gasket consisting of a hole in metal foil squeezed between the anvils. The gasketed specimen, which can be surrounded by a soft solid or a fluid, is subjected to less deviatoric stress than the ungasketed specimen and is thicker, providing more intense X-ray signals. Gaskets also distribute the local stresses at the edges of the anvils so that anvils survive to much higher pressures.

For single-crystal diffraction studies, the specimen may consist of a single crystal either surrounded by a fluid medium or grown in situ from a liquid [2, 3, 4]. DACs for single crystal analyses require a much wider range of angles of access and scattering and for that reason are often constructed from beryllium parts [5].

Diffraction patterns have been made of specimens at pressures up to and beyond that at the center of the earth (3.6 megabars or 360 gigapascals) [6] and optical observations have been made to even higher pressures [5]. Diffraction patterns of specimens under pressure in the DAC have been obtained at temperatures up to 1200°C by resistance heating [7, 8, 9] and up to several thousand degrees centigrade by laser heating [10, 11].

Diffraction data collected by directing an X-ray beam at right angles to the load axis of a DAC are especially useful for studying deviatoric stresses in the specimen [12].

6.1.2 Specimen Preparation for the Diamond Anvil Cell

Perhaps the most important tool for specimen preparation is a zoom stereo binocular microscope with both substage and superstage lights. A petrographic or biological microscope with a stage that can be dropped or a tube that can be raised to accept a diamond cell is also valuable, especially if it has provision for a camera. Next to the binocular microscope there needs to be an array of small tools, including sharpened spatulas; fine tweezers; fine reverse tweezers; an assortment of sewing needles mounted in pin vises; brushes; and dental tools. In addition, it is useful to have fine screw drivers; Allen wrenches; small pliers; wire cutters; a carbide tipped stylus; a hand-held grinding tool; emery paper; tin snips; wire cutters; and c-clamps. It is especially useful to have a can of compressed gas for blowing away unwanted dirt and excess specimen. It is very useful to have a variety of adhesives: a plastic cement such as Duco for room temperature runs and ceramic cements for high temperature runs. Two excellent cements are alumina cements manufactured by Zircar and Aremco. Also, Rutland black cement designed for use with wood-burning stoves can be used.

An ungasketed specimen is loaded by generously piling powdered specimen on the lower anvil face and then bringing the two diamond anvils together. As force is applied, the highest pressure develops at the specimen center. Alternatively, a polycrystalline specimen can be loaded into a gasket consisting of a hole drilled in a metal foil and squeezed between the diamonds. Several metals have been used for gaskets, including temperated 301 stainless steel, inconel, and rhenium. The stainless steels and inconel have the advantage of being relatively inexpensive, easy to handle, and fairly easy to drill. Rhenium, which is harder to drill, has the advantage of being stiffer and more chemically inert. Gasket thicknesses usually range from 0.005

to 0.010 in. (0.127–0.254 mm). The gasket hole, which is typically about half the diameter of the anvil face, can be drilled with a very fine drill in a precision drill press or by means of a focused, Q-switched YAG laser. However, a powerful enough laser is expensive and not normally warranted for this purpose alone.

A powdered specimen can be loaded directly into the hole of the gasket or it can be surrounded by a fluid in the hole. If the fluid is water, a drop can be hung from the bottom of the upper diamond and lowered until it runs into the hole and around the specimen already there. Argon or neon can be loaded into a gasket by lowering the entire diamond cell into a Dewar filled with liquid argon or neon and then closing the cell by means of a long rod that turns the screw [13]. Preservation of the centering of the hole with increased pressure depends on how well centered the hole was at the start, how parallel the diamond faces were, and how symmetrical the gasket was.

Growing a single crystal from a liquid encapsulated in a diamond anvil cell has made it possible to study a variety of samples from water [14] to hydrogen [15]. It is accomplished by trapping the liquid in a gasketed DAC, increasing the pressure until the liquid solidifies, and then slowly melting the solid by releasing pressure. When only one crystal remains, the pressure is again increased, causing the crystal to grow and fill the gasket hole with specimen all in one orientation.

6.1.3 Large Volume High Pressure–Temperature Devices

Large volume high pressure–temperature devices usually consist of more than two anvils that are driven by a hydraulic ram and slide system (Figure 6.2). Specimen size may be millimeters across. There are several styles of large volume presses. In one of these, the DIA system, six anvils truncated with square faces bear on the faces of a cube of X-ray transparent pressure medium that contains the specimen and heater (Figures 6.2 and 6.3). There are four vertical anvil gaps, convenient for incident and exiting X-rays, as well as thermocouple leads. A counter or solid-state detector is swung around the specimen for angle dispersion measurements [16, 17] or held at a fixed 2θ angle for energy dispersion measurements [18]. With polycrystalline sintered diamond (PSD) as anvil material, 1500°C and 150 kbar (15 GPa) have been achieved with 4×4 mm anvils [19] and 1000°C and 210 kbar (21 GPa) [20] have been achieved with 3×3 mm anvils.

6.1.4 Pressure Medium

The cube in the DIA system that the anvils bear on and that contains the specimen, the heater, and the thermocouple is made of the pressure medium (Figure 6.3). The pressure medium must have low X-ray absorption, low compressibility, low thermal conductivity, and stability under pressure and temperature. In addition, it must be able to deform and flow to some degree. The pressure medium that best satisfies these conditions is amorphous boron mixed with epoxy. A typical mixture consists of amorphous boron with a grain size of about 1 μm with epoxy in a weight ratio of 4:1 boron:epoxy.

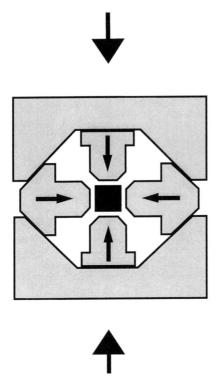

Figure 6.2. Arrangement of anvils in the DIA type of large volume press. Each anvil presses against the faces of a cube of pressure medium containing the specimen. X-rays enter and exit through the spaces between the anvils.

For making an assemblage with the specimen, heater, and thermocouple imbedded in the pressure medium, one should have a set of jigs and a hand pump press to form the mixture into a cube with the desired dimensions. Epo-Tek 457 ND epoxy works best. The mixture is loaded in the jig and compressed to 2000 psi for about a minute. After the pressure is released slowly, the cube is then heated to 170°C for 2–4 hr.

Heating for in situ diffraction studies is usually accomplished by passing electric current through a graphite or amorphous carbon heater that is in electric contact with the top and bottom anvils. When pressure–temperature conditions are high enough to convert the carbon heaters to diamond, causing loss of conductivity, TiC may be used as a substitute.

The millimeter size specimen has several advantages. For instance, the specimen and the calibrant can be separated thereby avoiding possible diffraction peak over-lap. The same can be done for different specimens where the investigator wants to make comparisons at the same pressure–temperature conditions without interfering diffraction peaks.

A set of tools similar to those used for diamond cell specimen preparation are needed: binocular microscope, pins, etc. The specimen chamber in the polyhedral

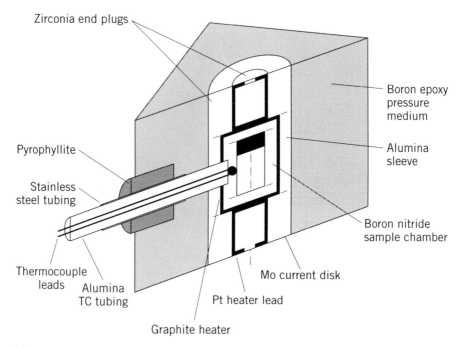

Figure 6.3. Typical construction of a specimen contained within the pressure medium cube in a DIA-type large volume press.

pressure medium is usually about 0.5–1 mm in diameter. The specimen in powdered form is usually funneled into the chamber from a folded piece of weighing paper with the aid of a fine probe. Sometimes prepressed pellets of specimen are used. Sometimes polyethylene or epoxy is mixed with the specimen to provide a more hydrostatic environment, especially if the run is to be made at room temperature. A liquid medium can also be used if the container is a Teflon capsule. A mixture of ethanol and methanol will permit hydrostatic pressures up to 100 kbar (10 gPa). A jig to hold the assembly until it is completed is very useful.

Some factors that affect the quality of the diffraction patterns are:

- Grain size.
- Specimen position and geometry.
- Recrystallization at high temperature.
- Nonhydrostatic stress.
- Absorption.

In the first two cases, exceptional care in specimen preparation and assembly is needed. In the third case, salts, which are usually the most likely materials to recrystallize, should be kept strictly dry before loading and in extreme cases may need to be mixed with another phase such as BN to prevent the formation of large crystals.

In the fourth case the use of a fluid or soft solid pressure medium can minimize the effects of deviatoric stress. In the fifth case there are various ways to calculate the effects of absorption as a function of energy when energy dispersive diffraction is being used. When angle dispersion is being used, it can be very advantageous to use the high energy X-rays available at synchrotron sources.

6.1.5   X-ray Diffraction Pressure Calibrants

If the compression curve, i.e., pressure versus volume, of a crystalline material is known with sufficient accuracy, then XRD measurements of its lattice parameters can be used to calculate pressure. One of the first and most commonly used pressure calibrants is NaCl, which also serves as a pressure medium [21]. Because it passes through a phase transition at about 300 kbar (30 GPa), its usefulness is restricted to pressures below that. Another material, gold, has been calibrated for both pressure and temperature [22, 23]. In other words, the full equation of state is reasonably accurately determined so that it can be used to measure pressure even during high temperature runs. Unlike NaCl, gold has a high atomic number making it desirable because a small quantity yields intense peaks. Because of its large elastic anisotropy, discrepancies among lattice parameters based on different d-spacings can be used to detect deviatoric stress. Other metals such as copper, molybdenum, palladium, and silver have been used, especially for making pressure measurements at very high pressures [24]. Numerous other materials have been used, e.g., MgO and CaF_2.

One of the principal advantages of the diffraction type of calibrants is that they utilize the analytical technique already in use for studying the specimen. One of the main disadvantages of the diffraction type of calibrant is that the diffraction peaks often interfere with diffraction peaks from the specimen.

6.1.6   High Pressure Work with Synchrotron Sources

Much of the research conducted by in situ XRD using high pressure apparatus has been done with conventional X-rays from sealed tubes as well as rotating anode sources. In nearly all of this work, molybdenum targets have been favored. The characteristic $K\alpha$ radiation from a molybdenum tube has sufficient energy (about 17 keV) to penetrate a diamond anvil or an enclosing gasket without serious loss of intensity. At the same time, it has a long enough wavelength (0.7 Å) so that a representative set of d-spacings from a crystalline specimen can be collected over a reasonable 2θ angle. Because the X-rays must exit from most high pressure devices through a rather restrictive passage, it is important to be able to collect a lot of information over a small 2θ range. Although energy dispersive X-ray diffraction has been in use for decades, it was not a very practical approach because of the relatively weak bremsstrahlung continuum spectrum from conventional or rotating anode tubes.

The availability of synchrotron radiation has had a profound influence on in situ XRD of specimens at high pressures and temperatures [25]. Not only is synchrotron radiation more intense, it offers a continuum spectrum of high intensity that covers a range of energy from 10 keV to 100 keV. The continuum spectrum makes energy dispersive diffraction a much more attractive alternative. It is possible to use synchrotron radiation in combination with energy dispersive analysis to obtain a full diffraction pattern at a fixed 2θ angle (typically $10°$ to $15°$). The resolution of a solid-state detector (intrinsic germanium is the favorite) is not as good as the angle dispersive method. Nonetheless, when energy dispersion is used, new phases are easily and quickly detected by the appearance of new peaks; structures can be identified if the crystal structure is simple; lattice parameters can be measured with almost as much accuracy as with angle dispersive data. Therefore, because of the convenience and rapidity of data acquisition, energy dispersion is generally favored when synchrotron radiation is used. There are several synchrotron sources available in the United States, Japan, England, France, and Germany. In the United States, major facilities exist at Stanford University, Brookhaven National Laboratory, Cornell University, and at Argonne National Laboratory.

6.2 SPECIMEN PREPARATION AND LABORATORY PROCEDURES FOR RESIDUAL STRESS AND STRAIN DETERMINATION BY X-RAY DIFFRACTION

Techniques of measurement of residual stress and strain by XRD are well established, having been in use since 1930. The method has been implemented on several types of diffractometers: Ω-type, ψ-type, types that permit specimen tilting about both Ω and ψ axes, and on a number of portable instruments. The Ω-axis is defined here as the $\theta/2\theta$-axis, and the ψ-axis as the axis normal to the θ-axis, and lying in the specimen surface in the diffraction plane, i.e., the plane containing the incident and diffracted beams and the specimen normal. (The ψ-axis should not be confused with the ψ-angle defined later). The instruments have utilized a variety of detectors: solid-state, position sensitive, film, scintillation and proportional counters with and without monochromators, and three types of X-ray optics: fixed receiving slit, parallel-beam and parafocusing. Several excellent accounts of the procedures and techniques of the field exist [26, 27, 28, 29, 30, 31]; in particular, the book by Noyan and Cohen [30], should be consulted frequently by the new practitioner and gradually assimilated as experience is gained in stress determination. Papers on more recent developments can be found in *Advances in X-ray Analysis* (the proceedings of the Annual Denver X-ray Conference), *Proceedings of the International and European Conferences on Residual Stress, Proceedings of the European Powder Diffraction Conference* and in various journals.

Residual stress measurements are critically dependent on being able to achieve high accuracy in the determination of peak positions, generally to within $\pm 0.02°2\theta$. Consequently any specimen preparation issues that have an impact on peak position, including specimen displacement and refraction error must be considered in detail, as well as specimen preparation per se. The main emphasis in this section will be on techniques involving the Ω-type powder diffractometer equipped with parallel-beam optics and a solid-state detector (Figure 6.4). These Ω-diffractometers are common in most X-ray laboratories and can be readily equipped in this way. The resulting

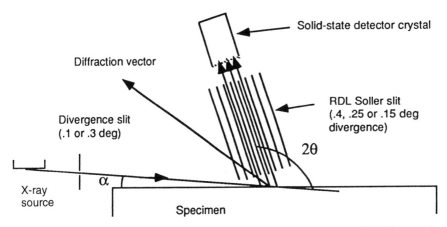

Figure 6.4. Typical pseudoparallel-beam optics arrangement on an Ω-diffractometer (schematic).

instrument can be used for most diffraction measurements (including strain) down to small incidence angles ($\alpha < 0.50°$) as well as for routine powder diffraction work, without requiring the optics to be changed back to the fixed slit configuration. The measurements at small α are of interest for thin films, surface treated materials, and for the measurement of depth profiles of most diffraction quantities without the need for layer removal.

6.2.1   Specimen Preparation for Stress Measurements

As a general rule it is best to assume that *anything* that is done to a specimen in the way of handling or preparation is likely to change the stress state. It is, therefore, advisable to disturb the specimen as little as possible, particularly the surface to be X-rayed, prior to making the X-ray measurements. This handling includes any handling or manipulation the specimen may receive prior to its arrival in the X-ray laboratory. Even mounting a specimen in the diffractometer specimen holder can "apply" stresses of significant magnitude to fragile specimens unless precautions are taken. Sectioning the specimen relieves the forces exerted by the portions removed and introduces stresses associated with the cutting process used. The result is a redistribution of the stresses in the specimen which may be quite different from that which was present before sectioning. Some sectioning is generally required to permit the specimen to fit in the holder but one should be aware of the effects. Measurements with a portable X-ray stress instrument, or at least the affixing of strain gages to the specimen prior to sectioning are often advisable to assess these effects.

In making measurements on as-received specimens, it is worth keeping in mind that the measured diffraction peak profile is the sum of the contributions from all regions in the diffracting volume, where the contribution from each region depends on its volume fraction and on the absorption of the beam along the beam path for each

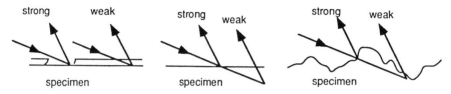

Figure 6.5. Effect of a discontinuous surface layer, beam penetration, and surface roughness on diffracted intensity (schematic).

volume. Thus, if the specimen is covered with a discontinuous, nonhomogeneous or nonuniform surface layer, e.g., cracked or scratched oxide, paint, or otherwise deposited implanted or diffused film, mill scale, setter residue, die lubricant, mold release, etc., areas where the layer is absent or less absorbing will contribute more strongly to the peak profile than covered areas. If the layer is absent (or uniform) the surface region of the specimen will contribute more strongly than a deeper region of the same thickness, and if the specimen surface is rough, asperities will contribute more strongly than depressions (Figure 6.5).

It is sometimes of interest to measure the stresses in the absence of surface layers. With metals the layers can be removed by light grinding and polishing followed by electropolishing to remove the grinding and polishing stresses. Etching is less desirable because it produces a rough surface and may change the stress state. Even electropolishing, if done incorrectly, may preferentially attack one phase of a two-phase alloy and change the microstresses near the surface. With ceramics, semiconductors, and inorganic glasses, grinding and polishing are usually sufficient but can be followed by light chemical etching for single-phase materials. If two or more phases are present, etching should be omitted because most etchants attack the phases nonuniformly as with metals. With hard organic polymers, grinding and polishing are often satisfactory, or alternatively surface layers can be removed with successive thin cuts on a microtome. With soft or rubbery polymers, the grinding and polishing or microtoming can be done at low temperature (below the glass transition). Chemical etching is inadvisable for polymers because etchants often swell the amorphous regions or attack them preferentially thereby changing the stress state.

With all the above materials except polymers, if the final step in layer removal leaves an altered stress state of sufficient thickness to affect the intended X-ray measurements, this layer can often be removed by sputtering without introducing significant chemical change. With polymers, conventional sputtering creates damage; however, if the sputtering is done using ^{3}He at grazing incidence this damage can be avoided [32]. A limitation of most sputtering systems is that the largest area that can be treated on the specimen is only about 1 cm^2.

6.2.2 Mounting Procedures

The objective with any mounting procedure is first not to introduce any stresses and second to locate the surface of the desired area to be irradiated so that the tilt axis of the diffractometer lies in that surface. If the surface is not flat (e.g., cylindrical or spherical), the irradiated area can often be reduced to an approximately flat region by collimating the incident beam or by masking the specimen surface. Lead tape is useful for the latter but contributes lead peaks to the diffraction pattern. Reducing the area decreases the number of diffracting grains and necessitates some type of small oscillatory motion of the specimen during exposure (or the summing of multiple scans at slightly different orientations) if the grain size is not sufficiently small. The absence of a sufficient number of diffracting grains is usually evident when the shape of a given diffraction peak differs significantly with small changes in specimen orientation or displacement.

If the specimen is a flat plate of appropriate size, it can usually be inserted directly into the specimen holder and held in place with spring pressure, or stuck to the specimen mount with rubber cement, depending on the type of goniometer. If the specimen is a thin or fragile plate, it can be backed with a thick plate and inserted or attached. If the specimen is too small for the spring-type holder, it is common practice to use a subholder which is designed to fit the main holder.

One type of subholder is shown in Figure 6.6. It is machined from a fully an-nealed aluminum or steel block, such that surfaces A, B, and C are coplanar within 0.013 mm (0.0005 in.) and parallel to the bottom surface D. The surfaces A, B, and C make contact with three indexing pins in the specimen holder, which define a plane containing the θ-axis of the diffractometer on a correctly aligned holder. If the specimen is reasonably rigid it can be mounted in the subholder as shown in Figure 6.6b such that surfaces A, B, C and the specimen surface S to be X-rayed are coplanar to within about 0.013 mm. This alignment is conveniently done with the aid of a basalt block about 30×20 cm $\times$ 8 cm thick, ground and polished flat on one face to within 0.013 mm. The specimen and subholder are then assembled as indicated in Figure 6.6b but with S a little above A, B, and C. The assembly is then inverted, placed on the polished face of the block and gently pressed down to make A, B, C, and S coplanar. This alignment is checked with a surface gauge placed on the block, by moving the gauge stylus from S to A, B, or C. The pressing down

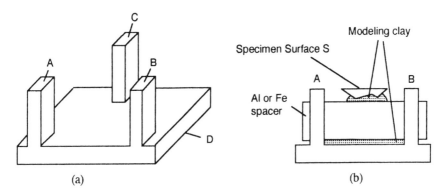

Figure 6.6. (a) Specimen subholder. (b) Subholder with specimen mounted.

step can be repeated if necessary and with a little practice the gauge readings can be made to agree within about 0.013 mm provided the modeling clay thicknesses are kept small (less than about 2 mm). If they are larger, visco-elastic springback in the clay becomes troublesome and makes the agreement between the gauge readings less precise as well as time dependent. Also, if the temperature where the specimen and subholder are assembled differs significantly from that in the diffractometer enclosure, the specimen may be displaced during X-ray measurement due to thermal expansion or creep of the clay. If the specimen is soft, the surface gauge stylus may cause deformation, in which case co-planarity should be checked optically. If the specimen is too fragile for the above procedure, it can sometimes be mounted on a suitable substrate (Si wafer, glass, or metal plate) using a liquid, room temperature curing, low shrinkage adhesive (epoxy resin or cyanoacrylate), or using Vaseline. With a three-pin type holder, shims or spacers are then needed to locate the surface of the specimen on the *ABC* plane.

6.2.3   Specimen Displacement Errors

Usually the main source of error in the measurement of diffraction peak positions is the displacement of the surface of the specimen away from the diffractometer axis. There are several other sources of error; these errors are discussed at length elsewhere [30, 33]. The effects of specimen surface roughness, transparency (i.e., the fact that diffracted intensity comes from a range of depths) and surface curvature (i.e., the fact that in general the specimen surface does not conform to the focusing circle) as well as physical displacement of the specimen, can all be considered effectively as specimen displacement errors. While physical displacement can be corrected, the other three contributors are difficult if not impossible to avoid with many types of specimens. The resulting error $\Delta 2\theta$ in the measurement of 2θ depends on θ, the tilt angle ψ, and the effective displacement Δx. The angle, ψ, is measured from the diffraction vector to the surface normal of the specimen and is defined as positive when in the same sense as 2θ; Δx is defined as positive when towards the X-ray source (Figure 6.7). For a conventional Ω-diffractometer with fixed-slit optics, $\Delta 2\theta$ in radians is given by [34]:

$$\Delta 2\theta = \frac{\Delta x \sin 2\theta}{R \sin \alpha} = \frac{\Delta x}{R} \frac{\sin 2\theta}{\sin(\theta + \psi)} \tag{6.1}$$

where R is the diffractometer radius, θ is the Bragg angle, and α is the incidence angle of the X-ray beam ($\alpha = \theta + \psi$). Equation (6.1) is obtained by applying the sine formula to triangle *DOB* in Figure 6.7 and neglects a small error due to defocussing of the diffracted beam [35]. It can be seen qualitatively in Figure 6.7 and quantitatively in Figure 6.8 that $\Delta 2\theta$ becomes large as α decreases, which effectively precludes stress measurement using small α angles (large negative ψ) with the fixed-slit Ω-diffractometer.

If parallel-beam optics are used, the displacement error is effectively eliminated as illustrated in Figure 6.9 [30, 34] in which the three types of optics are compared

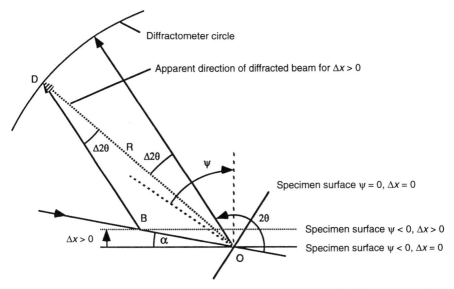

Figure 6.7. Effect of displacement error Δx on the 2θ error for an Ω-diffractometer with a fixed receiving slit at the diffractometer circle. Angle $OBD = 2\theta$.

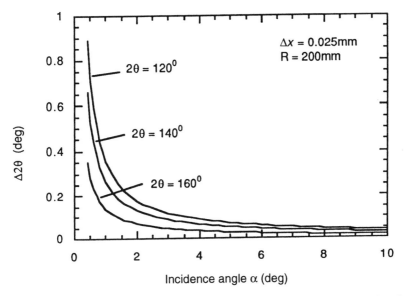

Figure 6.8. Rapid increase in 2θ error, with decreasing incidence angle. Specimen displacement $= 0.02$ mm, diffractometer radius $= 200$ mm.

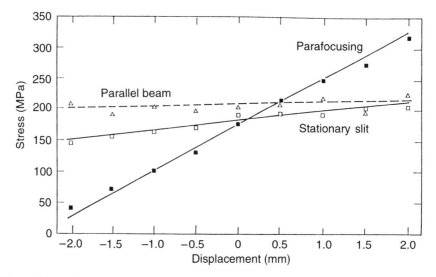

Figure 6.9. Effect of the type of optics used in an Ω-diffractometer on the error incurred in stress measurement as a result of specimen displacement from the diffractometer axis.

in terms of the effect of the error on the stress measured in a uniformly stressed steel specimen. The conditions used are as follows: the specimen is uniformly stressed, normalized 1045 steel, (211) reflection, Cr $K\alpha$ radiation. The solid lines were calculated using (6.1) for the fixed slit case and (6.1) with R replaced by

$$R' = R\cos(\psi + (90 - \theta))\cos(\psi - (90 - \theta))$$

for the parafocusing case. The data points are measurements.

6.2.4 Parallel-beam Optics

The simplest PBO setup on an Ω-diffractometer is shown in Figure 6.4. It consists of an X-ray tube set at a small take-off angle (about $6°$), a small divergence incidence beam ($0.3°$, $0.1°$, or less), a long, radial divergence limiting (RDL) Soller slit, and a solid-state detector, or a flat monochromator crystal plus scintillation counter. RDL Soller slits are available with divergences of 0.4, 0.25, or $0.15°$. (The divergence is the angle between the two most divergent rays passing through the gap between adjacent leaves). The arrangement shown in Figure 6.4 is more correctly termed "pseudoparallel-beam" because the incident and diffracted beams have nonnegligible divergence. The arrangement can be modified in several ways:

- By insertion of short, axial divergence limiting (ADL) Soller slits in both the incident and diffracted beams,
- By replacement of the incident slit with another RDL Soller (used with a larger take-off angle), by insertion of an incident beam conditioner to reduce

divergence, increase intensity and monochromatize the incident beam. This monochromatization can be done either with an asymmetrically cut monochromator crystal or a parabolic, graded multilayer optical device [36]

Displacement of the specimen merely moves the diffracted beam from the center region of the RDL Soller slit, toward one side or the other. It is only the angle between the leaves of the slit and the incident beam, i.e., the detector arm angle (2θ angle) that determines the direction of the rays transmitted to the detector. Another important advantage is that the width of the diffracted beam is largely determined by the divergence of the RDL Soller slit and is approximately independent of 2θ or α. Typical values of FWHM are about $0.3°$ and $0.15°$ for divergences of $0.4°$ and $0.15°$, respectively. Still narrower beams are obtained with a small mosaic spread, flat monochromator crystal plus scintillation counter in place of the solid-state detector [37] but at substantial cost in intensity. With fixed-slit optics, peak widths become very large at small α due to defocussing of the diffracted beam.

Although PBO are not sensitive to specimen displacement at large α, measurements at small α require accurate positioning of the specimen on the diffractometer axis or else the incident beam will miss the desired area or the diffracted beams will miss the RDL Soller slit [or will miss the detector crystal behind the slit if this crystal is narrower than the Soller slit (Figure 6.4)]. The latter effect can be used to advantage when one wishes to exclude unwanted diffraction peaks or background scattered from a substrate surface, provided the surface is at a different height than the specimen surface. There is also a small displacement error at small α as shown in Figure 6.10, which appears to be due to the divergence of the incident and diffracted

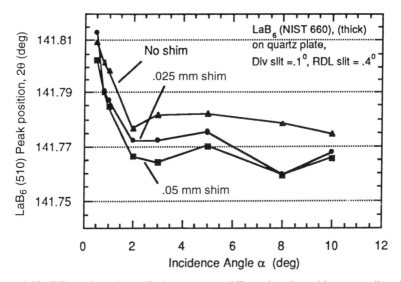

Figure 6.10. Effect of specimen displacement on diffracted peak position at small α. Displacements were produced by inserting shims of thickness shown and are negative.

beams. The measurement conditions were as follows: The specimen is LaB_6 (NIST 660) powder layer 0.05–0.1 mm thick on a quartz plate, (510) reflection, Cu $K\alpha_1$ radiation, peaks fitted to a pseudo-Voigt function. The NIST 2θ value for the (510) is 141.769°. In this and subsequent figures, the quartz plate was recessed slightly so that its surface was 0.025 mm below the ABC plane in Figure 6.6.

Figure 6.10 reveals another source of 2θ error at α less than about 2° with the asymmetric geometry, namely the refraction error. For an ideal optically continuous surface, this error $\Delta 2\theta$ in radians is given by [38, 39, 40]:

$$\Delta 2\theta = 2\theta_{meas} - 2\theta_{true}$$

$$= \frac{\delta}{\sin 2\theta_{true}} \times \left[2 + \frac{\sin \alpha}{\sin(2\theta_{true} - \alpha)} + \frac{\sin(2\theta_{true} - \alpha)}{\sin \alpha} \right] \qquad (6.2)$$

$$\simeq \frac{2\delta}{\sin 2\theta_{true}} + \frac{\delta}{\alpha} \quad \text{for small } \alpha \qquad (6.3)$$

where δ in SI units is given by [39]:

$$\delta = \frac{Ne^2\lambda^2}{4\pi\epsilon_o(2\pi mc^2)} \qquad (6.4)$$

Here $\eta_0 = 8.85 \times 10^{-12}$ F/m, λ is the wavelength, e and m are the charge and mass of the electron respectively, c is the speed of light and N is the number of electrons per unit volume irradiated. For a single element specimen:

$$N = \rho N_{av} \times (Z + \Delta f')/M \qquad (6.5)$$

where ρ is the mass density, N_{av} is the Avogadro number, Z is the atomic number, M is the atomic weight (kg/mole) and $\Delta f'$ is the real part of the dispersion correction to the scattering factor. $\Delta f'$ values are tabulated in the International Tables for Crystallography. Thus, using recommended values of the constants [41], equation (6.4) becomes:

$$\delta = 2.70087 \times 10^8 \times \rho \times \lambda^2 (Z + \Delta f')/M \qquad (6.6)$$

and, e.g., for Ag with Cu $K\alpha_1$ radiation, $\Delta f' = -0.6$ and $\delta = 2.895 \times 10^{-5}$. For a compound $A_x B_y$:

$$N = (\rho N_{av}/M) \times [x(Z_A + \Delta f_A') + y(Z_B + \Delta f_B')] \qquad (6.7)$$

where M is now the molecular weight.

There is some uncertainty as to whether or not a refraction error is present with a powder specimen [40]. The results shown in Figure 6.11 suggest that for the specimens investigated using the system shown in Figure 6.4, there appears to be a refraction error. The magnitude of this error decreases in the order: smooth solid surface, packed

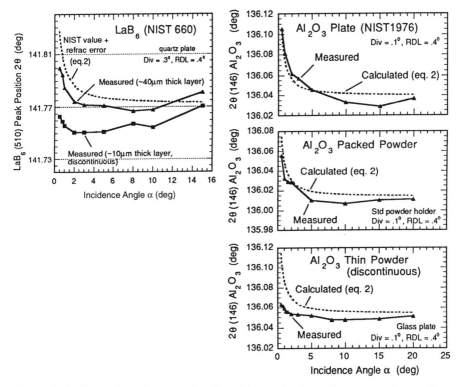

Figure 6.11. Comparison of measured peak positions and calculated ones assuming a refraction error, for LaB$_6$ (left) and Al$_2$O$_3$ (right). The specimens are LaB$_6$: (NIST 660) powder 1–20 μm size; α-Al$_2$O$_3$: (NIST 1976) plate 1–7 μm grain size (top); commercial 99% alumina powder about 5 μm size (middle); Buehler #1 polishing alumina 5 μm size, 0.8 mg/cm^2 (about 30% coverage) (bottom). 2 θ (NIST 660) (510) = 141.769°. Refraction error for the Al$_2$O$_3$ specimens was calculated assuming measured values at α = 20° were true. All peak profiles were fitted to a pseudo-Voigt function. Errors on the measured points are ±0.01°2θ. Calculations were based on equation 7.2.

powder surface, thin (discontinuous) powder layer, and presumably tends to zero for a specimen consisting of discrete individual crystallites on a substrate. The effect of specimen surface roughness on the refraction error for the Figure 6.4 geometry has not been reported to the authors' knowledge, but one might expect an effect analogous to the increased discretization of diffracting crystallites seen in Figure 6.11. Although a refraction correction of $\Delta 2\theta$ given by equations (6.2) or (6.3) is normally subtracted from measured peak positions at small α for continuous solid specimens, a correction which decreases with increasing surface roughness may be more accurate. The form of such a correction is not presently known. The refraction correction is not required for the symmetric geometry in which both the incident and diffracted beams are grazing to the surface [42].

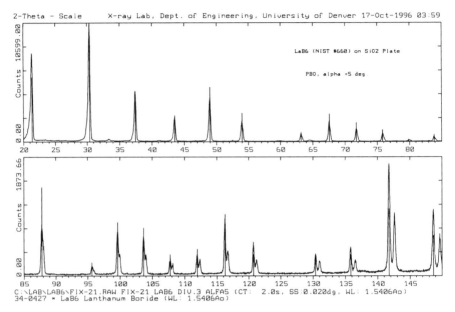

Figure 6.12. Diffraction pattern of a thin layer of LaB_6 (NIST 660) on a quartz plate substrate using the PBO arrangement.

A disadvantage of the system shown in Figure 6.4 is that diffraction peaks at small to intermediate 2θ angles are asymmetric, showing a tail on the low 2θ side as seen in Figure 6.12. This figure shows data obtained with the PBO arrangement shown in Figure 6.4. (Siemens D-500 diffractometer, Cu $K\alpha_1$ radiation, 30 kV, 25 mA, 2 s/step, 0.02°/step, $\alpha = 5°$, divergence slit $= 0.3°$, RDL Soller slit $= 0.4°$, Kevex Si(Li) detector: $E = 8.04$ keV, $W = 340$ eV). Such asymmetry is due to axial divergence [43] and can be removed by inserting ADL Soller slits in the incident and diffracted beams at considerable cost in intensity (an order of magnitude or more). Because stress measurements are usually made at $2\theta > 90°$, the ADL Soller slits are not essential. Even at smaller 2θ angles, the error in 2θ due to asymmetry is largely eliminated if a split function (e.g., split Pearson-7) is used to fit the diffraction peaks instead of a symmetric function as shown in Figure 6.13. The peak asymmetry is also reduced by using a smaller divergence RDL Soller slit [37].

6.2.5 Sequence of Procedures in Conducting Stress Measurements

Alignment Before undertaking stress measurements it is especially important to align the diffractometer well. There is no substitute for good alignment. The first step is to ensure that the specimen holder locates the surface of the specimen so that the θ-axis lies in that surface and that the incident beam passing through the narrowest divergence slit to be used is directed so that the θ-axis is centered in that beam. This step is normally done with the RDL Soller slit removed using standard alignment

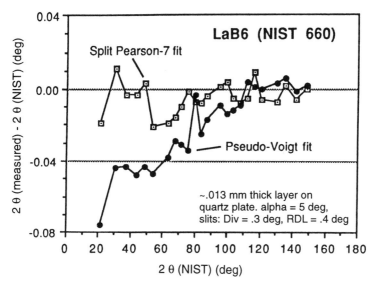

Figure 6.13. Effect of profile fitting function on the 2θ error obtained with PBO geometry, specimen LaB$_6$ powder about 13 μm thick on a quartz plate. Diffraction conditions are the same as those used for Figure 6.12.

procedures specific to the diffractometer. The RDL slit is then inserted and adjusted to give maximum intensity using the direct beam. If the solid-state detector crystal (Figure 6.4) is narrower than the RDL Soller slit, it is important that the location of the crystal be adjusted by moving the detector transverse to the beam so as to obtain the maximum diffracted intensity from a thin specimen located accurately on the diffractometer axis. At small α's, the width and location of this crystal determine the width and location of the irradiated area from which the diffracted beam is originating.

If measurements at small αs are planned, the $\alpha = 0$ (i.e., $\theta = 0$) position must be accurately determined so that α is known accurately. This is conveniently done using a long, 50 μm wide slit mounted in the specimen holder so that the θ-axis lies in the plane of the slit at its center. Such a slit is normally supplied with the alignment tools and is used in the beam alignment procedure above (if not supplied, it can be readily fabricated from a pair of flat glass or metal plates). This slit θ-rotated to the point where the intensity transmitted through the slit is a maximum determines $\theta = 0$ (and 180°) usually to within ±0.02°. In general the 2θ readings for the maximum transmitted intensity through the long slit at $\theta = 0$ and $\theta = 180°$ will not agree exactly even for a well-aligned specimen holder. This lack of agreement results in a beam-off-center error which, fortunately is independent of 2θ. The error $\Delta 2\theta$ is approximately equal to the angle γ between the incident beam and the true on-center beam.

Alignment Check   A good test of the alignment is to run a NIST [SRM660 (LaB$_6$) or SRM6406 (Si)] in the form of a thin layer about 13 μm thick on a low background substrate (quartz plate or silicon wafer) at, say, $\alpha = 5°$ over the whole 2θ range. The 2θ peak positions obtained by profile fitting to a split function should agree with the NIST certified values to within $\pm 0.03°$ (Figure 6.13). To check for ψ-tilt errors, select a peak(s) from the standard which is close to the peak(s) to be measured on the specimen and measure the effect of ψ-tilt on the peak position. The 2θ values should not vary by more than $\pm 0.01°$ (Figure 6.14) even down to small α values, except in the region where the refraction error becomes large (α less than about $2°$) (Figure 6.11).

Normally with the PBO geometry in Figure 6.4, more accurate results will be obtained using negative rather than positive ψ-tilts, particularly at large negative ψ (small α) because with positive ψ-tilts the irradiated area becomes small and an insufficient number of grains may be diffracting. To obtain positive ψ-tilts with respect to the specimen while applying negative ψ-tilts geometrically, the specimen is removed, rotated $180°$ about the normal to its surface and reinserted into the goniometer. The small irradiated area obtained with geometrically positive ψ-tilts can be used to advantage if high spatial resolution is desired.

Penetration Depth   Before a set of diffraction measurements is implemented, the 1/e penetration depth should be calculated. Assuming the specimen can be considered homogeneous (negligible gradients in composition or density in the depth direction), τ for an Ω-diffractometer is given by [44]:

$$\tau = \frac{\sin^2 \theta - \sin^2 \psi}{2\mu \sin \theta \cos \psi} \tag{6.8}$$

where μ is the linear attenuation coefficient of the irradiated volume. If α is small, of the order of α_c the critical angle for total external reflection, equation (6.8) is no

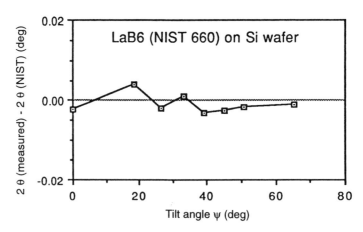

Figure 6.14. Variation of 2θ error with negative ψ-tilt for the (510) reflection of NIST 660 (30 μm thick layer on an silicon wafer) using PBO geometry, Cu $K\alpha_1$ radiation. Divergence slit $= 0.1°$, RDL slit $= 0.4°$, pseudo-Voigt fit to peaks.

longer accurate and τ is given by [45]:

$$\tau(\alpha) = (\lambda/4\pi)([((\alpha^2 - \alpha_c^2) + 4\beta^2)^{1/2} + \alpha_c^2 - \alpha^2]/2)^{-1/2} \qquad (6.9)$$

where $\beta = 4\pi/\mu\lambda$ and $\alpha_c = \cos^{-1}(1-\delta)$ and the α and α_c are in radians. Thus, e.g., with Cu $K\alpha_1$ radiation and $\alpha = 1°$, Ag has $\alpha_c = 0.436°$ and $\tau = 68.6$ nm, whereas Al (with $\Delta f' = 0.1$) has, $\delta = 8.404 \times 10^{-6}$, $\alpha_c = 0.2349°$, and $\tau = 1251$ nm. A plot of equation (6.9) shows that $\tau(\alpha)$ is an s-shaped curve which changes rapidly with α in the vicinity of α_c. Actually, equation (6.9) only considers absorption along the incident beam path; there is an additional absorption term for the diffracted beam path which can be neglected if α is sufficiently small and the diffracted beam is not at a small angle to the surface. In considering the significance of the stress measurements, it should be remembered that 63% of the diffracted intensity comes from a layer extending from the surface down to a depth τ and the remaining 37% (or $1/e$) comes from depths $> \tau$ assuming the specimen is homogeneous.

Mounting and Use of an Internal Standard   After the above steps are completed, the specimen can be prepared and mounted in the diffractometer as discussed previously. An internal standard in the form of a thin layer of standard powder, preferably SRM660 LaB_6 or SRM640b silicon, dispersed uniformly over the irradiated area on the specimen, is often useful as an additional check on alignment provided the peaks of the standard do not seriously overlap those of the specimen. The thickness of the standard powder is adjusted until the peaks from the specimen and from the standard have comparable intensity at the ψ-values used. The thickness has to be decreased drastically at small α to the point where the standard may no longer be useful. The measured peak positions for the specimen can be corrected using nearby standard peaks; however, note that the refraction correction cannot be performed in this way, as discussed above.

Use of an internal standard is much more important if fixed-slit optics are used instead of PBO, particularly for measurements at large ψ. In applying the correction to the specimen peaks it may be necessary to account for the fact that the standard layer is displaced from the τ_{50} depth of the specimen (the depth from which 50% of the diffracted intensity originates).

6.3   MEASUREMENTS ON SINGLE CRYSTALS

In the course of materials studies, occasions will arise when the orientation of a (sometimes large) single crystal is needed. The single crystal may exhibit natural growth faces, but for optical and electronic materials the surfaces are more likely to be cut and polished faces. Natural growth faces are indexed classically by optically determining the interfacial angles. The identification of cut and polished faces is often accomplished with front- or back-reflection Laue photographs. However, a powder diffractometer with electronically decoupled $2\theta/\theta$ or θ/θ motions can be used very effectively to determine orientation and other subsequent information about a crystal. The first step in the identification of crystal faces or crystal orientation is to look at the specimen crystal and assess its size, thickness, and whether or not it exhibits one or more parallel sets of faces. Then decide how much time is available for the analysis. If the specimen exhibits one or more parallel set of faces, is of a manageable size to fit in a powder diffractometer holder, the generator will allow very low tube currents, and time is short, consider using the powder diffractometer technique described below. If, however, time is not an issue, and some very aesthetic photographs exhibiting symmetry would enhance a talk or impress family and friends, use Laue photographs.

6.3.1 Identification of Crystal Faces

The identification of crystal faces using a Bragg–Bretano ($2\theta/\theta$ or θ/θ) powder diffractometer (decoupled $2\theta/\theta$ or θ/ω motions) requires that a set of opposite (natural or polished) faces are (very nearly) parallel. The "back" face of a pair of faces is used to define the crystal orientation on the goniometer, whereas the opposite face, the "front" face, is the face which is irradiated by X-rays and the face, therefore, which is identified. Small, thin crystals less than a few millimeters thick, such as crystals of $YBa_2Cu_3O_x$ or semiconductor wafers, can be mounted on an off-axis-cut "zero-background" quartz (ZBQ) substrate with a thin smear of petroleum jelly to hold the crystal in place on the substrate. *Note:* Be sure and check with the owner of the crystal as to the types of adhesive that can be used! In some cases, petroleum jelly is acceptable, but in other cases double-sticky tape may be preferred, depending on how the crystal or wafer will be cleaned subsequent to the analysis or what other analyses will be conducted on the crystal. In handling electronic wafers and optical or electronic crystals use "wafer" tweezers if possible; if the crystals must be handled, use gloves. If the breadth or length of the crystal is large enough, a ZBQ substrate is unnecessary. The crystal can be placed directly on the goniometer spring-loaded specimen-holder plate and under the lips or edge of the specimen holder that defines the center of the goniometer. If a ZBQ substrate is used, strips of index cards and/or strips of "Post-Its" on the edges of the ZBQ substrate can be used to compensate for the crystal thickness and set the front face of the crystal at the center of the diffractometer. Semiconductor wafers often have approximately the same thickness, especially if they are from the same manufacturer; consequently, once a ZBQ with the correct amount of edge height/thickness compensation has been prepared, it can continue to be used over and over again. If the crystal is more than a few millimeters thick, the crystal can be set in a deep-well specimen holder. If additional height compensation is still needed to bring the front face of the crystal to the center of the diffractometer, shim stock and/or pieces of wooden popsicle sticks can be used.

Caution is required when checking for diffraction from a crystal face. The diffracted intensity from single crystals is *very intense*, at least an order of magnitude greater than for powders. To initiate a search for a reflection from the face of a crystal, set the kV generator to the voltage usually used (for powder scans), which results in the same energy distribution of X-rays and, therefore, matches the pulse-height energy discrimination set on the detector electronics. Set the tube current to a low value, initially about 3–5 mA. If the generator does not allow this amount of control of the tube current, *do not go any further—do not try this experiment.* Access to very low tube currents must be possible or the detector could be damaged by excessive diffracted beam intensity! Use narrow divergence and receiving slits; a divergence angle of 0.7° and acceptance (receiving slit) angle of 0.07° are good values to try initially. If approximate lattice-matching in MBE-grown (molecular beam epitaxy-grown) electronic wafers is being checked, the angular values of the beam should be divergence of about 0.07° and acceptance of 0.02° (or less) to have sufficient resolution to observe δ_a/a values of 2×10^{-3} to 6×10^{-4}. Except for evaluating lattice-matching of surface layers on a substrate for which absolute maximum resolution is needed, narrow slits are used primarily to reduce the incident and diffracted beam intensity from a crystal face. The ultimate resolution achievable for diffraction from single crystals is the limiting resolution of the optics of the powder diffractometer, as most well-crystallized, well-ordered single crystals (especially optical or electronic materials) have inherent diffraction line widths far narrower than the few arc minutes resolution limit of powder diffractometers.

6.3.2 Orientation of Single Crystal Specimens

Determination of the orientation of the single crystal requires some thought about the physical shape of the crystal in relation to its symmetry and/or what the crystal is and the information desired. For example, the perpendicular to the surface of a silicon wafer of unknown orientation is likely to be [100], [111], [110], or a 3–6° off-axis variant of these orientations. Substrates InP or GaAs used for MBE and LPE growth are likely to be [100] or a 3–4° off-axis cut. High-temperature superconductors such as $YBa_2Cu_3O_x$, as well as the Tl- or Bi- containing phases such as $Tl_2Ba_2Ca_2Cu_3O_8$, form layered crystals with the c-axis perpendicular to the large face. For other types of optical or electronic materials, cut and polished faces can generally be assumed (as a first guess) to have low index faces. However, this condition will *not* be true for temperature-compensated cuts of quartz or some nonlinear optical materials. Although at what angle(s) the search for diffracted intensity begins is dependent on the crystal, the procedure for all types of single crystals is the same. Drive the detector to the expected 2θ value for an (hkl) potentially representative of that face; drive θ to $1°/2\theta$. With the tube current low (i.e., no greater than 5 mA) and with narrow slits in place, open the shutter while watching the ratemeter. If the ratemeter zooms off-scale, immediately shut the shutter and readjust the current down to 1–3 mA. Of course, should this high count rate occur, the orientation of the face has been identified! Further tweaking of θ can be done to improve the alignment of the crystal orientation relative to the goniometer, and then a scan of higher order reflections can be run to confirm the identification. Other types of scans appropriate for the desired information can also be run. More often, however, when the shutter is first opened, the ratemeter registers absolutely nothing. In this case, keep 2θ fixed and scan θ (ω) to both sides of the ideal, expected ω value. This motion is equivalent to doing a "rocking curve" about a fixed 2θ position. On a θ/θ diffractometer, ω and θ are shifted in a correlated manner to maintain the fixed 2θ value. If the amount of omega scan to either side of the expected value exceeds about five degrees, the front surface of the crystal is either not even approximately parallel to the specimen holder (check the crystal mounting) or the face has a different index [try a different (hkl) family and 2θ value]. If diffraction intensity is observed during the θ (ω) scan search, carefully scan θ (ω) to determine the optimum peak position. The (hkl) corresponding to the 2θ value at which diffraction is observed is, then, the family index of the crystal face.

6.3.3   Example of an Orientation Experiment

A rectangular-shaped crystal of tetragonal $(Sr, Ba)Nb_2O_6$ with four polished faces was provided for an analysis of the orientation of the faces. Due to ease of alignment and ease of data interpretation (i.e., time), the identification of the faces was carried out on a $2\theta/\theta$ diffractometer rather than by using back-reflection Laue. Inherent in this choice was the assumption that each set of opposite polished faces were (reasonably) parallel, i.e., the back face, which would be used to define the crystal orientation on the goniometer, would be equivalent and parallel to the front face, which would be irradiated by X-rays. The crystal was set in a deep-welled Plexiglas holder and held in place on the holder with Vaseline. Pieces of wooden popsicle sticks were inserted on the edges of the holder to compensate for the thickness of the crystal and bring the "front" face to the center of the diffractometer. As the cut and polished faces were expected to have low-angle indices such as (100), (110), (001), or (101), the initial search for diffracting intensity centered around the approximate angular positions expected for the (001), (400), and (330) reflections. (See ICDD PDF # 39-0265 for reference data for powdered $Sr_{0.5}Ba_{0.5}Nb_2O_6$ [46].) From the larger rectangular face a diffraction peak was found at a 2θ value approximately equal to that expected for a (400) reflection. The crystal was "rocked" in omega about the observed 2θ value until a maximum was found. To maintain focusing conditions during a $2\theta/\theta$ coupled scan, the numerical value of omega was reset to account for the offset from the ideal omega value. Using the same procedure, the polished face at $90°$ to the large rectangular face [now identified as a (100) face from the observation of a (400) reflection] was found to diffract (very strongly) at a 2θ value expected for the (001) reflection. The observed omega offsets could represent off-axis cutting and polishing errors, but are equally (or, perhaps more) likely due to a tilted or warped bottom in the Plexiglas holder. (Tilting of the crystal would not have occurred with a zero-background quartz substrate, but this crystal was too large for use with such a substrate.) To confirm the identification of the faces, scans were recorded over narrow angular ranges in which diffraction from multiples of the face indices would be expected to occur. The results of these scans confirmed the identity of the polished faces. The larger polished faces were found to be (100), the smaller polished faces (001). The (110) direction should be at $45°$ to the (100) direction when rotated about the (001) direction. Note that this system is tetragonal with 4 mm symmetry.

Further experimental details were as follows. The (100) face, with an omega offset of $-0.16°$ was scanned at 1 sec/0.01° step over the 2θ ranges of 27–29.5°, 42–45°, and 75–79° with a divergence angle of 0.72° and acceptance angle of 0.072°. The (001) face, with an omega offset of $-0.13°$ was scanned at 1 sec per 0.005°

step over the 2θ ranges of 21.5–23.5° and 44.8–46.8° with a divergence angle of 0.143° and acceptance angle of 0.036°. Scans were obtained using nickel-filtered copper radiation with tube settings of 45 kV and 1 mA on a Scintag PAD V $2\theta/\theta$ powder diffractometer equipped with a liquid nitrogen cooled solid-state intrinsic germanium detector set to discriminate for Cu $K\alpha$. (Nickel filtering was used to decrease the incident-beam intensity.) The radius of the goniometer was set at 220 mm; both incident and receiving Soller slits were in place to improve resolution and limit axial divergence.

6.3.4 Final Comments

If Laue photographs are selected for orientation characterization because the crystal is simply too large to be managed at all on a powder diffractometer (or aesthetic appeal is a high priority), crystallography texts, such as McKie and McKie [47] and Amoros, Buerger, and Canut de Amoros [48], may be consulted for a discussion of this technique. If the specimen crystal contains light atoms (atoms with low atomic numbers) and is less than about 1.5 mm thick, or contains heavy atoms but is less than about 1 mm thick, front-reflection (transmission) Laue photographs can be obtained using unfiltered radiation from a tungsten or molybdenum X-ray tube. In this case, the crystal is affixed with an amorphous adhesive to a glass fiber which is mounted on a brass pin; the brass pin with fiber and crystal is subsequently mounted on a goniometer head. Techniques for mounting small single crystals can be found in [49]. A transmission Laue photograph can be recorded using a special Laue camera (see, for example, Figure 3, Chapter 5, in [48]) or by using a precession camera with all settings at zero and the gears disengaged. In the case of large and/or thick crystals, the type of Laue photograph recorded will of necessity be back reflection. The incident X-ray beam is collimated through a hole in the film and film holder and impinges on the crystal face perpendicular to the beam. Backward scatter from the face irradiated (high-angle diffraction) is recorded on the film. For an example of a back-reflection Laue photograph see Figure 6, Chapter 5, in [48]. Large crystals can be mounted for back-reflection Laue using a standard goniometer head (if the camera track being used has an appropriate attachment), but the crystal can also be mounted using a Griak holder (very effective for large wafers) or by using a Bond barrel holder. Commercial holders are available that will fit on standard Laue camera tracks that can also be directly transferred to a crystal saw without loss of orientation. From Laue photographs the symmetry of the crystal direction parallel to the beam is determined; tables containing the Laue projection symmetry for crystallographic directions in all six crystal systems can be found in Table 3, Chapter 6 [48] or in Table 3.7.2, *International Tables*, Vol. I [50]. For those cases in which the projection symmetry is unique for a given direction, the face which is perpendicular to that direction will be defined, i.e., if the crystal is known to be tetragonal and the Laue photograph exhibits 4 mm or 4 symmetry, then the direction of the beam is parallel (001) and the face which is perpendicular to the beam is, therefore (001). For those cases in which the projection symmetry is not unique, resorting to indexing the Laue photograph and recording photographs at other orientations will be required. For example, if a crystal is known to be tetragonal and exhibits 2 mm symmetry in a Laue photograph, an ambiguity exists as to whether the beam direction is parallel to (100) or (110). At this point, putting the crystal back on a diffractometer may be a very attractive alternative!

6.4 UNUSUAL TRICKS WITH UNUSUAL SAMPLES

During the course of research for this book, many inquiries were made to experienced diffractionists all over the world asking for suggestions on handling unusual situations. Many responses were received. Most of the techniques that were proposed are derivatives of methods described elsewhere in this book. This section will describe a few of the more unusual situations. Individual contributors will not be identified because of the probability of forgetting some of the contributors and because several of the ideas have roots with multiple individuals or even in the forgotten past.

6.4.1 Isolation of Material

The concentration and isolation of sufficient material for a diffraction specimen mount is sometimes the principal obstacle to successful analyses. It has been emphasized already that the time spent isolating a specimen may well be returned with interest (time) during the analysis. The surest method of isolating the desired phase is by handpicking under a binocular microscope. Many times the particles are imbedded in a matrix and need to be released. Dental picks or a vibratory engraving tool may be needed to release the grains. A film of thick mineral oil will prevent the particle from disappearing when it is released. Alternatively, the surface may be etched with an appropriate reagent such as an acid that preferentially dissolves the matrix, leaving the loose particles on the surface. The particles may then be collected by dusting into a collecting dish, pressing a piece of Magic Mending Tape on the surface and removing, or by placing a cellulose in ethyl acetate solution over the etched area and removing the film after the solvent has evaporated. Another example of concentration is the removal of sulfur from organic preparations by dissolution with C_2S. Other properties such as density, magnetism, and electrostatic behavior can be exploited also to achieve a concentration.

   An example of separating a phase is isolating the solid film or slag on a hot liquid such as a molten metal. If the specimen is allowed to cool, the film is too intimately adhered to the mass to collect. Bubbling an inert gas through the molten metal will cause the film to lift and froth, so it can be collected free of the substrate.

6.4.2 Reactive Materials

Handling reactive materials is always a problem. The goal is to retain a barrier between the specimen and the atmosphere which is still X-ray transparent. Some specimens may be collected in an oil bath, and the oil film left on during the diffraction experiment. An example are oxide-coated filaments which may be studied by opening the lamp while they are immersed in oil and retaining the oil coating throughout the experiment. Some reactive particles may be protected sufficiently by preparing a slurry with a binder or cement (epoxy, liquid rubber) and then trimming the composite to fit a specimen holder. Cutting off the surface layers of such a composite may also give the benefit of reduced preferred orientation. Even asbestos fibers may be prepared by forming a composite and removing the surface layers.

Handling hygroscopic materials in normal diffraction experiments requires the maintenance of a dry atmosphere in the specimen chamber if not the whole room. One could do the experiments in a desert location. Alternatively, the specimens must be sealed in capillaries or special holders with Mylar windows that are loaded in a dry box. Mylar may be sealed to ordinary metal specimen holders using a hot-glue gun or sealed to a thin metal ring which is held over the specimen by a layer of vacuum grease. The Mylar should be retained in a stretched state until the glue sets to assist in keeping the specimen surface flat. One simple method for enclosing hygroscopic LiH employs Mylar tape with adhesive on one side. Windows were prepared in pieces of tape by removing the adhesive with a cotton swab and solvent, clearing an area about 1 cm^2. Powder was dusted over the cleared area, and a second windowed tape was then placed over the powder. This specimen could be used in transmission or reflection. This packet would remain dry for several hours. Another approach is to prepare a normal cavity mount in the dry box, then paint the surface with sprayed acrylic lacquer. Metallographic mounts may be treated the same way to prevent oxidation.

6.4.3 Toxic Materials

Toxic samples, in contrast to reactive samples, may be handled using normal procedures, but the material requires personnel safety. Dispersion of the material must be prevented and the specimen must be recovered for disposal. Many of the above procedures will encapsulate the material adequately; however, the specimen holder may become contaminated. It is advisable to use disposable specimen holders. One simple, inexpensive support is to use thin wafers of silicon to support the specimen, then discard the silicon wafer along with the specimen. Silicon in diameters up to 12 inches, which may be easily cut into appropriately sized rectangles using a diamond scribe, may be obtained from the electronic industry. Ideally, a (511) or (510) orientation is desirable, but any orientation may be used by ignoring the region where the silicon diffracts.

6.4.4 Radioactive Materials

Radioactive specimens pose severe problems. There are additional concerns when handling radioactive materials, especially safety of personnel, full recovery of the specimen, and preventing the radioactivity from interfering with the measurements. Special shielded enclosures are usually employed with the diffraction equipment either in the enclosure or attached. However, low-level radioactive specimens may be handled safely outside such enclosures. Usually, there is very little sample with which to work, so the capillary method is employed. Silica capillaries are loaded in a glove box, a silica rod is inserted to prevent the expulsion of particles during removal from the box, and the capillary is then sealed with a flame or hot wire. This capillary can be used in normal diffraction equipment. For a diffractometer, the specimen may be made up of several capillaries placed in the specimen plane oriented parallel to the diffractometer axis. Handling in a capillary or as a composite with epoxy or other cement may be necessary for α-emitters. Metal capsules may be required to attenuate the radiation from γ-emitters. For Cu $K\alpha$ radiation, nickel works very well for many elements. Radioactive metals, such as the transuranic elements, require further protection to prevent oxidation as well as for radiation safety. Some of the methods mentioned under reactive specimens are applicable for radioactive specimens, but the radiation will induce changes in the protective layer, so the user should determine how long the layer will remain coherent before doing experiments at the open diffractometer.

For high-level radioactive specimens, special instruments with atmosphere enclosures and vacuum transfer devices may be necessary. An excellent review of the handling of radioactive materials is given by Schiferl and Roof [51]. Further information may be obtained from the *Plutonium Handbook* [52].

6.4.5   Flowing Materials

Some special devices have been designed that allow a continuous slurry of particles and liquid or gas to agitate during the experiment. These containers have a thin window on one side for reflection diffractometry. Continuous agitation helps minimize the effect of preferred orientation. Cells allowing solid-gas reactions will allow following the kinetics of phase modifications during the experiment. Where the sample is not reactive but monitoring of a continuous feed stream is desired, the cell may be designed to allow a slurry to flow past the diffractometer. One example of this usage has the diffractometer in the horizontal position, and a steady stream of particles free-falls along the diffractometer axis during the monitoring.

6.4.6   High Temperature Diffraction

High temperature experiments require specimens supported on unreactive, thermally stable substrates. Quartz capillaries are usable up to about 1000°C and are loaded by techniques already described. Diffractometer heating attachments are usually of the metal ribbon design where the metal ribbon is both heater and specimen support. Vertical diffractometers are most convenient for hot-stage studies because the specimen is less likely to fall out of the support. The best way to load the hot stage with the specimen is to prepare a slurry of the material in xylene or isopropyl alcohol. The slurry is applied to the heater stage with a fine camel's hair brush or a very fine medicine dropper so that thin layers are formed and dried before the next layer is deposited. Intimate particle contact with the heater and other particles is critical for even heat distribution. Wet application of the specimen is the best method to achieve this condition. Because the temperature is usually measured by attaching a thermocouple to the underside of the ribbon heater, the actual temperature of the diffracting surface is usually not known well. If the specimen is thin, and most specimens only diffract from the upper hundred micrometers, then the specimen temperature may be close to the temperature of the heater.

6.4.7 Low Temperature Diffraction

Low temperature diffraction also requires intimate contact of the specimen with the cold finger that creates the cold environment. This condition is usually achieved by using a conductive cement or binder to attach the specimen to the copper block. If a block specimen is used, the cement should contain finely divided powdered copper. A good grease film will also suffice for many specimens. Most greases thicken at the low temperatures, but do not crystallize, and retain their adherence properties. Vaseline and vacuum grease work very well. Low temperature epoxies are also available. Cyanoacrylates also work, as does a silicophosphate dental cement. Specimens slurried in collodion and painted on the cold finger also work well.

6.4.8 Organic Materials

Most organic materials can be prepared by the same methods described for inorganic materials. If the sample is in solid form, the softness of the material may allow a specimen to be cut to a usable shape using a razor blade. Many organic materials cannot be pulverized in a mortar and will require cutting and dicing to reduce the particle size. Chopping with a razor blade on a glass surface will accomplish this goal. Because of the high transparency of organic compounds, the depth of X-ray beam penetration is considerable, which allows many crystallites to be in the beam assisting the crystallite statistics. This condition is beneficial for intensity measurements, but when accurate d-spacing measurements are desired, a thin specimen must be used. Some organic phases which are soft and deformable at room temperature may be successfully ground under liquid nitrogen. Specimens may be prepared from fibrous materials by winding the fibers around a rectangular or circular support. The windings may be either deliberately parallel or completely random. If the support is hollow in the center, the same specimen may be used in transmission geometry as well as in reflection geometry.

6.4.9 Bulk Specimens for the Diffractometer

Although it is not very common today, it is possible to use rather large bulk specimens directly in the diffractometer. Although there are portable diffractometers that may be transported to the piece to be examined, many facilities are not so equipped and need to use the piece directly in the laboratory instrument. The problem today is the need to defeat the safety interlocks to remove the specimen chamber to allow for the large dimensions. Where the diffractometer is also enclosed in a shielded enclosure, personnel safety is not sacrificed with the specimen chamber removed. The main problem becomes positioning the specimen so that the X-ray beam can see the desired area, and that the rest of the specimen does not interfere with the movements of the diffractometer. Even curved, convex, or concave surfaces may be used for study. For highly curved surfaces, it may be necessary to add some lead foil as collimation to narrow the beam so that only the center of the curved specimen is illuminated. These foils need to be placed around one inch above the specimen surface so they are out of focus for the diffractometer optics. Pieces up to several inches may be examined in laboratory diffractometers with the specimen chamber removed.

Even the interior surface of lengths of pipe may be examined if the pipe is cut in half and aligned with the long dimension perpendicular to the diffractometer axis so that the X-ray beam may see the interior wall. Irregular shapes may be used if the surface to be examined is reasonably flat and supported so that the surface is in the proper position on the diffractometer. Two problems with bulk pieces are the specimen displacement, which can be large because there is no good location for the diffractometer axis, and the surface irregularities and curvature leading to much of the diffracted beam being out of focus. Judicious use of lead foil can minimize these problems by confining the incident and diffracted beam to a small area on the specimen.

6.5 PREPARING SPECIMENS FOR THE SYNCHROTRON

Synchrotron features including geometries, properties, and effects are covered at some length in recent review articles ([53], [54], [55]). This section will illustrate considerations which affect specimen preparation.

6.5.1 Effects of Data Recording

- *Time Effects*   These effects are not important, because all data points are normalized to incident beam monitor counts.
- *Peak Shifts*   These shifts could occur (e.g., unstable beam orbit, heating effects in monochromator, radiation damage to specimen), but no evidence for this problem has been observed at X7A and similar instruments are NSLS.

6.5.2 Intensity Values

- *Relative*   Normal mode of measurement.
- *Absolute*   This measurement is reasonably easy to do with a reference sample such as silicon or CeO_2 provided absorption term is known. It can also be done if all instrumental parameters are known (e.g., incident intensity, solid angle seen by the detector, detector efficiency, etc.; see [56]).

6.5.3 Importance of Proper Alignment

The importance of proper alignment depends on diffraction geometry; in particular when using parallel beam geometry, the crystal-analyzer geometry is tolerant of small specimen displacements, say 0.2–0.4 mm. Normal alignment procedure is carried out with a transit and involves:

- Centering the horizontal goniometer axis in incident beam with a tapered pin at the specimen position.
- Centering specimen capillary in beam (the Blake flat-plate holder is designed to need no adjustment).[1]
- If necessary, alignment of receiving slit by centering on diffracted beam from strong reflection of reference specimen at correct 2θ value.

6.5.4 Statistics Applied to Samples and Sampling

A typical strategy is to count for short times at low angles, longer at high angles (e.g., 2–10 sec) and long enough to give several thousand counts at the strongest peak, at step intervals sufficiently small to give at least 5 points over the FWHM of the sharpest peak.

6.5.5 Methods for Commuting Powders

Most users employ standard grinding techniques for hard and brittle samples. Minimum grinding is recommended, just enough to eliminate any obviously gritty texture, preferably under an organic liquid (e.g., acetone, cyclohexane), in order to minimize line broadening effects and thus take full advantage of the instrumental resolution. If the original sample is a single crystal, crush lightly and sieve out the 30–50 μm fraction. This particle size may cause problems due to incomplete randomization, and a smaller size may prove necessary.

6.5.6 Preparing Specimens for Simple Routine Analysis

For quick checks of specimen quality, peak widths, etc., either recessed flat plates or capillaries are used.

6.5.7 Use of Standards in X-ray Diffraction Analysis

The standards mainly used for wavelength calibration include Si or CeO_2 as compacted disks. Typical procedure is to scan five strong peaks (111), (220), (311), (331), (422), determine peak positions from a fit to a psuedo-Voigt function, and use these positions to fit lambda and the zero 2θ offset. An important point here is to use an appropriate asymmetry function [57]. Typical estimated standard deviations for λ are 0.0001 Å.

Equipment errors will most likely show up in abnormally low integrated intensities or unusually large peak FWHMs.

---

[1] Available from Blake Industries, Inc., Scotch Plains, NJ. USA.

6.5.8 Preparing Specimens for Accurate *d*-spacing Measurements

One of the major advantages of the synchrotron is the very high instrumental resolution that can be achieved, $0.006°$ at 0.7 Å and $0.01°$ at 1.2 Å in the low-angle region. One can, therefore, detect line broadening at about the 20–30% level, corresponding to a particle size of 3 μm or so, and care is needed to avoid degradation of the specimen during the grinding procedure. In fact, very few real materials can be prepared with this kind of crystallite size, and in any case, strain effects may dominate at higher angles.

It is by no means clear how one obtains an *ideal* reference specimen, i.e., one with negligible line broadening. Crushing a single crystal wafer of silicon under acetone and sieving out the 270–400 fraction (33–51 μm) works well (see [58]). When loaded into a capillary and rotated at 1 Hz, this specimen gives close to the calculated instrumental resolution function and reasonably good intensities (5% or better). This technique should work with other crystals, e.g., Ge, CeO_2, Al_2O_3, and maybe CaF_2 and BaF_2. Annealing close to the melting point is recommended for the optimum specimen.

6.5.9 Preparing Specimens for Accurate Intensities

Reducing the Particle Size Ideally, one would probably like spherical crystallites about 0.5–1 μm in diameter in order to avoid line-broadening effects, but if the structure is not too complex, some broadening can be tolerated if the size were reduced further. Replicate measurements on highly crystalline reference specimens such as Si and LaB_6 (i.e., scanning the same peak several times) give standard deviations of 3% or so. With less crystalline material, 2% may be attainable, but from then on, it would take a lot of work to do better.

Specimen Mounting

1. *Flat Plates in Reflection* A recessed specimen holder made from aluminum or from zero-background quartz is used depending on whether the *infinite-thickness* criterion applies. Top load the specimen using a microscope slide to produce a smooth surface (e.g., press hard and rotate slightly with fingertips). Back- or side-loading tends to give material that is too loosely packed and very likely to fall out at higher angles while being rocked or rotated! This problem can also happen with top-loaded specimens, but a very thin smear of Vaseline applied to the recess will usually do the trick. If a thin pellet is available, it can be placed in a specifically machined recess. Another possibility is to use an organic binder, but binders contaminate the material and may give unwanted background scattering. For thin-film specimens, asymmetric geometry (towards grazing incidence) is a useful option.

2. *Flat Plates in Transmission* This type of specimen is useful for transparent specimens (organics or low atomic number), especially if it is possible to press a reasonably robust pellet sufficiently thin to satisfy the criterion $\mu^*d < 1$. If the attenuation is too high, or the amount of available material too small, the

specimen can be distributed uniformly between two plastic films or beryllium disks glued to a rigid "window-frame," whichever is less objectionable in terms of extraneous scattering. To avoid problems, attenuation losses should be acceptable for 100 μm wafers of beryllium in the 0.7 to 1.2 Å region. For these transmission measurements, it is important to set a rough measurement of the packing density in order to make absorption corrections in the subsequent data analysis. Occasionally, specimens may be dusted on sticky Kapton tape—this method requires less than a milligran of material, but requires the use of a position sensitive detector to get reasonable intensities.

3. *Capillary Specimens* Capillary specimens provide the geometry of choice for the most reliable intensity measurements, as they appear to eliminate or greatly diminish preferred orientation effects. Rotation or rocking through 10–20° also seems to result in proper powder averaging even for 30–50 μm crystallites. Capillaries are also very useful for reactive, toxic, and radioactive materials, and need only a few milligrams of sample. The disadvantages are (1) significantly reduced intensities, often an order of magnitude less than for flat-plate specimens and hardly feasible for highly absorbing materials with $\mu > 100$ cm^{-1}, say, except with a PSD (then it is possible to work with 0.1 to 0.2 mm capillaries); (2) unwanted diffuse scattering from the capillary. As above, it is important to determine the packing density from the weight and dimensions in order to apply absorption corrections if necessary (for $\mu^* R < 1$, the correction may be modeled as a negative temperature factor.)

REFERENCES

[1] Bassett, W. A., Takahashi, T., and Stook, P. (1967), "X-ray diffraction and optical observations on crystalline solids up to 300 kilobars," *Rev. Sci. Instrum.*, **38**, 37–42.

[2] Weir, C. E., Block, S., and Piermarini, G. J. (1965), "Single crystal X-ray diffraction at high pressures," *J. Res. Natl. Bur. Stand.* (U.S.) **C69**, 275–281.

[3] Merrill, L. and Bassett, W. A. (1975), "The crystal structure of CaCO$_3$ (II), a high-pressure metastable phase of calcium carbonate," *Acta Cryst.*, **B31**, 343–349.

[4] Hazen, R. M. and Finger, L. W. (1982), *Comparative Crystal Chemistry*, Wiley: London, 20–38, 231 pp.

[5] Xu, J. A., Mao, H. K., and Bell, P. M. (1986), "High-pressure ruby and diamond fluorescence: observations at 0.21 to 0.55 terapascal," *Science*, **232**, 1404–1406.

[6] Ruoff, A. L., Xia, H., Luo, H., and Vohra, Y. K. (1990), "Miniaturization techniques for obtaining static pressures comparable to the center of the earth: X-ray diffraction at 416 GPa," *Rev. Sci. Instrum.*, **61**, 3830–3833.

[7] Schiferl, D., Fritz, J. N., Katz, A. I., Schaefer, M, Skelton, E. F., Qadri, S. B., Ming, L.-C., and Manghnani, M. H. (1987), "Very high-temperature diamond anvil cell for X-ray diffraction: application to the comparison of the gold and tungsten high-temperature high-pressure internal standards," in *High-Pressure Research in Mineral Physics*, M. H. Manghnani and Y. Syono, eds., Terra Scientific/AGU: Tokyo/Washington, D.C. 75–83.

[8] Ming, L.-C., Manghnani, M. H., and Balogh J. (1987), "Resistive heating in the diamond anvil cell under vacuum conditions," in *High-Pressure Research in Mineral Physics*, M. H. Manghnani and Y. Syono, eds., Terra Scientific/AGU, Tokyo/Washington, D.C., 69–74.

[9] Bassett, W. A., Shen, A. H., and Bucknum, M. (1993), "A new diamond anvil cell for hydrothermal studies to 2.5 GPa and from −190 to 1200°C," *Rev. Sci. Instrum.*, **64**, 2340–2345.

[10] Boehler, R., von Bargen, N., and Chopelas, A. (1990), "Melting, thermal expansion, and phase transitions of iron at high pressures," *J. Geophys. Res.*, **95**, 21731–21736.

[11] Bassett, W. A. and Weathers, M. S. (1987), "Temperature measurements in a laser-heated diamond anvil cell in high-pressure research," in *Mineral Physics*, M. H. Manghnani and Y. Syono, eds., Terra Scientific: Tokyo/Washington, D.C., 129–133.

[12] Kinsland, G. K. and Bassett, W. A. (1977), "Strength of MgO and NaCl polycrystals to confining pressures of 250 kbar at 25°C," *J. Appl. Phys.*, **48**, 978–985.

[13] Mao, H. K. and Bell, P. M. (1980), "Design and operation of a diamond-window, high-pressure cell for the study of single crystal samples loaded cryogenically," *Carnegie Institution of Washington Year Book*, **79**, 409–411.

[14] Block, S., Weir, C. E., and Piermarini, G. J. (1965), "High-pressure single-crystal studies of ice VI," *Science*, **148**, 947–948.

[15] Mao, H. K., Jephcoat, A. B., Hemley, R. J., Finger, L. S., Zha, C. S., Hazen, R. M., and Cox, D. E. (1988), "Synchrotron X-ray diffraction measurements of single-crystal hydrogen to 26.5 gigapascals," *Science*, **239**, 1131–1134.

[16] Shimomura, O., Utsumi, W., Taniguchi, T., Kikegawa, T., and Nagashima, T. (1992), "A new high pressure and high temperature apparatus with the sintered diamond anvils fo synchrotron radiation use," in *High Pressure Research: Applications to Earth and Planetary Sciences*, Y. Syono and M. H. Manghnani, eds., Terra Scientific: Tokyo/ Washington, D.C., 3–11.

[17] Zhao, Y., Parise, J. B., Wang, Y., Kusaba, K., Vaughan, M. T., Weidner, D. J., Kikegawa, T., Chen, J., and Shimomura, O. (1994), "High pressure crystal chemistry of neighborite (NaMgF$_3$): an angle-dispersive diffraction study using monochromatic synchrotron X-radiation," *Amer. Mineral.*, **79**, 615–621.

[18] Vaughan, M. T. (1993), "In-situ X-ray diffraction using synchrotron radiation at high P and T in a multi-anvil device," in *Short Course Handbook on Experiments at High Pressure and Applications to the Earth's Mantle*, R. W. Ruth, ed., Min. Assoc. of Canada, Edmonton, Alberta, 95–130.

[19] Weidner, D., Vaughan, M. T., Ko, J., Wang, Y., Leinenweber, K., Liu, X., Yeganeh-Haeri, A., Pacalo, R. E., and Zhao, Y. (1992), "Large-volume high pressure research using the wiggler port at NSLS," *High Press. Res.*, **8**, 617–623.

[20] Utsumi, W., Yagi, T., Leinenweber, K. Shimomura, O., and Kikegawa, T. (1992), "High pressure and temperature generation using sintered diamond anvils," in *High-Pressure Research: Applications to Earth and Planetary Sciences*, Y. Syono and M. H. Manghnani, eds., Terra Scientific: Tokyo/Washington, D.C., 37–42.

[21] Decker, D. L. (1971), "High pressure equation of state of NaCl, KCl, and CsCl," *J. Appl. Phys.*, **42**, 3239–3244.

[22] Jamieson, J. C., Fritz, J. N., and Manghnani, M. H. (1982), "Pressure measurement at high temperature in X-ray diffraction studies: gold as a primary standard," in *High*

Pressure Research in Geophysics, Adv. Earth Planet. Sci., **12**, S. Akimoto and M. H. Manghnani, eds., Center for Academic Publications, Tokyo, 27–48.

[23] Heinz, D. L. and Jeanloz, R. (1984), "The equation of state of gold calibration standard," *J. Appl. Phys.*, **55**, 885–893.

[24] Mao, H. K., Bell, P. M., Shaner, J. W., and Steinberg, D. J. (1978), "Specific volume measurements of Cu, Mo, Pd, and Ag and calibration of the ruby R1 fluorescence pressure guage from 0.06 to 1 Mbar," *J. Appl. Phys.*, **49**, 3276–3283.

[25] Bassett, W. A. and Brown, G. E. (1990), "Synchrotron radiation: applications in the earth sciences," in *Annual Reviews of Earth and Planetary Sciences*, G. W. Weatherill, ed., Palo Alto, CA, 18, 387–447.

[26] Barrett, C. S. and Massalski, T. B. (1966), *Structure of Metals*, 3rd ed., McGraw-Hill: New York.

[27] *Residual Stress Measurement by X-ray Diffraction* (1971), Soc. Automotive Eng. Handbook Supplement, SAE J784a, 2nd ed.

[28] Klug, H. P. and Alexander, L. E. (1974), *X-ray Diffraction Procedures*, 2nd ed., Wiley: London.

[29] Cullity, B. D. (1978), *Elements of X-ray Diffraction*, 2nd ed., Addison-Wesley: New York.

[30] Noyan, I. C. and Cohen, J. B. (1987), *Residual Stress: Measurement by Diffraction and Interpretation*, Springer: New York.

[31] Noyan, I. C. and Cohen, J. B. (1991), "Residual stresses in materials," *American Scientist*, **79**, 142–153.

[32] King, D. E. and Czanderna, A. W. (1993), "ISS, XPS and Grazing Incidence FTIR of organized molecular assemblies," paper presented at the 41st Ann. Symp. of the AVS, Orlando FL. November 18, 555 pp.

[33] Cohen, J. B., Dolle, H., and James, M. R. (1980), "Stress analysis from powder diffraction patterns," in *Accuracy in Powder Diffraction*, NBS Special Publication 567, p. 453.

[34] James, M. R. (1977), "An examination of experimental techniques in X-ray stress analysis," Ph.D. Thesis, Northwestern Univ.

[35] James, M. R. and Cohen, J. B. (1979), "Geometric problems with a PSD employed on a diffractometer, including its use in the measurement of stress," *J. Appl. Cryst.*, **12**, 339–345.

[36] Schuster, M. and Goebel, H. (1995), "Applications of graded multilayer optics in X-ray diffraction," *Adv. X-ray Anal.*, **39**, In press.

[37] Eatough, M. O. and Goehner, R. (1994), "The effects of using long Soller slits as parallel beam optics for GIXRD, on diffraction data," *Adv. X-ray Anal.*, **37**, 167–173.

[38] James, R. W. (1963), "The dynamical theory of X-ray diffraction," *Solid State Phys.*, **15**, 53.

[39] Schwartz, L. H. and Cohen, J. B. (1987), *Diffraction from Materials*, 2nd ed., Springer: London, 112 pp.

[40] Hart, M., Parrish, W., Bellotto, M., and Lim, G. S. (1988), "The refractive index correction in powder diffraction," *Acta Cryst.*, **A44**, 193–197.

[41] Cohen, E. R. and Taylor, B. N. (1990), "The 1986 recommended values of the fundamental physical constants," *Physics Today*, Aug., p. BG9.

[42] Shute, C. J. and Cohen, J. B. (1991), "Strain gradients in Al-2% Cu thin films," *J. Appl. Phys.*, **70**(4), 2104–2110.

[43] Cheary, R. W. and Cline, J. P. (1995), "An analysis of the effect of different instrumental conditions on the shapes of X-ray powder line profiles," *Adv. X-ray Anal.*, **38**, 75–82.

[44] Wolfstieg, U. (1976) "The Psi goniometer," *Harterei Techn. Mitteil.*, **31**, 19–22.

[45] Parratt, L. G. (1954), "Surface studies of solids by total reflection of X-rays," *Phys. Rev.*, **95**, 359–369.

[46] The Powder Diffraction File (1996), International Centre for Diffraction Data (Newtown Square, PA).

[47] McKie, D. and McKie, C. (1974), *Crystalline Solids*, Wiley: New York.

[48] Amoros, J. L., Buerger, M. J., and Canut de Amoros, M. (1975), *The Laue Method*, Academic Press: New York, 375 pp.

[49] Stout, G. H. and Jensen, L. H. (1989), *X-ray Structure Determination*, 2nd ed., Wiley: New York, 453 pp.

[50] *International Tables of X-ray Crystallography*, Vol. I (1969). Kynoch Press: Birmingham, England, 558 pp.

[51] Schiferl, D. and Roof, R. B. (1979), "X-ray diffraction on radioactive materials," *Adv. X-ray Anal.*, **22**, 31–42.

[52] *Plutonium Handbook*, Wick (1980).

[53] Finger, L. W. (1989), "Synchrotron powder diffraction," in *Reviews in Mineralogy*, D. E. Bish and J. E. Post, eds., **20**, 309–331.

[54] Cox, D. E. (1991) "Powder Diffraction" in *Handbook on Synchrotron Radiation*, 155–200, Vol. 3, G. S. Brown and D. E. Moncton, eds., Elsevier: Amsterdam.

[55] Cox, D. E. (1992), "Powder diffraction and structure determination in synchrotron radiation crystallography," P. Coppens, ed., Academic Press: New York, p. 188.

[56] Suortti, P., Hastings, J. B., and Cox, D. E., (1985), "Powder diffraction with synchrotron radiation. I. Absolute Measurements," *Acta Cryst.*, **A41**, 413–416.

[57] (see Finger et al. JAC submitted).

[58] van Berkum, J. G. M., Sprong, G. J. M., de Keiyser, Th. H., Delhez, R., and Sonneveld, E. J. (1995), "The optimum standard specimen for X-ray diffraction line profile analysis," *Powd. Diff.*, **10**, 129–139.

CHAPTER 7

SPECIMEN PREPARATION FOR CAMERA METHODS

CONTRIBUTING AUTHORS: ERICKSSON, JENKINS, RASMUSSEN, RENDLE, SMITH

7.1 THE DEBYE–SCHERRER CAMERA

7.1.1 Geometry of the Debye–Scherrer Camera

The Debye–Scherrer camera [1] is one of the earliest powder diffraction devices to be perfected into its modern form. The later addition of the Gandolfi attachment [2] has not changed the Debye concept in any way; it has only modified the manner in which the specimen is presented to the X-ray beam. Figure 7.1 shows the essential geometric features of the camera. The X-ray beam from a spot source is collimated to 0.3 to 0.5 mm diameter. The specimen is a fine powder placed in this beam. The undeviated beam is collected in a beam stop designed to intercept the entire bundle without causing any scatter to other parts of the camera. The diffracted beams are detected by a cylindrical film placed with its axis perpendicular to the incident X-ray beam at the position of the specimen. The film is held in a cylindrical cassette whose diameter may be 57.29 mm, 114.59 mm, or some other value that allows easy conversion of the film measurements from millimeters to degrees. The whole assembly is a light-tight unit that may be transported easily between the workbench, the darkroom, and the generator.

In older Debye–Scherrer designs, the specimen was typically a rod-shaped bundle of the powder with a diameter of 0.2 mm to 0.5 mm with the axis of the rod perpendicular to the X-ray beam. The specimen is usually rotated on the camera axis to assure multiple orientations of the grains in the powder during the exposure. Techniques to prepare these specimens will be described in section 7.2.

7.1.2 Preparing Mounts for the Debye–Scherrer Camera

The goal of the Debye–Scherrer mount is a thin, cylindrical rod-shaped specimen which may be mounted on the camera rotation mechanism. The mechanism is usually a pulley driven spindle with a cup on the inside of the camera into which is placed

A Practical Guide for the Preparation of Specimens for X-Ray Fluorescence and X-Ray Diffraction Analysis, Edited by Victor E. Buhrke, Ron Jenkins, and Deane K. Smith. ISBN 0-471-19458-1 © 1998 John Wiley & Sons, Inc.

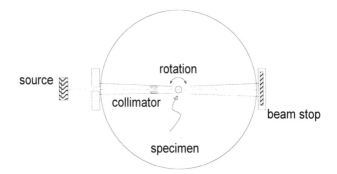

Figure 7.1. Geometry of the Debye–Scherrer camera.

a metal pin holding the specimen rod. Because of the optics of the Debye–Scherrer camera, the recorded line width on the film is affected by the size of the specimen. It is necessary to keep the specimen diameter as small as possible while getting as many grains as possible into the X-ray beam. Experience suggests that the diameter be 0.2 to 0.3 mm. The specimen should be around 8 mm long with powder covering that part of the rod which is in the X-ray beam, usually at least 2 mm of the length. The support may either be an X-ray transparent tube such as glass or quartz containing the powder or an X-ray amorphous rigid rod onto which powder has been cemented. Some methods for making appropriate specimens follow.

7.1.3 Self-supporting Specimens

Some materials may be prepared as stand-alone rod-shaped agglomerates suitable for mounting directly into the camera. Metal wires are ideal examples. Self-binding powders may be extruded as rods through hypodermic needles or small orifices. Some powders will wet with a cement and roll into a rod shape between microscope slides or one's fingers. In the most common cameras, the specimen is held in a metal pin with a cup in one end. If the cup is filled with cement or Tacki wax, one end of the specimen is placed in the wax to support it in the camera.

7.1.4 Rod-Mounted Specimens

The most common rod-mounted technique uses glass as the supporting material. Either soft glass or Pyrex may be used. Using the appropriate burner for the type of glass, thin rods may be prepared by softening the glass with the burner and stretching the heated segment to thin it. For example, start with a 7 mm outer diameter glass tube. Have a second person ready to help you stretch the heated tube. Heat the center part of the rod to its softening point. Have your assistant take one end of the tube and pull until the glass gets firm. Continue the tension to prevent sagging until the glass hardens. When the thin region is cool enough to touch, break it into 2 to 4 in. segments. These segments are tubes with a small capillary hole down the center. Keep only the thinnest of these pieces stored in a test tube for future use. The best rods will be just thick enough so they do not sag under their own weight when held at one end yet thin enough so they may be bent into a loop of one inch radius without breaking.

The following items are needed: a small mortar (B4C) and pestle; metal spatula; pin vise with needle; cement—collodian in ethyl acetate; cow gum or rubber cement in benzene, gum arabic in water; Tacki wax or Plasticine; and a small glass plate (petri dish).

To mount powder on the rods, a polished surface is needed, about 1 to 2 in. in diameter and preferably 1 to 2 in. high. A small diameter petri dish will serve the purpose. Grind the specimen to a fine powder in a mortar and place a small pile of the powder on the polished surface (near the right edge if you are right-handed). Place a drop of cement on the surface near the powder and quickly start the next step. Hold the end of the glass rod between your thumb and forefinger so you can roll it without having the rod flip around. Place 2 to 8 mm of the tip of the rod into the cement. Remove the rod and roll the wet end against the glass surface to remove any excess glue. Then, roll the tip into the pile of powder. Repeat the process until you have developed a powder film on the tip of the rod that has a diameter of 0.2 to 0.5 mm. Break the rod so that the tip section is about 7 to 10 mm long and mount the specimen in the pin that goes into the camera.

Some cements remain tacky for longer periods of time than others, which affects the ease of getting the particles to adhere to the rod. Airplane cements, for example, usually skin quickly on the surface, and the particles refuse to adhere. Cow gum diluted and gum arabic in water, on the other hand, remain tacky for some time, allowing many particles to be picked up and minimizing the need for multiple dipping into the cement source. Collodian thinned 10:90 in ethyl acetate makes a good cement. The low viscosity of this mixture allows a thin layer to adhere to the rod each time the rod is dipped, and multiple dippings to build up a good layer of powder on the rod. As the solvent evaporates, the cement will get stiff and pull the prepared mount off the rod if the user does not stop in time. A 50:50 mixture of collodian in ether used to be an excellent cement until ether was shown to be dangerous in a laboratory.

Some alternates to the glass rod are silk or nylon thread, nitrocellulose, and gelatin. The thread will have to be held in both hands when cementing the powder and should be kept under tension until the glue has thoroughly dried. Glass rods may be obtained from fiber-optic cables. The cable should be cut into segments before the fibers are collected. Because the fibers are coated, it is necessary to remove the coating with an acid bath prior to use. A blank rod should be run in the camera to confirm its cleanliness.

When only a small amount of specimen is available, it is necessary to concentrate the powder into a small ball mounted on the tip of the glass rod. When a highly polished boron carbide mortar is not available, grinding a small specimen can be accomplished by placing a glass cover slip over the material on the microscope slide and pressing with a flat plate on the cover slip. If the material is not so hard that it causes the glass to shatter, the fragmentation can be confined to a small volume on the glass. The original grains could also be encapsulated in a collodian film before crushing. This technique prevents the crushed particles from disbursing large distances but may require the removal of the collodian before use. For more detail on this approach, see section 7.2.2.

A needle in a pin vise can be used under the microscope to collect the powder and combine it into a ball using a small amount of cement. This ball is then cemented to the tip of the fiber. Use the needle to be sure the ball is free of the glass, then touch the tip of the support rod to a drop of cement and to the ball. The capillary will suck up a trace of cement and make it easier to get the ball to stick at the tip of the rod. If the rod is already mounted in the camera pin and the pin is held in a pin vise, it will be easier to manipulate the mount and eliminates the need to handle the individual rod independently. If the specimen is a shaving or flake that cannot be ground, the fragment may be mounted directly on the end of the rod support. Be sure the fragment is free on the microscope slide before trying to cement it to the rod.

Small amounts of specimen may also be mounted on a gelatin point. To make the point, cut a gelatin capsule into thin triangles about 8 mm long. Mount the base end into the pin mount that goes into the camera. Holding the pin in a pin vise, wet the tip of the triangle with water to soften it and touch the sticky tip to the powder. Collect as much powder concentrated at the point as possible.

7.1.5 Capillary Mounts for the Debye–Scherrer Camera

Commercial capillaries are available which are useful for specimen mounts. The capillaries are very thin walled and may be made of silica glass ("quartz") or soft glass. Sometimes you may be able to get Lindemann glass (which is supposed to be a boron glass). These capillaries are prepared in the manner described above, starting with 4 mm very thin-walled tubes. The overall length is around 10 cm with about 1 cm of the original tube at one end to act as a funnel and the thin other end sealed into a tiny bead. Capillaries of many diameters from 0.2 to 2.0 mm are available.

Commercial sources are mentioned in the Appendix.

Supplies These supplies include capillary tubes of appropriate diameter; a mortar and pestle; spatula; glass tube 1 cm OD × 1.0 m long; and a small triangular file.

Loading a capillary is not trivial. The best diameter for Debye–Scherrer use is 0.3, although 0.2 and 0.5 mm are useful. The powder must be ground very fine, so no clumps will wedge in the funnel and block the flow of powder to the thin bottom. Using the spatula, a small amount of powder is inserted into the funnel end. (The end result should have about 5 to 8 mm of powder in the capillary.) A small triangular file or similar object is then used to vibrate the tube to get the powder to flow out of the funnel into the lower tube. Rub the file against the top of the funnel, and the powder should disaggregate and move downwards. When sufficient powder is in the capillary, the funnel is cut off using a match or hot wire. Next, get a 10 mm inner diameter glass tube around 1 m long. Place one end on the table surface and drop the capillary down the open end. The capillary should free-fall down the tube and bounce off the table back up the tube. This procedure tamps the powder into the small end of the capillary. If all goes well, three to six drops should be sufficient to densify the powder into the bottom 5 to 8 mm of the capillary. Now use the hot wire or a match to cut the capillary into a final length of no more than 1 cm. Handle the capillary only with your fingers. Tweezers will shatter the capillary like picking up an egg with pliers.

A hot wire can be prepared by using nichrome or platinum wire across terminals connected to a variable power source. Very low voltage (< 10 V) are required to reach red heat which is sufficient to melt the capillary. This device may be used in glove boxes where open flames would not be allowed.

An alternative method for tamping the particles into the tip of the capillary is to use ultrasonic vibrations to keep the powder moving from the funnel to the narrow tip [3]. Place the capillary with the powder into a small, 50 ml Erlenmeyer flask containing water up to the level of the funnel and put this assembly into a sonic cleaner bath. Densification usually occurs in less than five minutes. Further densification can be achieved by using the tube dropping method described above.

7.1.6 Preparing Debye–Scherrer Specimens for Nonambient Measurements

For loading capillaries in a glove box, it is usually convenient to have means to handle the small capillaries with other than the bulky gloves. Test tube racks with test tubes fitted with one-hole stoppers (and a pressure relief hole) [4] or a frame with plastic tubing of appropriate diameter [5] that can be transferred through the box portal assist considerably. If the hole is just the size of the funnel, the capillaries are fitted snugly in the stopper, and the test tube becomes the handle in the box. The capillaries are loaded in the controlled atmosphere and sealed before removal from the controlled environment. If possible, the powder should be induced to the narrow portion below the funnel by tamping or vibration, so the capillaries can be sealed in the box. However, the capillaries may be sealed with a wax plug in the funnel end, removed from the box and then processed. A hot wire may be used to seal the capillaries in the box if an open flame is not compatible with the atmosphere. Once sealed, the capillaries may be removed from the box and treated as above to densify the powder into the narrow tip. Toxic and hazardous materials may be handled in this manner, but exceptional care must be taken not to break the capillary in subsequent processing. Capillaries prepared in this manner are also usable for diffractometry and other diffraction experiments.

7.2 THE GANDOLFI CAMERA

The Gandolfi attachment replaces the single-axis rotation mechanism with a device that rotates about two axes which intersect at the center of the camera. A cluster of grains is supported on a thin fiber held in a flexible pin so that the axes of rotation all intersect at a point within the volume of the specimen which is also in the center of the camera in the X-ray beam. The concept of the Gandolfi attachment is that a smaller number of grains will be necessary to achieve the random orientation of crystallite orientations on which the powder method is based. The motion covers most of the orientations in space, but not all. The device was initially designed to achieve a powder pattern from a single crystal fragment, but because of a shadow region, a minimum of four grains are advisable to get total coverage.

7.2.1 Gandolfi Mounts

Preparing mounts for the Gandolfi camera is different from the Debye–Scherrer camera because the motion requires the specimen to be concentrated into a spherical shape which can be centered in the beam path at the intersection point of the rotation axes. As for the Debye–Scherrer camera, the smaller the diameter of the specimen, the sharper the diffraction lines. Because of the two-axis rotation mechanism in the Gandolfi camera, there are more severe limitations on where the specimen must be mounted on the supporting pin. Consequently, it is usually advisable to prepare the mounting assembly prior to collecting the powder.

Most Gandolfi designs use a flexible pin with a cup in the end to hold a glass rod. Only a few millimeters of length leeway are allowed. Cement the support rod into the cup and cut off the length to meet the dimensional requirements. Tacki wax makes a good cement. Hold the pin assembly in a pin vise for easy manipulation.

Prepare the powder and add a drop of cement. With the tip of a needle, form a small ball of cemented powder. Bring in the pin assembly and cement the ball to the tip of the glass rod. It is best if the ball is on the very end of the rod, but it can be on the side. The ball should be no more than 0.5 mm in diameter and smaller is preferable. If only a few grains are available, use the mount assembly to pick up a drop of cement and then the grains.[1]

7.2.2 Handling Very Small Specimens in the Gandolfi Camera

A very special technique for mounting nanogram quantities has been developed by the McCrone Research Institute [6]. The technique applies to both the Debye–Scherrer method and the Gandolfi method. The goal is to optimize the diffracted X-rays while minimizing the X-rays from all other sources.

First, it is necessary to mount the glass rod into the camera pin and cut the tip to the appropriate length for the camera used. Cement the rod permanently in place. Fisher High Pyseal works well. The pin should then be clamped in a pin vise. A hot-wire microforge should be constructed with a platinum or nichrome wire controlled by a low-voltage power supply and a guide block for the pin vise. Working under a stereo microscope, the wire should be heated to orange heat. The pin vise assembly is then moved to touch the tip of the glass rod to the hot wire. The very tip will melt and adhere to the wire. Withdrawing the pin vise assembly will draw a short, sharp tip

---

[1]Glass fibers work best if they have the tip cut off cleanly as a flat end. This condition may be achieved most easily if the fibers are broken under tension rather than by twisting. Grab the two ends and pull until the fiber separates somewhere. Usually there is a defect that initiates the break, and the tension fracture will be perpendicular to the fiber length.

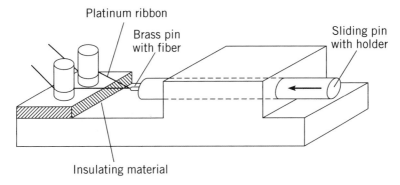

Platinum ribbon

Brass pin
with fiber

Sliding pin
with holder

Insulating material

Figure 7.2. Sharpening the tip of a glass fiber with a microforge.

on the glass rod (Figure 7.2). The thickness and length of this tip depend on the temperature of the wire and the rate of withdrawal of the pin holder. Now cut the sharpened tip to the length required by the camera (between 5 mm and 7 mm for one camera). The very tip of this support should be as small in diameter as to minimize the amount of material and still be rigid enough to support particles. A 3 μm to 5 μm diameter is ideal.

Prepare a solution of collodian 10:90 in ethyl acetate or amyl acetate as a cement. Place a drop of cement on a clean microscope slide. Use the needle to move the grain(s) into the center of the drop and spread out the drop with the needle. After the solvent has evaporated, there should be a very thin film of collodian on the slide with the grain(s) included. Using the needle, cut the particles loose from the glass retaining a small amount of the collodian film. Pick up this composite of particle(s) and film on the sharpened tip of the mount. It should adhere easily. Then, to make the mount permanent, hold the tip over the open mouth of the solvent (ethyl acetate) so that the fumes soften the collodian film and make it attach firmly to the support rod. Particles down to 2 μm may be examined by this method. Thirty micrometer diameter particles are routinely handled by this technique.

The technique of imbedding a particle in a film of collodian is an excellent method for archiving particles and for shipping them to another facility for analysis. For example, particles from air filters can be picked off with a needle and transferred to a collodian drop on a microscope slide. When dry, the location of the particle should be indicated by alignment scratches drawn with the needle in the film. The microscope slide may then be mailed or stored in a box or case especially designed for slides.

For the Gandolfi technique, the particles may be polycrystalline or single crystals. For successful Debye–Scherrer results, the specimen must be polycrystalline. Very small particles may be crushed successfully using two clean microscope slides with the particles in between. Firm pressure will shatter the grain which will usually not disburse very far due to the confinement of the slides. Further confinement may be accomplished by first encapsulating the particle in a film of collodian as mentioned above. The fragments are then transferred to the tip of the mount. Successful Debye–

Scherrer results have been obtained from nanogram quantities before the introduction of the Gandolfi attachment, but the use of the Gandolfi device whenever possible is recommended.

The secret to achieving good photographs is to center perfectly the specimen in the camera and X-ray beam and to minimize all other sources of X-ray scatter. The centering is required to get the diffracted X-rays to produce the sharpest, most intense lines. If the particle is not centered, it behaves like a larger particle with a low concentration and produces broader, weaker lines. With a little experience and practice on noncritical specimens, centering can be accomplished in a short time using the fingers to move the flexible joint while looking in the optical alignment device even when using inverting optics. However, some designs for micromanipulators have been developed to assist the less dexterous operators.

Several modifications of the camera assist in reducing the background. Evacuating the atmosphere is the most effective step. Collimator modification may also help. Collimators should come close to the specimen to minimize the atmosphere path, and they should be tapered to create the minimum shadow zone at low and high diffraction angles. A smaller diameter film cassette insert may also be used to shorten the exposure time; however, a 57.28 mm diameter camera produces good results when all the other factors are optimized.

7.3 THE GUINIER CAMERA

The pretreatment of specimens for Guinier techniques is no different from that of other powder diffraction methods. Normally, the specimen is first examined with a binocular microscope and, if it is transparent, also with a polarizing microscope. This preliminary examination can be most helpful for estimating the homogeneity of the specimen and sometimes for selecting specific parts of a heterogeneous specimen, e.g., fibers when searching for asbestos.

The preparation of a specimen depends on the physical and chemical properties of the sample and on the amount available. When several grams of a stable, inorganic material or a mineral are available, ball milling may be appropriate. In commercial laboratory centrifugal mill systems, the balls in the bowl can reach accelerations up to 20 times that of the gravitation, dependent of the size of the bowl. A useful system is the Planetary Micro Mill Pulverisette 7.[2] Bowls are marketed in sizes of 12, 25 and 45 ml volume, and bowls and balls are offered in the following materials: agate, sintered corundum, tungsten carbide, stainless steel, and zirconium oxide. Small samples may be ground by hand in small mortars of agate or, for very hard samples, of boron carbide. The choice between wet or dry grinding is made on the basis of the nature of the sample. Wet grinding with alcohol is often useful for getting homogeneous specimens and the alcohol is easily removed by evaporation. When enough sample is available ball milling or grinding may be followed by sieving, but

---

[2]Marketed by Fritsch, Industriestrasse 8, D-6850 Idar-Oberstein, Germany.

it should be realized that sieving might result in partial separation of multicomponent specimens.

Samples of metals and intermetallic compounds are often not suitable for grinding. In such cases powders can be prepared by filing the sample with a very fine-grained diamond file.

Organic crystals are often soft and easily deformed by grinding as are malleable metals and fibrous and platy compounds. Cooling mortar, pestle, and sample to solid carbon dioxide temperature or lower may increase the brittleness of the samples in some cases.

7.3.1   Mounting the Specimen in the Guinier Camera

As we are dealing with a transmission method, specimens normally have to be supported by something, most often a foil, which, ideally, should be both highly transparent to the radiation employed and completely amorphous. The glue that fixes the powder to the foil should have the same characteristics. Another condition that must be fulfilled is that foil and glue do not react chemically with each other and neither foil nor glue may react chemically with the specimen. As always in experimental work, perfect conditions are hard to find, and supporting foils and glues must be chosen in accordance with the chemical nature of the specimens. If no suitable foil and/or glue can be found, the powder may be loaded into a thin-walled glass capillary as described below.

In general, the foil is mounted on a metal support which ensures that the foil is kept flat and that the specimen can be mounted in a camera or on a diffractometer in a well-defined manner. In some cameras several specimens can be exposed simultaneously. A diagram of three exposures on one film is shown in Figure 7.3.

7.3.2   Foils and Glues

Mylar foils of thickness 4–6 μm are readily available.[3] A foil 6 μm thick transmits 99.9% of Cu $K\alpha$ radiation. When covered by a layer of "carpenter's glue" it transmits 98.9% Cu $K\alpha$ radiation. Many adhesive tapes are also applicable as supports, e.g., Scotch Magic 3M which transmits 95% of Cu $K\alpha$ radiation. It might be tempting

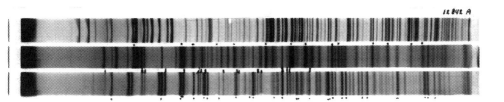

Figure 7.3. Three exposures on one film.

[3] E.g., From Spectrofilm, Somar Laboratories, P. O. Box 5234, Grand Central Station, NY 10017.

to use a metal foil as support, e.g., to attenuate fluorescence from the specimen, but very strong lines from the metal foil may disturb the diagram. Carpenter's glue,[4] dissolved in water in the ratio 1:10 by volume is excellent, together with Mylar foils for substances that do not react with water. Mineral powders are normally well dispersed in the glue and form a thin, fairly evenly distributed layer of powder on the supporting foil after the glue has dried. For substances which react with water, shellac dissolved in acetone may be useful when Mylar is used as a support. The necessary amount of specimen can roughly be estimated as that which can be held on a 4–5 mm broad nickel spatula on the outer 1–2 mm of the tip. For quartz this volume corresponds to 1–2.5 mg.

The glue on the 3M Magic Mending tape is, however, the most universally applicable one that we have encountered so far. The main difficulty of using adhesive tape is that of distributing the powder uniformly on the tape. Another slight drawback of adhesive tape as compared to Mylar foil is the fact that the tape gives a slightly higher background than Mylar and that the background is a little more structured.

7.3.3   Moving the Specimen in the Guinier Camera

The flat specimen used with the Guinier method is normally very thin. In order to get a good, statistical distribution of diffracting crystallites, it is necessary to make the area of the specimen bigger than the rectangular cross-section of the incident beam and to move the specimen in such a way that all crystallites in the specimen get into the beam path for some fraction of the time of exposure. In some cameras and diffractometers, the specimen undergoes a translational movement during the experiment and in other devices the specimen is spun. The spinning movement is preferable, as it can to some extent counteract some preferred orientation that could be incurred while mounting the powder on the foil. Preferred orientation caused by particle shape cannot, however, be eliminated by spinning the specimen.

7.3.4   Capillary Techniques with the Guinier Camera

Specimens that react with the ambient atmosphere or with available foils or glues may be enclosed in thin-walled capillaries (Lindemann tubes[5]) made from a borosilicate glass of low absorption. Preferred orientation effects may be somewhat reduced by mounting the specimen in a capillary whether it be air sensitive or not. Spinning the capillary around its axis whirls the powder around, provided it is not packed too densely, and in this way more orientations of the crystallites come into diffracting position than is the case for flat specimens in reflection or transmission geometry. The capillary technique employed in a Guinier instrument is practically equivalent with the classical Debye–Scherrer technique, the main difference being the use of a convergent beam and a pure $K\alpha_1$ radiation. The use of the capillary technique

---

[4]E.g., Cascol from Casco Nobel AB, S10061, Stockholm, Sweden.
[5]Supplied by Fa. Hilgenberg, D-3509, Germany.

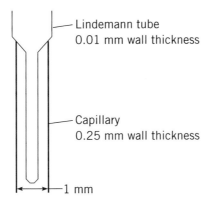

Figure 7.4. A Lindemann capillary with protective capillary for loading.

presupposes that the powder is dry and finely divided. Otherwise it may be difficult to load the capillary uniformly with specimen. Lindemann tubes of internal diameters of 0.3–0.4 mm are fairly easy to fill, whereas thinner capillaries can be difficult, if not impossible, to load. The following procedure has proved useful for loading a capillary: The Lindemann tube, which is very fragile, is placed in another capillary of 1 mm inner diameter and of 0.25 mm wall thickness as shown in Figure 7.4. Some specimen is placed in the broad, upper end of the Lindemann tube and the outer capillary is set in vibrations by gentle strokes with a coarse file or a similar device; the powder then descends rather easily down the Lindemann tube, which can be sealed, e.g., by the flame of a small candle, once a suitable degree of fill has been attained.

7.3.5 Nonambient Temperatures—High Temperatures

The problems of specimen mounting become increasingly difficult with increasing temperature. The laws of high temperature chemistry state that everything reacts with everything at high temperatures and the higher the temperature, the more rapid the reactions. It is not difficult to build furnaces for diffraction at high temperatures; the difficulty is normally that of finding a suitable support for the specimen. Quartz capillaries can be used up to $1000°$–$1100°C$ with specimens that do not react with SiO_2 at these temperatures. Capillaries made of β-filter nickel foils may be usable if they are unreactive and protected from oxidation by a surrounding atmosphere. Another possibility is that of mounting the powder on a platinum net by pressing the powder into the holes in the net. Preferred orientation may be an additional problem with this latter technique. Again, chemical reactions may be a limiting factor for this kind of mounting. Most metals will alloy or otherwise react with platinum at elevated temperatures as will many alkaline substances, e.g., $BaCO_3$. The Bragg–Brentano geometry is better suited than the Guinier geometry for very high temperature work. Furnaces for Guinier techniques are commercially available from the companies Stoe and Huber.

7.3.6   Nonambient Temperatures—Low Temperatures

The problems associated with diffraction at low temperature is not that of chemical reactions but of heat transfer. If the specimen is mounted in a capillary, a stream of dry, cold air or of nitrogen can be blown over the capillary and temperatures down to 90K can be reached. For lower temperatures, cooling by helium gas may be used. Temperatures down to 10K can be reached by closed cycle refrigerators. Guinier instruments based on those kinds of cryostats have been described by various authors [7, 8]. Still lower temperatures demand the use of liquid helium.

REFERENCES

[1] Debye, P. and Scherrer, P. (1916), "Interferenzen an regellos orientieren teilchen im Röntgenlicht," *Nachr. Kgl. Gies. Gottingen Math.-Physik*, **K1**, 1–15.

[2] Gandolfi, G. (1967), "Discussion upon methods to obtain powder patterns from a single crystal," *Mineral. Petrogr. Acta*, **13**, 67–74.

[3] Foit, F. F., Jr. (1982), "A technique for loading glass capiliaries used in X-ray powder diffraction," *J. Appl. Cryst.*, **15**, 357.

[4] Schiferl, D. and Roof, R. B. (1979), "X-ray diffraction on radioactive materials," *Adv. X-ray Anal.*, **22**, 31–42.

[5] Lange, B. A. and Haendler, H. M. (1972) "A capilliary support apparatus for use in glove bags and dry boxes," *J. Appl. Cryst.*, **5**, 310.

[6] McCrone, W. C. and Delly, J. G., *The Particle Atlas*, vol 1., 2nd ed., pp 119–129. Ann Arbor, MI.

[7] Ihringer, J. (1982), "An automated low temperature Guinier X-ray diffractometer and camera," *J. Appl. Cryst.*, **15**, 1–4.

[8] Rasmussen, S. E. and Larsen, F. K. (1986), "Conversion of a high temperature Guinier camera to a low temperature device," *J. Appl. Cryst.*, **19**, 413–414.

CHAPTER 8

SPECIMEN PREPARATION EQUIPMENT

CONTRIBUTING AUTHORS: BUHRKE, HAMILTON, JENKINS, SMITH, TAGGART

8.1 INTRODUCTION

The earlier chapters of this book have been devoted to the procedures involved in the preparations of specimens for various types of X-ray analysis. This chapter will emphasize equipment, including potential sources and maintenance procedures. The vendors mentioned in this book are based on suggestions of the many authors who have contributed to the content and are not meant to imply that the product is the only option available. Because manufacturers of equipment often change products, change names and locations, or go out of business, any listing of sources is only current at the time of preparation. Regardless, the authors of this book feel that the information will still be useful for several years after publication.

This chapter is arranged to start with the equipment for processing the coarse sample and to follow the stream to the final small particle aliquot used in the measurements. Some sections are divided into specific equipment for diffraction or fluorescence. Only vendor names (no address) are listed within the main sections. Sometimes a specific model number of a particular piece of equipment is indicated. Additional sections are devoted to special topics, including the care of platinum ware. Section 9.5 in the next chapter contains a separate list with names and addresses and sometimes a phone number of all the vendors cited in earlier sections.

This chapter does not go into any detail about how to use the equipment. The reader is advised to consult manufacturers' instructions and previous sections in previous chapters for these details.

8.2 GRINDING AND PULVERIZING EQUIPMENT

Disaggregating a sample into a form that allows a proper aliquot to be obtained for analysis is the first step in most specimen preparation procedures. Necessary equipment may include everything from massive crushers and pulverizers to micro-

A Practical Guide for the Preparation of Specimens for X-Ray Fluorescence and X-Ray Diffraction Analysis, Edited by Victor E. Buhrke, Ron Jenkins, and Deane K. Smith. ISBN 0-471-19458-1 © 1998 John Wiley & Sons, Inc.

mortars, depending on the types of samples encountered. The initial sample might be 200 pounds of large heterogeneous blocks or a few milligrams of fine powder. Processing the large bulk sample poses the most problems. The challenge is to pulverize the sample to produce fine particles without contamination or fractionation, homogenize the product, and split it into representative aliquots for study. The operation may involve crushers, splitters, mills, and mortars. Each of these groups will be considered.

In the comminution of samples, one must remember there are two mechanisms that are involved, compression and shear. For XRF specimen preparation, it usually does not matter which mechanism is involved. However, for XRD specimens, the shear motion has more potential for damaging the crystallites which comprise the specimen, affecting the diffraction measurements. Especially in the later stages of preparation, one should use only equipment which does not involve shear.

The goal of the crushing step is to comminute the bulk sample to one with particle sizes no larger than a few millimeters. Pulverizers reduce the particle size and homogenize the sample with a binder.

8.2.1 Large Crushing and Pulverizing Equipment

The XRF analysis of a powder specimen requires that particles be of a uniform size range and with small enough grain size to yield chemical homogeneity without absorption masking. The correct size range depends upon the analyte and matrix elements. XRD analysis requires uniform particles packed in a random orientation. The particles should be no larger than 75 μm and preferably smaller than 10 μm without introducing a fraction of submicrometer-sized particles and without inducing lattice damage to the particles. The crushing, pulverizing, and milling equipment listed here may be used to achieve these goals. The mortar and pestle is not a good alternative (see comments in the mortar and pestle section) and should be used for XRF only when there is a very limited amount of sample. When employing a pulverizer, hydraulic press, surface grinder, and a milling machine, a user should follow the manufacturer's recommendations for the best procedure to obtain consistent and quality results.

Pulverizer, V-Belt, #24253X3UA (242-53x3) from Bico Braun.

Jaw Crusher, Chipmunk #2413400VD, VD V-Belt from Bico Braun.

Jaw Crusher, Model #LC-36 from Bico Braun.

Jaw Crusher, Model #LC-36 from Gilson.

Mini jaw crusher for small samples from Sepor.

Fritch pulverisette jaw crusher, Model #1 from Sepor.

Additives are sometimes useful for proper grinding of powders. Some sources of additives are listed below.

Grinding aids of various types from Spex Certi Prep Industries, Somar Laboratories, Chemplex Industries, and Buehler.

Wax tablets, Copolywax #91, #F-Code 289 from Cargille Tab-Pro.

Mix tablets, #220 from Somar Laboratories.

Mix tablets, #625 from Chemplex Industries.

Metallurgical Binder/Grinding Aid, Copolywax R91 Tablets and Powder from Cargille Tab-Pro.

Once the sample has been crushed, it is often necessary to separate out the too coarse fraction for further treatment. Sieves are used for this step. One must be careful in any quantitative analysis procedure, XRF or XRD, not to physically separate the final specimen by properties. Some sources for sieves are listed below.

Sieve, stainless cloth, No. 100, #V8SF#100 from Gilson.

Sieve, U.S.A. Standard, #100, Cat. CS-8100 from Soiltest.

Sieves of various types from Spex Certi Prep Industries, Chemplex Inc., Somar Laboratories, Gilson Company, and Soiltest.

In XRF analysis, special specimen cups are necessary. Some sources are listed below.

Specimen cups of various types from Spex Certi Prep Industries, Somar Laboratories, and Chemplex Industries.

Specimen cap, pre-flared, 31 mm, Cat. #3619A from Spex Industries.

Aluminum dish, disposable, Model #8-732 from Fisher Scientific.

Dewar flask, 5 liter, Cat.#5028-M25 and Dewar caddie from Thomas Scientific Instruments.

Although there are many ovens on the market, the following have been used in several X-ray laboratories.

Oven, forced draft type, Model L-24 from Soiltest.

Oven, Blue-M drying, #MO-1450A from VWR Scientific.

Dust benches for protection from harmful dusts are available from many sources.

Hamilton Industries

Thomas Scientific Instruments

Herzog

Buehler

Rocklabs

Angstrom

8.2.2 Splitters

Once the sample is obtained in a pulverized state, the next step is to reduce the size of the supplied material to a quantity suitable for laboratory analysis. The problems of sampling were discussed in section 1.5. Riffle dividers or splitters are used to split larger samples into smaller aliquots which are representative of the bulk sample. Some of the suitable devices are listed below.

Riffler, spinning type, Accu-Max, Model #SP-245 from Gilson.

Splitter, Stainless Steel, Model #CL-286 from Soiltest.

Universal sample splitters from Sepor.

8.2.3 Mills

There are two basic types of mills, those that use rods and those that use balls. Mills come in all sizes from those that can handle continuous flow of crushed fragments in a slurry at rates of tons per day to those that process a few grams of sample for the laboratory experiment. This section will only consider the latter category.

The goal of milling the laboratory sample is to reduce the particle size to below 10 μm. Various mechanical grinders are available for automatic grinding. There is the old motorized mortar and pestle which applies more shear motion than compression. Modern grinders include the McCrone micronizing mill, the Brinkman "Retsh" grinder, the Pitchford grinder, and the Spex shatter box. These devices employ methods that emphasize crushing. Grinding may be done dry or in a liquid. Using a liquid is usually more successful because the finest particles are flushed off the coarser ones that still need to be crushed. Liquid nitrogen may be used also to help minimize the particle damage.

Mixer Mill #800, swing mill type from Spex Certi Prep Industries.

Puck-and-ring grinder, Model #HSM-F from Herzog.

Puck-and-ring grinder, Shatter box, Model #8500/10 from Spex Certi Prep Industries.

Freezer/Mill, Cat.#6700 from Spex Certi Prep Industries.

Micronizing Mill, Cat. #232 from McCrone Accessories & Components.

Brinkman Mill, from Brinkman Instruments.

Pitchford grinder, from Cianflone Scientific.

8.2.4 Mortars

The final stage of specimen preparation may involve such a small amount that the crushing must be done in a mortar. Alternatively, the final product of milling may need to be mixed with a standard and blended. The mortar and pestle are common equipment for this step for XRD samples but not for XRF samples. The following discussion applies only to XRD.

Mortars and pestles may be manufactured out of many different materials. Steel is used for the larger types whose goal is to crush fragments up to 2 cm into finer particles. A useful type is the Plattner mortar (from any instrument supply house), which consists of a base plate with a 1.25 in. diameter cavity, a sleeve with a 1 in. inside diameter to contain the fragments, and a 1 inch rod which fits into the sleeve. A heavy hammer is used to apply the force to the rod. Repeated crushings and sievings can produce a product which passes a 200 mesh screen for brittle samples. This technique does minimal damage to the actual crystallites.

There is a problem with grinding samples that contain particles of markedly different hardness in a mortar, which affects samples for both XRF and XRD. Very soft particles will often collect on coarser hard particles and lubricate the grinding action so that the hard particle is not reached by the pestle. The use of a mortar and pestle for this situation will probably not produce a specimen with particles that are homogeneous in composition and size. Samples with clay minerals and many commercial cements are examples of materials which cause problems in grinding. Usually the puck-and-ring crusher is more efficient in achieving particle size homogeneity. The reader is advised to consult the appropriate chapters for further discussion of this problem.

Mortars and pestles may also be made from agate, alumina, mullite (porcelain), and boron carbide. Cavity dimensions range from several inches to less than one-fourth inch in diameter. Grinding in a mortar involves mostly shear motion and does considerable damage to the crystallites of soft materials. Also the action of the pestle against the mortar surface results in abrasion of the surfaces and contamination of the sample. Use of the harder mortar types, boron carbide, and highly polished cavities and pestle faces helps minimize this contamination. Unfortunately, the polish degrades with use, and all mortars must be replaced after hard use. Most mortars are available through chemical supply houses. The boron carbide mortars are available from Spex Industries.

8.3 SPECIAL EQUIPMENT

Several techniques mentioned in the preparation sections require special equipment. Some of these items are considered in this section.

8.3.1 Presses

The powder specimen for XRF analysis has to fit into the specimen chamber of the X-ray spectrometer. It is desirable, in many cases, to have the specimen the shape of a round, flat disk. The vendors listed can supply presses and dies for this purpose. Automatic presses are available to apply the pressure and release it in a reproducible motion.

Chemplex Industries
Spex Certi Prep Industries
Fred S. Carver
Graseby Specac

Herzog

Angstrom

Foundry Testing Equipment

8.3.2 Fluxers

Samples which do not go into solution easily and tend to remain heterogeneous are often treated by the technique of flux fusion. There are only a few suppliers of automated and manual fusion equipment.

Spex Claisse Fluxer from Spex Certi Prep Industries

The Spectroflux from Alfa-Aesar

Anhydrous lithium bromide, 7550-35-8 from Alfa-Aesar

Bis crucible #7130 from Spex Certi Prep Industries

8.3.3 Aerosol Suspension Devices

1. 4-liter aspirator bottle.

   4-liter aspirator bottle, VWR Scientific Co., Stock No. 16331-368.

   47-mm open faced anodized aluminum filter cassette, VWR Scientific, Stock No. 28144-404 (also as Nalgene composition, CMS No. 125-443).

   2–3 inch flat rubber gasket—local hardware store (or can be cut from tire tubing or sheet rubber from hobby stores).

   Pliobond rubber cement (to fix gasket to aspirator bottle mouth)—local hardware store.

2. 250-ml aspirator bottle.

   250-ml aspirator bottle, VWR Scientific, Stock No. 16331-288.

   25-mm open faced filter Delrin cassette, VWR Scientific, Stock No. 28144-302.

   1-in. flat rubber gasket—local hardware store (or can be cut from tire tubing or sheet rubber from hobby stores).

   Pliobond rubber cement (to fix gasket to aspirator bottle mouth)—local hardware store.

3. Tubular Aerosol Suspension Chamber 1. (This list applies to the "System A" TASC; Systems B and C are modular fabricated and available from Davis Consulting.)

   1 500-ml gas burette (this is cut in half, providing two complete units), Fisher Scientific, Stock No. 03-790B (glass stopcock), or 03-789-5B (plastic stopcock).

   1 10-ml distillation receiver, VWR Scientific, Stock No. 26344-049.

   1 Drierite cell (optional, but important in humid climates without humidity control), VWR Scientific, Stock No. 26668-007.

1 flow meter, Gilmont (0.2–14 liter/min range), VWR Scientific, Stock No. 29894-080.

1 37-mm or 47-mm open-faced anodized aluminum filter cassette, VWR Scientific, Stock No. (37 mm: 28144-403, 47 mm: 28144-404).

1 Rotary vane pressure/vacuum pump (GAST – 35 lpm), VWR Scientific, Stock No. 54906-001.

1 25-mm in-line filter cassette (for pump protection), VWR Scientific Co., Stock No. 28144-255.

1 long-stem loading funnel, VWR Scientific, Stock No. 30230-928.

1 support stand (24-in. height), VWR Scientific, Stock No. 60110-244.

2 3-finger extension clamps, VWR Scientific, Stock No. 21572-800.

2 round jaw extension clamps, VWR Scientific, Stock No. 21570-109.

4 auxiliary clamps, Fisher Scientific, Stock No. 05-754.

1 lb 0.5-mm diam. spherical glass beads (for approx. 50 loadings), Stock No. A-050; Potters Industries, available only in minimum 50-lb lots.

3 ft. 3/8 in. i.d., 3/32 in. wall tygon tubing, VWR Scientific, Stock No. 63010-133.

47-mm Whatman GF/C glass fiber filters; VWR Scientific, Stock No. 28497-696.

4. Microhood (see section 4.8.4). All of the items listed may be purchased at a hardware store.

4-in. (10.2 cm) square cross-section tapered plastic container.

2-in. (5.1 cm) PVC conduit.

2 small eye hooks (1/2-in. long) or round-head screws (to attach elastic bands holding cassette).

1 2-in. (5.1 cm), 1/8 to 1/4 in. wide flat rubber gasket (between PVC conduit and cassette).

Silicon rubber cement (attach chamber to PVC conduit).

8.3.4 Liquid Specimen Holders

The coupling distance from the window of the X-ray tube to the surface of the liquid being analyzed must remain constant during analysis or else large random errors will occur. This distance can change if the liquid expands or contracts during analysis. Holders are available specifically for liquid samples from the sources listed below.

Chemplex Industries

Spex Certi Prep Industries

8.3.5 Metallic Specimens

Processing metallic specimens may involve procedures other than disaggregation into discrete particles. Many metal and alloy samples must be processed as a solid button usually mounted in a cylindrical briquette and then polished to present a surface for both XRD and XRF analysis. The whole process of metallography is beyond the scope of this section, but some sources for metallographic processing equipment are listed.

1. Equipment for cast iron and steel preparation.

    Ladle bowl, model NB20 (7.8 kg iron capacity), from Canadian Thermix.

    Sample mold from Herzog, according to specifications.

    Bench vice and hammer: ordinary supply.

    Sample grinder: manual (HS 200) and semiautomatic (HTS 200) surface grinders for cast iron or steel disk specimen, are sold by Herzog. Equivalent devices (a bench top polishing machine such as Leco AP-600, or surface grinder model HT 350 from Herzog Corp.) equipped with a grid 60 polishing paper can also be used. Use grinding disk as supplied or recommended by the manufacturer of the grinder. Usually, it is a 37 grit Al_2O_3 stone wheel. For a large sample throughput, an automatic surface grinder such as model HB or HVTS from Herzog should be used.

    Vibratory polisher, such as Vibromet I from Buehler (to be used only if carbon is to be analyzed).

    Desiccator: with silica gel, or activated alumina drying agent.

2. Milling machines for preparation of nonferrous metals. (The milling machine must be equipped with diamond cutter(s). Isopropyl alcohol should be used for cooling and lubrication during machining.)

    Automatic, such as model HN-FF (with two milling heads) or HN-SF (with cutting blade and milling machine) available from Herzog. Depending on the needs, the following modules can be combined with the milling machines: (1) input and output sample magazines: turntable for 10 samples, chain magazine for 20 samples, spiral magazine for 100 samples; (2) sample transport systems: pneumatic tube systems, transport belts, pneumatic linear transport systems, servo driven linear transport systems, robots.

    Manual, such as model Polycut-E from Reichert Jung.

8.4 PREPARING X-RAY DIFFRACTION SPECIMEN HOLDERS

Specimen holders for XRD are usually supplied by the manufacturer of the diffraction equipment. Every diffractometer manufacturer uses different dimensions for their specimen holders, and even with a single manufacturer, different holders are used for different mechanisms. Sometimes, though, the user will want to prepare a special type of holder. This section illustrates the thinking necessary for making one specific holder that can be applied to any other design.

8.4.1 Specimen Holders for Large Specimens

1. Insert a round glass (48 mm diameter, 3 mm thick) on the rotating sample changer of the diffractometer.
2. Mark the position of the glass (referred to the sample changer) with a permanent marker, for subsequent uses.
3. Expose the glass to the same register conditions that you will use for your specimens, for example: aperture slit = 1 mm, divergence slit = 1 mm, and detector slit = 0.2 mm. Intensities are collected by continuous scanning from 5 to 70°2θ with a step size of 0.05°2θ and a counting time of 15 seconds for each step. A total of five and one-half hours are needed for each sample.
4. When the glass zone exposed becomes dark, using a ruler, draw a rectangle around the dark zone (approx. $2 \times 40 \times 1$ mm^3).
5. Using the diamond milling cutter, make a small rectangular hole (approx. $2 \times 40 \times 1$ mm^3).

8.4.2 Specimen Holders for Small Specimens

1. Insert a round glass (48 mm diameter, 3 mm thick) on the rotating specimen changer of the diffractometer.
2. Mark the position of the glass (referred to the sample changer) with a permanent marker, for subsequent uses.
3. Expose the glass to the same register conditions that you will use for your specimens, for example: a microdiaphragm with a diameter of 1 mm inserted into the aperture diaphragm, divergence slit = 1 mm, detector slit = 0.2 mm, and scintillation detector. Intensities are collected by continuous scanning from 5 to 70°2θ with a step size of 0.02°2θ and a counting time of 20 seconds for each step. A total of 18 hours is needed for each sample.
4. When the glass zone exposed becomes dark, with an irregular perimeter, using the diamond milling cutter, make a small hole (approx. $2 \times 4 \times 1$ mm^3).

8.5 RECOVERING DAMAGED PLATINUM WARE

8.5.1 General Hints on the Use of Platinum Crucibles

Failure of platinum crucibles is invariably due to the formation of low melting point eutectic mixtures. The worst examples include arsenic, silicon, phosphorus, and sulfur. When these materials are fully oxidized there is generally not a problem. Difficulties often occur when reducing conditions are present. A good example is carbon and silicon together. If the silicon is present as SiO_2 there is no problem, but under reducing conditions platinum silicide is formed, which forms a eutectic melting at about 800°C compared to 1800°C for pure platinum. Similarly, platinum phosphide forms a eutectic with platinum which melts at 588°C. A further problem is that Pt_5P_2 separates on the grain boundaries and is very brittle. Thus problems occur in the analysis of materials such as grain, corn, etc., where both carbon and phosphorus are present. In this instance it is better to use a gold–platinum (90% gold:10% platinum) crucible, even though this material has less mechanical strength and melts at a lower temperature than Pt-based crucibles. Note too that rhodium sulfide melts at 925°C; thus platinum–rhodium crucibles are not good for the fusion of P and S containing materials under reducing conditions.

Rules for Use of Platinum Crucibles

- Do not touch with bare hands.
- Always use platinum-tipped tongs.
- Minimize mechanical damage.
- Never place hot platinum on a wooden surface.
- Avoid use of reducing flames.
- Never clean a platinum crucible by heating. The "muck" may seem to disappear—in actual fact it has probably diffused into the platinum.
- When removing carbonaceous material keep the temperature as low as possible.
- Never place platinum crucibles on SiC trays.
- Keep platinum crucibles clear of SiC heating elements.
- Ensure oxidizing conditions for ferro-alloys and carbides.
- Keep an old crucible for potential problem samples.

In general, use platinum for simple fusions; platinum–5% gold to minimize wetting of the crucible; use gold–10% platinum if unoxidized phosphorus is unavoidable; never use platinum–rhodium if sulfur is present. Note that copper and nickel form alloys with platinum in reducing conditions.

8.5.2 Discussion of the Problem

Because of the elevated temperatures at which alkali borate fusions are performed, platinum crucibles and molds offer many advantages, most notably strength and the ability to be fabricated into precise forms that yield superior sample discs. Conversely, platinum alloys are costly. While the investment in the platinum metal can be recovered, the cost of refabrication of the laboratory ware is significant. If cracks can be repaired, the savings are obvious. The following method shows how cracks may be soldered.

8.5.3 Repair Procedure

An appropriate solder must have a low enough melting point so that the need to use an extremely hot torch doesn't damage the platinum. When solder is molten, it must combine with the platinum, yielding an alloy that resists elevated temperatures and chemical attack. Gold meets these requirements. Gold melts at 1064°C; therefore, it is very likely that the fluxing equipment itself can be used to melt the gold repair.

Because of the high temperatures involved, it is safer and more convenient if the patch can fill the crack without close observation by the technician. This repair is accomplished by cutting a small rectangular patch from an 0.015 in. thick sheet of gold, using a pair of office scissors. The patch is then pressed on the repair site at room temperature with the thumb or a pencil eraser until it takes the curvature of the object.

As shown in Figure 8.1 balance the patch in the inside curve of the repair site and orient the platinum ware (*a*) so that when the gold melts it is contained over the crack, instead of flowing downhill to another site. When molten, capillary action will cause the gold to wick into the crack, following the dendritic pattern until the entire crack is filled (*b*). The gold and platinum are allowed to remain at the high temperature, so they will continue to alloy. After being safely cooled, the repair should be inspected. On the back side of the soldered area the crack should be easily visible as a completely filled gold-colored tracing of the crack. The front side should show a gold outline of the patch on the platinum. If a gold trail leads away from the patched spot, it means the patch was too thick, or the platinum was tilted too much. With continued use at high temperatures, the alloying of the gold with the platinum will continue, until the gold color disappears and the crack is difficult to locate.

8.5.4 Special Problems and Hints

It is tempting to assume that if a small piece of gold works, more should work better. It might even seem that treating the molten gold as if it were a liquid glue and using it to coat a large spot on the crucible would be the ideal solution. It is not. Different platinum-gold alloys appear to have different rates of thermal expansion. Greater differences in expansion rates between the section with the excess of gold and the remainder of the platinum are more likely to cause a crack to form around the repair. This new crack will be impossible to repair effectively by adding more gold solder. The thickness of the gold patch and attempts to patch the crack should be kept to a minimum. The first repair has the greatest likelihood of lasting for the long term.

It is not necessary to pay for the precise fabrication of thin (0.015 in.) sheets of gold. Any scraps of gold that are available may be melted into a single button in

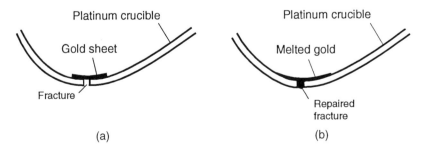

Figure 8.1. Repairing platinum ware.

a porcelain crucible (at this temperature the crucible can only be used once). The button can then be hammered thin, using a flat piece of steel as an anvil.

If attempts at patching are not successful because the crack is too wide, another try can be made by peening the immediately surrounding area. This expands the metal, thus bringing the sides of the crack closer together, after which another solder patch may be successful. The day will inevitably come, however, when further repair is unsuccessful, and the platinum will have to be refabricated.

8.6 SPECIMEN PREPARATION KIT FOR X-RAY DIFFRACTION

The list of equipment and supplies in this section have proven useful for preparing specimens for general diffraction analysis. Most of the items are available from chemical supply houses or local stores or have been mentioned elsewhere in this chapter. A second list of sources for some items is presented in Section 9.5.

8.6.1 Kit for Preparing Specimens

1. General

    Microscope (binocular 8x to 40x)
    d-spacing standards
    Intensity standards
    Profile-shape standards
    Crystallite-size standards
    Desiccators

2. Mortars with pestles

    Plattner (steel, 1 in.)
    Agate or Diamonite (2 in.)
    Boron carbide (1 in. and/or larger)
    Mortar covers (washers or plastic)

3. Tools

    Pliers (for crushing)
    Pin vises (for holding needles and drills)
    Small screwdriver
    Magnet
    Sharpening stone (fine)
    Tweezers (medium to very fine tip)
    Diamond scriber
    Scissors
    Hemostats

Spatulas
Vibratory engraver
Awl
Files.

4. Expendable supplies

Sieves (small diam. 10, 35, 60, 80, 100, 200, 325, 400 mesh)
Razor blades (single edge)
Microscope slides (petrographic)
Vaseline or Apiezon grease
Brushes (small)
Tacki wax
Acrylic spray (clear)
Small funnels
Sample vials (screw top and cork)
Petri dishes (glass and plastic)
Medicine droppers
Mylar film (1/4 mil)
Tape (Magic-Mending)
Weighing papers
Filter paper (Whatman #20)
Markers (for samples and glass vials)
Needles (medium sewing)
Drills (steel, carbide and/or diamond)
Glazed ceramic spot plates (black and white)
Confined spot filter paper
Radiation paper
Glass plate (> 18 in. square)
Scalpel with replaceable blades
Glove bags
Pipe cleaners
Clear nail polish

5. Chemicals

Acetone (in wash bottle)
Alcohol (in wash bottle)
Collodion (flexible)
Ethyl acetate
Amyl acetate
Ether (anhydrous—very dangerous)

> Silica gel or cornstarch

6. Optional laboratory equipment

> Micronizing mill
> Spray drier
> Pellet press
> Suction systems
> Sonifier
> Glove box (controlled atmosphere)
> Knurled or serrated plate
> Oven
> Flexible shaft grinder
> Aerosol deposition system

7. Diffractometer

> Specimen holders for diffractometer
> Glass or quartz or silicon slides
> Cardboard or fine filter paper
> Clips

8. Debye–Gandolfi camera equipment

> Camera sample pins
> Glass tubing (drawn thin)
> Glass tubing ($>$ 8 mm ID, 1 m long)
> Glass rods to fit in capillaries
> Capillaries
> Hot-wire device
> Mortar (boron carbide 1 cm.)

9. Guinier camera

> Sample holders for camera
> Metal or glass rod (flat end)

8.6.2 Sources for Kit Items

The list of equipment needed to prepare an XRD specimen may be quite extensive, as many devices could practically do a similar job [1, 2, 3]. In addition, a device recommended in connection with aluminas, for example, may also be used for many other materials. Based on our personal experience and contacts with various laboratories, the following are recommended. (See section 8.8 for addresses.)

1. Thin-walled capillaries, quartz or glass, 0.3 mm ID, Part No. GCT 100; 0.5 mm ID, Part No. GCT 102. Pantak, W. Mueller, Charles Supper, Blake Industries.
2. Cow gum rubber solution adhesive. BTR Graphic Products, local office supply stores.
3. Gum arabic (makes a water-based glue). Product No. G 9752, Sigma Chemical.
4. Mylar film (0.25 to 6 μm thick polyethylene terephthalate). Product No. 250, Chemplex Industries, Graesby Specac, and Spex Certi Prep Industries.
5. Vaseline (petroleum jelly) or Apiezon grease. Apiezon grease available from Apiezon Products, Vaseline available from local drug store.
6. Plasticine (modeling clay). Available from toy stores or florists shops.
7. Glass fibers (for supporting specimens). See section 7.1 for instructions for laboratory fabrication.
8. Thick-walled glass capillary tubes (for annealing metallic filings). Melting point capillary tubes (soda glass) supplied in 100 mm lengths, 100 tubes per package. Code No. 6119/02 Catalogue No. MFB-210-5381, Fisons Plc.
9. Muffle furnace (for annealing metal filings or slivers), Product No. F88-256, Model CSF 12/7, Carbolite furnace, Philip Harris Scientific.
10. Silicon or quartz single crystal substrates. The Gem Dugout.
11. Spray mount adhesive. 3M Company.
12. Pin vises. Starrett No. 0 or 1, local hardware store.
13. Tacki wax Catalogue No. 11444-000, Boekel Industries. Order from a chemical supply house that orders from Boekel Industries.
14. Calcium hexametaphosphate. Local store as Calgon or other washing additive.
15. Ceramic plates for clay analysis. Soil Moisture Equipment, Monarch Tile, Robertson Mfg.
16. Millipore filters. Millipore Laboratory Products.
17. Silver membrane filters. Selas.
18. Polyvinylchloride (PVC), filters, Nuclepore filters. Nuclepore.

8.7 EQUIPMENT FOR PREPARATION OF ZEOLITE SPECIMENS

Zeolite specimens have some unique properties that often require special procedures and equipment. This section contains a list of equipment and supplies that have proved useful in laboratories that routinely prepare specimens of zeolites or related materials.

8.7.1 Equipment Suppliers, Zeolite Preparation

McCrone Micronizing Mill. Cat. No. 232, McCrone Accessories & Components. Approximate cost $4000.

Glass desiccator, 200 mm ID, (Cat. No. 3751-H20) with porcelain plate insert, 190 mm diam. (Cat. No. 3760-L30), Thomas Scientific Instruments. Approximate cost $230.

Glass Slides, Frosted one side, 25 × 75 mm, Cat. No. 6685-N30, Thomas Scientific Instruments. Approximate Cost $13 per package.

Magnesium Nitrate 6 Hydrate, ACS Reagent. Cat. No. C469-E92, Thomas Scientific Instruments. Approximate Cost $37 per 500 g.

Bico Braun UA Pulverizer V-Belt. #24253X3UA (242-53x3) Bico.

Bico Braun Chipmunk crusher. #2413400VD V-Belt jaw crusher, Bico.

No.100 sieve stainless cloth, stainless frame. V8SF#100, Gilson.

U.S.A. standard sieve. #100 CS-8100, Soiltest.

800 Spex Mixer/Mill. Spex Certi Prep Industries.

Swing mill. Spex Certi Prep Industries.

Forced Draft Laboratory Oven Model L-24. Soiltest.

BLUE-M Drying Oven MO-1450A. VWR Scientific Co.

Metal bonded diamond grinding wheel. #40-4468, Buehler.

J.T. Baker Propylene Glycol "Baker Analyzed" 99.5% min. #JTU510-7, VWR Scientific Co.

Jaw Crusher Model #LC-36. Bico.

ACCU-MAX Spinning Riffler. Model #SP-245, Gilson.

Stainless steel sample splitters (riffles). Model #CL-286, Soiltest.

31 mm Pre-Flared SpecCap. #3619A, Spex Certi Prep Industries.

TAB-PRO Copolywax tablets 91 #F-Code 289, Cargille Tab-Pro.

Somar-Mix tablets. #220, Somar Laboratories.

X-ray mix tablets. #625, Chemplex Industries.

Spex-Claisse Fluxer Crucibles, Bis crucible. #7130, Spex Certi Prep Industries.

Spex-Claisse Fluxer-Bis. #7100 Bis-10, Fluxer Spex Certi Prep Industries.

Spectroflux 100. #12078, Alpha-Aesar.

Johnson Matthey Lithium bromide, anhydrous, 99%. #7550-35-8, Alpha-Aesar.

8.8 VENDOR ADDRESSES

The following information is offered to assist the user in locating specific equipment. The addresses are correct as of the date of preparation of this book, but may change with time. The authors waive responsibility for any future changes and possible inaccuracies that may have been included in this list. More current and alternate vendor information may be located by consulting the Thomas Registry in your local library or calling a purchasing agent at a local college.

Alfa-Aesar, 30 Bond Street, Wall Hall, MA 01835-8099, Phone: 1-508-521-6300.

Angstrom, Inc., 6245 S. Oak Park Ave., Chicago, IL 60638, Phone: 1-312-229-8484.

Apiezon Products, Ltd., 8 York Road, London, UK, or James Biddle Co., Plymouth Meeting, MA 09462.

Bico Braun International, 3118 Valhalla Drive, Burbank, CA 91510-1235, Phone: 1-800-893-3490.

Blake Industries, 660 Jerusalem Road, Scotch Plains, NJ 07076, Phone: 1-908-233-7240.

Boekel Industries, Inc., 509 Vine Street, Philadelphia, PA 19106, Phone: 1-215-627-1611.

Brinkman Instruments, No.1 Cantiague Rd., Westbury, NY 11590, Phone: 1-516-334-7599.

BTR Graphic Products, Ltd., Eastbourne Road Trading Estate, Slough, Berkshire SL1 4SF, UK.

Buehler, Ltd., 41 Waukegan Rd., P.O. Box 1, Lake Bluff, IL 60044-1699, Phone: 1-800-BUEHLER.

Canadian Thermix, Inc., 226 Industrial Parkway North, Unit 2A, Aurora, Ontario L4G 3G8, Canada.

Cargille Tab-Pro Corp., 4-T East Frederick Place, Cedar Knolls, NJ 07927-1894, Phone: 1-201-267-8888.

Carpco, Inc., 4122 Haines Street, Jacksonville, FL 32206, Phone: 1-904-353-3681.

Champion Products, 2975 Airline Circle, Waterloo, IA 50703, Phone: 1-319-232-8444.

Chemplex Industries, 140 Marbledale Road, Tuckahoe, NY 10707, Phone: 1-914-337-4200.

Charles Supper Co., 15 Tech Circle, Natick, MA 01760, Phone: 1-508-655-4610.

Cianflone Scientific, 228 RIDC Park West Drive, Pittsburgh, PA 15275, Phone: 1-800-569-8400.

CMS Corp., 20100 E. 35th Drive, Aurora, CO 80011-8154, Phone: 1-303-375-8989.

Davis Consulting Co., 4022 Helen Ct., Rapid City, SD 57702.

Eberbach, P.O. Box 1024, Ann Arbor, MI 48106, Phone: 1-313-665-8877.

Enraf-Nonius, Roentgenweg 1, P.O. Box 483, 2500 AL Delft, the Netherlands.

Fisher Scientific, Midwest Distribution Center, 1600 W. Glen Lake Road, Itasca, IL 60143, Phone 1-630-773-3050.

Fisons Plc., Bishop Meadow Road, Loughborough, Leicestershire, LE11 ORG, UK, Phone: 01509-321166.

Foundry Testing Equipment, Inc., 9190 Roselawn, Detroit, MI 48204, Phone: 1-313-491-4680.

Fred S. Carver, Inc., P.O. Box 428, Menomonee Falls, WI 53051, Phone: 1-219-563-7577.

Garlock, Inc., Plastomer Products, 23 Friends Lane, Newton, PA 18940, Phone: 1-800-618-4670.

Gilson Company, P.O. Box 677, Worthington, OH 43085-0677, Phone: 1-614-548-7298.

Glenn Mills, Inc., 395 Allwood Rd., Clifton, NJ 07012.

Graseby Specac Ltd., River House, Lagoon Road, St Mary Cray, Orpington, Kent BR5 3QX, UK.

Hamilton Industries, 1316 18th St., Two Rivers, WI 54241, Phone: 1-414-793-1121.

Herzog Corp., 16600 Sprague Rd., Suite 400, Cleveland, OH 44130.

Herzog Machinfabrik Gmbh, Osnabruck, Germany D-4500.

Huber Diffraktionstech, Sommerstrasse 4, D-8219 Rimsting/Chiemsee, Germany.

McCrone Accessories & Components, 850 Pasquinelle Dr., Westmont, IL 60559.

Millipore Laboratory Products, 80 Ashby Rd., Bedford, MA 01730, Phone: 1-617-275-9200

Miners, Inc., P.O. Box 1301, Riggins, ID 83549.

MMM (3M) United Kingdom PLC, 3M House, Bracknell, Berkshire RG12 1JU, UK, Phone: 01344-858000.

Monarch Tile Co., P.O. Box 2041, San Angelo, TX, Phone: 1-915-655-9193.

Nuclepore Corp., 7035 Commerce Circle, Pleasanton, CA 94566, Phone: 1-510-463-2230.

Optical Instrument Laboratories, P.O. Box 608, Bellaire, TX 77401, Phone: 1-713-772-7294.

Pantak Ltd., Unit 30, The Robert Cort Industrial Estate, Britten Road, Reading RG2 OAU, UK.

Philip Harris Scientific, 618 Western Ave., Park Royal, London, WE 0KTE, UK, Phone: 0181-992-5555, Fax: 0181-993-8020.

Potters Industries, Inc., Brownwood Div., Highway 279 North, HC 30, Box 20, Brownwood, TX 76801, Phone: 1-915-752-6711.

Reichert Jung, Postfach 112, Heild, Germany.

Robertson Mfg. Co., S. Pennsylvania Avenue, Morrisville, PA, Phone: 1-610-295-1121.

Rocklabs, Ltd., 187 Morrin Rd., P.O. Box 18-142, Auckland, New Zealand.

SCT Sales, Inc., 2305 E. Arapahoe Rd., Suite 115, Littleton, CO 80122, Phone: 1-303-730-0084.

Selas Corporation of America, Flotronics Division, Phone: 1-215-646-6600.

Sepor, Inc., 718-T Fries Ave., P.O. Box 578, Wilmington, CA 90748 Phone: 1-310-830-6601.

Sigma Chemical, Sigma-Aldrich Co., Ltd., Fancy Road, Poole Dorset BH12 4QH, UK, Phone: 0800-447788, Fax: 0900-378785. Overseas customers call 01202-733114 and reverse charges.

Soil Moisture Equipment Corp., P. O. Box 30025, Santa Barbara, CA 93105, Phone: 1-305-964-3525.

Soiltest, Inc., 86 Albrecht DR., P.O. Box 8004, Lake Bluff, IL 60044-8004.

Soiltest, Inc., Materials Testing Div., 2205 Lee St., Evanston, IL 60202.

Somar Labs., 115 Main St., Tuckahoe, NY 10707, Phone: 1-914-961-1400.

Spex Certi Prep, Inc., 203 Norcross Ave., Metuchen, NJ 08840, Phone: 1-732-603-9647.

Stoe & CIE Gmbh., Hilperstrasse 10, P.O. Box 4110, D-6100 Darmstadt, Germany.

Struers, Inc., 26100 First Street, Westlake, OH 44145, Phone: 1-216-871-0071.

Sturtevant, Inc., 3 Clayton St., Boston, MA 02122, Phone: 1-800-992-0209.

Sweco, 7120 New Buffington Road, P.O. Box 1509, Florence, KY 41022, Phone: 1-800-849-3259.

Swiss Precision Instruments, Inc., 2425 S. Eastern Ave., Los Angeles, CA 90040, Phone: 1-213-721-1818.

Tab-Pro Corporation, (see Cargille Tab-Pro Corp.) Techmar, 7143 E. Kemper Rd., Cincinnati, OH 45429, Phone: 1-513-247-7000.

The Gem Dugout, 1652 Princeton Drive, State College, PA 16803 Phone: 1-814-238-4069.

Thomas Scientific Instruments, Inc., P.O. Box 99, Swedesboro, NJ 08085, Phone: 1-609-467-2000.

Van Aken International, Rancho Cucamonga, CA 91730.

VWR Scientific Co., P.O. Box 39396, Denver, CO 80239, Phone: 1-303-371-0970

VWR, P.O. Box 66929, O'Hare AMS, Chicago, IL 60666.

W. Mueller, D-1000, Berlin 27, Germany, Phone (030) 4316172.

REFERENCES

[1] McMurdie, H. F., Morris, M. C., Evans, E. H., Paretzkin, B., and Wong-Ng, W. (1986), "Methods of Producing Standard X-Ray Diffraction Powder Patterns," *Powder Diffr.*, **1**, 40–43.

[2] Esteve, V., Justo, A., and Amigo, J. M. (1994). "X-Ray diffraction analysis of airborne particulates collected by a cascade impactor sampler. Phase distribution versus particle size," *Materials Science Forum*, 166–169, pp. 705–710.

[3] Esteve, V. and Amigo, J. M. (1994). "Chemistry and phase determination of individual atmospheric aerosol particles in a Mediterranean coastal site," *Air Pollution Research Reports: Physico-chemical behaviour of atmospheric pollutants*, pp. 718–723.

CHAPTER 9

USE OF STANDARDS IN X-RAY FLUORESCENCE ANALYSIS

CONTRIBUTING AUTHORS: BUHRKE, BABIKIAN, CROKE, FERET

9.1 STANDARDS—AN OVERVIEW

9.1.1 Introduction

History   Standards were developed as the need became obvious in analytical laboratories associated with mining, metallurgy, ceramic, and the earth sciences. Quality assurance and analytical departments of industries wanted the communication between scientists concerning quantity and property of materials to be reliable and accurate. To accomplish effective communication with meaningful measurements, they questioned the efficiency of the analytical methods [1]. That inquiry brought about in 1901, the beginning of the National Bureau of Standards (NBS), renamed in August 1988 to the National Institute of Standards and Technology (NIST).

Inaccurate determinations and analyses have created and are still creating important and sometimes very serious economic consequences. When a company is investing millions of dollars to acquire mineral deposits and plan a new production plant for the next ten years, the chemical analyses for such a decision must be accurate. The determination of a meaningful measurement, e.g., $SiO_2 = 20.2 \pm 0.1\%$ by XRF, is a complex process and sound decisions can be made only when the information is reliable. In our competitive market, reliable information, meaningful measurements, and accurate chemical analysis have become the *raison d'être* of analytical laboratories.

Availability   The available reference standards used in XRF spectrometers are sometimes taken for granted, and they are not used as often as they should be. There are three thousand reference standards listed in catalogues and hundreds more in preparation. These standards include ores, metals and alloys, rocks, minerals, waste products, and dusts collected by electrostatic precipitators. Environmental problems have increased the demand for standards for tailings, industrial wastes, and isotopic materials.

A Practical Guide for the Preparation of Specimens for X-Ray Fluorescence and X-Ray Diffraction Analysis, Edited by Victor E. Buhrke, Ron Jenkins, and Deane K. Smith. ISBN 0-471-19458-1 © 1998 John Wiley & Sons, Inc.

9.1.2   Definitions

The well-known scientific and technical organization founded in 1898, the American Society for Testing and Materials (ASTM), defines standard as *a document that has been developed and established within the consensus principles of the Society and that meets the approval requirements of ASTM procedures and regulations.* The term *"standard" serves in ASTM as an adjective in the title documents, such as test methods or specifications, to connote specified consensus and approval* [2]. In this chapter the word "standard" does not refer to standardization documents but to analytical reference materials that are packed in bottles, vials, and jars.

Standards or analytical reference materials have accepted reference values. These assigned values are based on the experimental work of some national or international organization, e.g., NIST.

The International Organization for Standardization (ISO) Guide 30-198 uses the following definitions in "Terms and definitions used in connection with reference materials":

1. *Reference Material* (RM) A material or substance one or more properties of which are sufficiently well established to be used for the calibration of an apparatus, the assessment of a measurement method, or for assigning values to materials [3].

2. *Certified Reference Materials* (CRM) A reference material one or more of whose property values are certified by a technically valid procedure, accompanied by or traceable to a certificate or other documentation which is issued by a certified body [3].

3. *Standard Reference Materials* (SRM) Certified reference materials issued by NIST. These SRMs are well-characterized materials produced in quantity to improve measurement science. SRMs are certified for specific chemical or physical properties and are issued by NIST with certificates that report the results of the characterization and indicate the intended use of the material. They are prepared and used for three main purposes:

   - To help develop accurate methods of analysis (reference methods).

   - To calibrate measurement systems used to: (a) facilitate exchange of goods, (b) institute quality control, (c) determine performance of characteristics, or (d) measure a property at the state-of-the-art limit.

   - To assure the long-term adequacy and integrity of measurement quality assurance programs [3].

4. *Setting up Samples* (SUS) Materials which have an uncertified analysis and are intended to be used as a guide. These materials are assigned values for each element during calibration of the instrument with CRMs and RMs. These assigned values are used for day-to-day or shift-to-shift adjustment of the instrument. This process is often called *drift correction, standardization of calibration,* or *restandardization.* Instrument users are advised to obtain a

reasonably long-term supply of each SUS to avoid having to recalibrate the instrument when the current SUS is exhausted [4].

5. *Geostandards* Geochemical reference materials are CRMs prepared by the United States Geological Survey. Although several of their initial reference materials are exhausted, a great effort is being made for a second generation of these reference materials to be available.

9.1.3 Catalogues

The five main sources of reference materials are listed below. The catalogues are published approximately every two years, the price lists are printed separately as a supplement, and upon your request, they are sent free of charge. The price of the standards range from moderate to expensive but when the information available in the documentation accompanying them is taken into consideration, the prices are more than justifiable.

1. National Institute of Standards and Technology (NIST), Standard Reference Materials Catalogue. United States Department of Commerce, Gaithersburg, MD 20899. Tel.: (301) 975-6776; FAX: (301) 948-3730.

2. Brammer Standards Company, Quality Analytical Reference Materials from Around the World. 14603 Benfer Road, Houston, TX 77069-2895. Tel.: (713) 440-9396; FAX: (713) 440-4432.

3. United States Geological Survey, Rock standards–Geostandards Newsletter. P.O.Box 25046, Denver, CO. Tel.: (303) 236-2454; FAX: (303) 236-3200.

4. Analytical Reference Materials Inc., P.O. Box 2246, Evergreen, CO 80439-2246. Tel.: 1-800-421-9454.

5. Geostandards, C.R.P.G. 15, Rue Notre-Dame-des Pauvres, B.P. 20, 54501 Vandoeuvre-lès-Nancy Cedex, France. Tel.: 83.51.22.13; FAX: (33)83.51.17.98

Besides the catalogues, an essential tool of absolute importance for any analytical laboratory is the Special Issue of *Geostandards Newsletter* issued in July 1994. It is published by the International Working Group "Analytical Standards of Minerals, Ores and Rocks" with the assistance of Centre de Recherches Pétrographiques et Géochimiques (CNRS). The editor-in-chief, K. Govindaraju, calls this special issue a " rich bank of geostandards." It consists of four appendices giving values for both major, minor and trace elements with descriptions and references.

9.1.4 Origin and Types

A reference material can be a liquid, a gas, or a solid. It can be pure, mixed powders, natural, or synthetic materials. Relevant information concerning physical character-istics such as color, source, and location; method of extraction from the previous environment, the producing company name, and references should be mentioned in the certificates [5].

9.1.5   Preparation Procedure

The reference material in the final packaged bottles, vials, and jars has to meet strict requirements for homogeneity and stability [5]. The preparation procedure will be covered in detail later in this chapter.

9.1.6   Certificate Evaluation

Each standard shipped is accompanied by a certificate reporting the chemical or physical properties and their uncertainties. Different programs of round robins are established to determine these properties. The accumulated data is processed, the mean, the standard error of estimate, and hence the uncertainties, are calculated.

Round Robins   These programs may be one of the following:

1. An interlaboratory program, which is the most common procedure, includes ten or more selected laboratories using the analysis method of their choice.
2. A small number of laboratories using the same well-known method of analysis.
3. A single laboratory using a specific noncomparative (nonrelative) method of analysis. The requirement for this program is that an assumption is made:

   - The laboratory involved has superior staff expertise.
   - The analysis method used has high precision.
   - The analysis method used is free of bias [5].

Pitfalls

1. *Outliers*   The value that appears to deviate markedly from other values in the set in which it occurs is called the *outlier*. During the preparation of a reference material, an outlier could be due to:

   - An obvious human error.
   - A loss in calibration of the analytical instrument used.
   - An undetected change of the analytical instrument.

   An investigation could clarify the reason for the deviation, but if the doubtful data falls within the random variability in the normal Gaussian distribution, the value is retained and processed with the rest of the data. The Huge Error formula or Dixon's Q should be used to determine an actual outlier [6].
2. *Caveat*   There are reference materials with values given that are described using terms like "best," "usable," "proposed," and "information value." These figures should not be used with utmost confidence but with some reluctance [5]. If the data and the mean values are reported, the practical rule of the Huge Error can be applied again. Depending on the situation, the mean value can either be improved or be drastically wrong.

Validation

1. Before using a reference material:

   - Visually check for color changes.
   - Visually check for segregation of the particles.
   - Check instructions for handling and storage.
   - Check expiry date.

2. Once the bottle is opened:

   - It can be reused provided it is kept in an environment where its stability is retained and not contaminated.
   - The moisture content and the loss on ignition (LOI) are points to deal with caution. When the determined LOI value, which is the mass loss of the specimen heated at a specific temperature to remove mainly structural water and carbonates, is much higher (generally accepted $\pm 0.1\%$) than the certified value, even though the storage instructions are followed and the sample has been kept in a desiccator, the standard has to be dried at the specific temperature to remove the adsorbed water or carbon dioxide. The LOI determined after the drying process should correspond to the certified value. The correct determination of the loss on ignition is the first step in recognizing the applicability and validity of a reference material. If the LOI cannot be determined accurately, the possibility of having a valid chemical analysis is minimal. LOI is a basic test for laboratories using XRF; it indicates the total volatiles (water, carbon dioxide, chlorine, sulfur, fluorine, hydroxyl groups, and organic matter) which adds up with the major oxides to give a total of 100% [7]. When there is a gain in weight instead of a loss, it indicates an oxidation reaction. e.g.,

$$4Fe_3O_4 + O_2 \rightarrow 6Fe_2O_3$$

In some cases the LOI is the difference between two reactions, a gain in weight and a loss in weight, e.g.,

$$FeCO_3 \rightarrow FeO + CO_2 \text{ (loss)}$$

$$4FeO + O_2 \rightarrow 2Fe_2O_3 \text{ (gain)}$$

9.1.7 Applications

Calibration–Correlation   XRF analysis uses calibration curves to determine concentrations; it is a relative analysis or a comparative method of analysis. The process of producing calibration curves for a specific XRF spectrometer includes the preliminary steps of deciding what are the unknown specimens and which standards to use to analyze those specimens. The availability of the appropriate and accurate standards is the major requirement for such a calibration. The selection of these standards will result in a significant reduction in systematic errors and have an impact on the final quantitative analysis results.

There are three major methods to provide standards for calibration. Buying the pure chemicals and using specimens analyzed by a noncomparative method are the only alternatives when SRMs, CRMs, and RMs are not available.

Buying the Pure Chemicals (Ultrex, Puratronic) and Preparing the Required Concentrations Although theoretically feasible, the difference in the particle size of the components will greatly affect the homogeneity of the sample. This inhomogeneity could be partly overcome by grinding the mixture in a shatter box for a few seconds and saving it in an airtight container for future use. The chemical compound ratios shown in Table 9.1 facilitate the calculations involving the synthetic standard preparations of Table 9.2. For the major and minor elemental analysis, the recipes in Table 9.2 were calculated using a computer and a spreadsheet.

Using Samples That Have Been Analyzed in Triplicate by a Noncomparative Method, e.g., Wet Methods The details of these analyses—sample preparation, procedure used, accumulated data, and statistically accepted data from which the mean has been calculated—should be recorded and be traceable for quality control purposes.

Purchasing SRMs and CRMs Is the Most Reliable Method to Acquire Standards The standards chosen should have a high degree of similarity, or ideally, they should be identical to the compared samples. For raw materials analysis in the cement industry, a set of nineteen reference materials used for the calibration are listed in Table 9.3. The covered concentrations (percentages) on loss free basis for five major and five minor elements are listed in Table 9.4 [8].

The set in Table 9.3 is flexible. It can be modified to include the analysis of Mn_2O_3 and SrO. The covered ranges can be changed by adding more standards or deleting standards. These international reference materials are also used to make up standards that are not available. Glass beads can be made by using 50% by weight of one ignited reference material and 50% by weight of another ignited reference material. The beads prepared are very homogeneous due to the dissolution in the flux. In comparison, when powders are mixed, this homogeneity cannot be attained. Extreme

TABLE 9.1. Chemical Compound Ratios

Compound	Formula Weight (g)	Ratio of ingredients
$CaCO_3$	100.09	56.03% CaO:43.97%CO_3
K_2SO_4	174.26	54.06% K_2O:45.94% SO_3
Na_2SO_4	142.04	43.64% Na_2O:56.36% SO_3

TABLE 9.2. Synthetic Standard Preparations

Pure	Weight (g)	Oxide	Concentration (%)
		Mix A	
$CaCO_3$	30.000	CaO	32.33
SiO_2	10.000	SiO_2	19.23
Al_2O_3	3.000	Al_2O_3	5.77
Fe_2O_3	5.000	Fe_2O_3	9.62
MgO	2.000	MgO	3.85
K_2SO_4	1.000	K_2O	1.04
Na_2SO_4	1.000	Na_2O	0.84
		SO_3	2.14
TOTAL	52.000	CO_2	25.37
		Mix B	
$CaCO_3$	15.000	CaO	16.16
SiO_2	20.000	SiO_2	38.46
Al_2O_3	4.000	Al_2O_3	7.69
Fe_2O_3	6.000	Fe_2O_3	11.54
MgO	3.000	MgO	5.77
K_2SO_4	2.000	K_2O	2.08
Na_2SO_4	2.000	Na_2O	1.68
		SO_3	4.28
TOTAL	52.000	CO_2	12.68

TABLE 9.3. Calibration Standards

Source	Code	Oxide
NIST	SRM 99a	Feldspar
	SRM 688	Basalt rock
	SRM 1884	Portland cement
	SRM 635	Portland cement
	SRM 1c	Argillaceous limestone
Brammer	BCS 313-1	High purity silica
	BCS 368	Dolomite
	BCS 393	Limestone
	BCS 309	Sillimanite
	BCS 376	Potash feldspar
	ECRM 777-1	Silica brick
	ECRM 878-1	Slag
	SARM 1	Granite
	SARM 3	Lujavrite
	SARM 5	Pyroxenite
	CAN SY/2	Syenite
	IPT 51	Burnt refractory
	IPT 44	Limestone

TABLE 9.4. Covered Ranges on Loss Free Basis

Oxide	Range
SiO_2	1–99.8%
Al_2O_3	0.2–61.2%
TiO_2	0–2.2%
P_2O_5	0–0.5%
Fe_2O_3	0–12.7%
CaO	0.2–97.9%
MgO	0.2–39.2%
Na_2O	0–8.6%
K_2O	0–11.2%
SO_3	0–7.3%

caution is still important when concentrations are calculated and while weighing for the preparation of the mixed bead, e.g., as indicated in Table 9.5.

Check Standards and Quality Assurance No quantitative result is of any value unless it is accompanied by an estimate of the error. This analysis of error applies not only to XRF analysis, but to any other numerical experimental results. First, the precision, the reproducibility, the random error, or the imprecision of our data, which falls on both sides of the average, is calculated. Next, the systematic error or bias, which is the error always in the same direction, either always high or always low, should be investigated.

1. *Precision* Measuring precision is a pure statistical control that does not necessarily require reference materials. A homogeneous material that is often analyzed in a laboratory could be used to calculate the precision of the chemical determinations. However, if a SRM or CRM is used, both precision and bias are investigated simultaneously. The following example below shows a set of data and the basic statistical analysis.

TABLE 9.5. Composition of Mixed Bead[1]

	Certified SiO_2 %	Certified CaO %	LOI %
BCS-393	0.7	55.4	43.4
IPT-51	55.0	0.06	0.16
	On loss free basis:		
BCS-393	1.24	97.88	0.0
IPT-51	55.09	0.06	0.0

[1]Beads made of 50% by weight BCS-393 and 50% by weight IPT-51 will have $SiO_2 = 28.16\%$ and CaO = 48.97%.

Determinations of SiO_2, Al_2O_3, Na_2O, and K_2O, using an ARL 8420+ spectrometer.

Reference Material: NIST-278 Obsidian Rock Reference values:

$SiO_2 = 73.05\%$
$Al_2O_3 = 14.15\%$
$Na_2O = 4.84\%$
$K_2O = 4.16\%$

The data have been accumulated by analyzing two glass beads every two months, as shown in Table 9.6. Assuming the data follows the known Gaussian distribution, also called normal distribution, then:

Approximately 68% of the data will be within $\pm 1\sigma$.
Approximately 95% of the data will be within $\pm 2\sigma$.
Approximately 99.7% of the data will be within $\pm 3\sigma$.

In the example shown, at 95% confidence limit for a single datum:

$SiO_2\% = 72.98 \pm 1.0$
$Al_2O_3\% = 14.23 \pm 0.18$
$Na_2O\% = 4.81 \pm 0.10$
$K_2O\% = 4.22 \pm 0.08$

TABLE 9.6. Data Collected Bimonthly on Two Glass Beads

Month	Bead	$SiO_2\%$	$Al_2O_3\%$	$Na_2O\%$	$K_2O\%$
February	1	72.73	14.20	4.75	4.18
	2	72.32	14.10	4.71	4.17
April	1	73.33	14.28	4.83	4.25
	2	73.59	14.35	4.83	4.26
June	1	73.87	14.39	4.90	4.28
	2	73.83	14.35	4.89	4.28
August	1	72.72	14.26	4.83	4.21
	2	72.96	14.23	4.84	4.23
October	1	73.00	14.23	4.81	4.23
	2	73.05	14.26	4.84	4.24
Mean (x)		72.98	14.23	4.81	4.22
Measurements (n)		10	10	10	10
Standard Deviation(s)		0.50	0.09	0.05	0.04
Rel Standard Deviation (RSD)		0.69	0.65	1.05	0.86

For instrumental methods in general, the RSD of $< 2\%$ is acceptable, while $> 5\%$ indicates serious problems. The lower the RSD, the better is the precision of the method. When comparing different specimen preparation procedures, either glass beads or pressed powder, the RSD is a good indicator; the preparation with the lowest RSD has the highest precision. With any homogeneous material, as in the example, the precision of the test could have been calculated, but whether or not the analysis is biased could not be determined.

2. *Bias*   Bias or systematic errors on any test can be determined only if you have a standard reference material and the test method is actually under a statistical control—the accumulated data should be a minimum of six determinations. The Student t-test is mostly used by the analytical chemist to determine whether the test value is significantly different from the reference value. For example: Is the difference between the mean (x) SiO_2 analysis value of 72.98% and the certified value (μ) of 73.05% significant? When the t-value is calculated from equation (9.1):

$$|t| = (x - \mu) \times \sqrt{n/s} \tag{9.1}$$

and it is less than the critical value, then there is no evidence of systematic error. If the t-value is greater than the critical value, then the data obtained significantly differs from the reference value.

For $SiO_2 - |t| = (72.98 - 73.05)\sqrt{10/0.50} = 0.443$.
For $Al_2O_3 - |t| = (14.23 - 14.15)\sqrt{10/0.09} = 2.811$.
For $Na_2O - |t| = (4.81 - 4.84)\sqrt{10/0.05} = 1.897$.
For $K_2O - |t| = (4.22 - 4.16)\sqrt{10/0.04} = 4.743$.

From t-distribution tables found in most statistical books, the mathematical model is compared to the above figures. At 95% confidence $(P = 0.05)$ and 9 degrees freedom (df), the critical value t is 2.262. Because the experimental $|t|$ values of SiO_2 and Na_2O are less than the critical value 2.262. The differences obtained from the reference values in SiO_2 and Na_2O, are not significant. This comparison does not mean that there are no systematic errors, but there is no mathematical evidence that systematic errors do exist. As for Al_2O_3 and K_2O, the experimental $|t|$ value is higher than the critical value t; therefore there is mathematical evidence of systematic error [9].

3. *Accuracy*   The term "accuracy" is used for the overall exactness, which includes both precision and bias.

$$Accuracy = Precision + Bias$$

When the number of measurements is ≥ 30 and the mean is considered to be the true value, then the term "accuracy" refers to bias only. To cover the legal aspect of analytical chemistry and prove the reliability and accuracy of the data, it is wise and advisable to analyze SRMs and CRMs simultaneously with the unknown samples. In a quality assurance program, the long-term reliability and accuracy of the produced analyses are assured only by using SRMs and CRMs.

ASTM Qualification   According to ASTM, XRF analysis is a rapid method of analysis; therefore, in order to be qualified for the manufacturer's certificate, the XRF spectrometer should meet the specifications set up by the organization. These specifications and the fine print should be read carefully and followed systematically.

Seven NIST Cements Analyzed for SiO_2   To be qualified for the manufacturer's certificate, six of the seven analyzed cements are allowed to have the differences mentioned below and only one of the seven is allowed to have no more than twice the differences mentioned below [10].

Requirement A

- Maximum difference between duplicates analyzed on different days = 0.16

Requirement B

- Maximum difference of the average from SRM certified values = ±0.2

The difference between the duplicates of day one and day two are:

SRM Black	0.01
SRM Ivory	0.07
SRM Turquoise	0.16
SRM Cranberry	0.06
SRM Brown	0.08
SRM Purple	0.13
SRM Grey	0.05

The above figures meet Requirement A; if the figure above for SRM Turquoise were 0.33 ($> 2 \times 0.16$), the manufacturer's certification would not be granted.
    The difference of the average from SRM value:

SRM Black	+0.07
SRM Ivory	−0.06
SRM Turquoise	+0.02
SRM Cranberry	+0.03
SRM Brown	−0.25
SRM Grey	+0.05

The second decimal figure is not defined in Requirement B, indicating that the difference could be between 0.20 and 0.29; therefore, the seven cements meet the requirement. If the figure above for SRM Purple were 0.50 ($> 2 \times 0.2$), the manufacturer's certification would not be granted. This procedure should be repeated for all the elements analyzed by XRF to have the spectrometer approved by ASTM specifications.

9.1.8 Conclusion

SRMs and CRMs used in an XRF calibration provide reliable and accurate analytical results; they are well suited to establish a statistical control over the results produced by the spectrometer. As for ASTM approval, NIST Standards are unconditionally required. Both International Reference Materials and NIST Standards should be available to laboratory staff at a convenient location, thereby encouraging frequent use.

9.1.9 Monitor Samples

The monitor samples should be chosen so that they correct the intensities for drift over a large range, matching or even exceeding the range covered by the calibration for as many different materials and programs (calibrations) as possible. In order to reduce the time needed to correct the instrument drift, the aim is to keep the total number of monitors as small as possible.

Examples are given in this section on how to design and fabricate general purpose monitors [12]. The examples involve two glass disks [12, 13], a Fe-based disk, and a Ni-based alloy disk [12]. An additional (fifth) Teflon disk may serve in the drift correction for only one element: fluorine. The disks are 40 mm in diameter and 3–7 mm thick (alloy disks and glass disks, respectively). The Teflon disk is 26 mm thick. They were specifically designed to cover the light elements encountered in the aluminum industry but may serve as highly stable drift correction monitors in the X-ray fluorescence analysis of many materials. They cover the major, minor, and trace elements in most routine analyses in the areas from mining to fabrication. They can also be used to compare instrumental performance and to set up instrument quality control practices. *The first four monitors are practically indestructible.*

Glass Disks   The first two samples were cast as Si-based glass disks by Analytical Reference Materials International, a U.S.-based supplier of standard reference materials. They were manufactured according to the specification provided by Alcan, using the compounds given in Table 9.7.

The powdered constituents were mixed and heated in a graphite crucible, melting at approximately 1100°C. Quartz alone melts at about 1750°C to an extremely viscous liquid, but if some sodium carbonate is added to the pulverized quartz, it can be fused at a temperature easily obtained in a muffle furnace. Quartz combines with metal oxides to form silicates. For successful casting of a quartz-based glass, the SiO_2 content must be at least 35–38%. Therefore, production of a single glass disk saturated with all elements of interest would be impossible. An initial glass gullet was produced by melting previously dehydrated, weighed, and mixed powdered components (the target composition of the two glass disks is given in Table 9.7).

TABLE 9.7. Data for Casting of the Iron-based Alloy

Constituent	Disk 1 (%)	Constituent	Disk 2 (%)
Na_2CO_3	13.5	Na_2CO_3	10
MgO	7	Na_2SO_4	3
Al_2O_3	10	SiO_2	43
SiO_2	42	$Na_4P_2O_7$	10
$Na_4P_2O_7$	5	V_2O_5	3
K_2O	1	Cr_2O_3	3
CaO	3	$Mn_2P_2O_7$	5
TiO_2	4	CO_2	3
V_2O_5	2	$SrSO_4$	3
Cr_2O_3	2	MoO_2	2
MnO	2	Nb_2O_5	2
Fe_2O_3	6	$Cd_3(PO_4)_2$	3
ZnO	1	Sb_2O_3	2
Ga_2O_3	0.5	$BaSO_4$	2
ZrO_2	1	PbS	3
		Bi_2O_3	3

The glass was then ground to -200 mesh and homogenized. Approximately 30 g of the homogenized powder were "slumped" to produce each 40 mm glass disk. Slumping is a process in which a material placed in a mold or a crucible is taken to a temperature above the softening point of the glass. Typically this temperature would be in the 1050°C to 1150°C range. The glass simply flows (or slumps) to fit the mold. The glass disks were then annealed, lapped, and polished to an optical finish. Annealing is a controlled cooling process (slow cooling) designed to reduce the thermal stress in the glass to an acceptable level. The lapping process took the glass disks to approximately two light bands of flatness, and the polishing step (using cerium oxide) was then the final step. The glass disks are exceptionally hard and have extremely high mechanical resistance.

Fe-based Disk A sample of the second type was made by S. Kuyucak of Metals Technology Laboratories of Energy, Mines and Resources Canada in Ottawa, Ontario. First, a high sulfur master alloy was prepared to circumvent the variable recovery resulting from pure sulfur addition. In a 25 kg capacity air induction furnace, 17.4 kg steel billets were melted. Sulfur flowers weighing 2.4 kg wrapped in a total of 0.080 kg aluminum foil in small portions, were stirred into the melt. The furnace was tapped to a ladle at 1580°C, and the melt was poured into 1 in. diameter CO_2 bonded sand molds to produce FeS sticks. Analyzed sulfur turned out to be 9.1%, resulting in a 75% recovery.

The final melts were prepared in a Balzers vacuum furnace, using a 2.8 kg steel (0.4 L) capacity crucible. Table 9.8 shows the charge composition and mass of each constituent added for the total of 2.4 kg.

TABLE 9.8. Composition of Glass Disks, wt %

Element	Charge Material	Mass (Kg)	Concentration (%)
	Furnace Charge		
Fe	Fe bar	0.95	Balance
S	FeS sticks	1.08	4.1
Ni	Ni crowns	0.24	10.0
Cu	Cu chips	0.024	1.0
Mo	FeMo (68%)	0.014	0.40
	Deoxidation and Early Additions		
Al	Al chips	0.014	0.60
Nb	FeNb (67%)	0.016	0.44
V	FeV (79%)	0.012	0.40
	Late Additions		
Al	Al foil	0.010	0.40
Ti	Ti sponge	0.012	0.50
Ta	Ta powder	0.011	0.46
Ce	Ce (95%)	0.013	0.51

As seen in Table 9.8, alloying additions were made in two batches. Early additions were those with moderate affinity for oxygen (FeNb, FeV), and late additions were those with high affinity for oxygen (Ce, Ti, Ta). To improve the corrosion resistance, nickel and copper were added, whereas molybdenum was added to increase hardenability. Aluminum was added for complete deoxidation, prior to the alloy additions. Although manganese and chromium addition would improve the sulfur stability, they were not added.

Two cylinders 100 mm long and 55 mm in diameter were cast. They were lip poured into a CO_2 bonded sand mold within the vacuum chamber. They were machined to the final diameter of 40 mm and sliced into 5 mm thick disks. A few individual disks display a limited microporosity in the center, but this porosity does not affect the disk performance in the X-ray measurement or its stability.

Ni-based Alloy Disk A sample of the third type was produced by melting a complex Ni-based alloy, followed by die-casting to form a cylindrical ingot. It was made by A. Palmer of Metals Technology Laboratories of Energy, Mines and Resources Canada in Ottawa, Ontario.

First, a master alloy of low-melting metals, fused in a crucible below 530°C, was prepared in a muffle furnace. The elements bismuth, lead, tin, and zinc were melted together in the proportion 4:5:3:2 (below 530°C) to produce a 3 kg bar which could be cut up for addition to the nickel-based melt. This master alloy had a fine, white crystalline structure which was easily broken. Next, an electrical furnace was used to manufacture the final sample. The furnace had six compartments for additives, which were loaded in a predetermined order:

Loaded in the crucible	Ni	1.68 kg
	Co	0.12 kg
	Cu	0.06 kg
	Cr	0.06 kg
	Fe	0.03 kg
	Si	0.02 kg
	Mg	0.06 kg
	Mo	0.02 kg
No. 1 compartment:	Ti	0.06 kg
	+Mn	0.06 kg
No. 2 compartment:	Zr	0.09 kg
No. 3–6 compartments:	master alloy divided into four roughly equal portions	0.84 kg

The vessel was pumped down and back filled with argon at about 70 kPa, while the initial charge was melted. The power was switched off during each addition; in this way splashing and fuming were kept to a minimum. The metal was seen to boil gently when all of the elements had been added. It was not possible to get a temperature reading. The melt was held at about 45 kPa for 10 minutes after all of the additions had been made, in an attempt to remove dissolved gases.

The alloy was poured into the mold made from silicate-bonded quartz-sand and treated with a graphite wash. A small "hot-top" was provided on its upper part, in order to delay freezing of that part of the ingot and to minimize shrinkage porosity in the main body. The final ingot showed no porosity, but there was some center-line shrinkage, which was unavoidable in a narrow cylindrical shape. The 120 mm cylinder, 50 mm in diameter yielded 18 slices, each a few mm thick. One surface of each disk (the analytical surface) was polished.

Teflon Disk   A commercial Teflon rod was scalped to 40 mm diameter and sliced into 26 mm thick disks. It is a high performance material, machinable with ordinary metal-working tools. The disk is white in color, and is nonwetting, exhibits zero porosity, and unlike ductile materials, will not deform and will not out-gas in vacuum environments.

Ausmon Setting-up Sample   A setting-up sample called Ausmon and made as a borate based disk was purchased from Philips. The disk contains the following 13 elements as majors: fluorine, sodium, magnesium, aluminum, silicon, phosphorus, sulfur, chlorine, potassium, calcium, titanium, manganese and iron. In addition, the disk also contains approximately 2000 ppm of 34 different elements. No long-term stability test was performed for this disk, so it is difficult to speculate on its value as drift correction monitor.

Low Setting-up Samples   The monitors described so far can mostly serve as high setting-up samples in spectrometer drift correction procedures. Two low setting-up samples, for low intensity measurement, were also examined. A quartz disk was purchased from Philips. A nickel cylinder (SUS-RNI 11/6) was obtained from Brammer Standard Company. It was sliced, polished, and analyzed.

9.1.10   Procedure

Choice of Elements   Table 9.9 lists the intensities measured for elements present in each monitor, obtained on ARL 8480 spectrometer equipped with the rhodium end window tube. All modern software requires the use of a certain peak intensity to correct for drift. Background measurement is not necessary. Therefore, only peak intensities obtained in the analysis of each element are given. It has to be noted that

TABLE 9.9. Intensities of Elements in Monitor

	Intensity (K c/s)						
Element	No. 1 Brown Glass	No. 2 Blue Glass	No. 3 Fe Disk	No. 4 Ni Disk	No. 5 Teflon Disk	Recommended Monitor High	Low
C	1.56	1.55	0.43	0.56	1.37	5	1
O	5.70	5.25	0.28	0.20	—	1	4
F	0.28	0.28	2.71	0.96	60	5	1
Na	40	39	0.58	2.26	—	1	4
Mg	6.5	0.45	0.46	0.25	—	1	4
Al	72	62	2.46	1.36	—	1	4
Si	322	322	3.18	0.92	—	1	4
P	47	48	0.66	0.31	—	1	4
S	0.76	6.85	152	0.77	—	3	4
Cl	0.29	0.48	0.25	0.39	—	2	3
K	40	8.3	0.72	0.65	—	1	4
Ca	83	4.3	0.81	1.56	—	1	4
Ti	43	2.1	9.22	58	—	1	2
V	39	37	18	5.33	—	1	4
Cr	45	71	6.24	160	—	4	3
Mn	49	54	8.67	121	—	4	3
Fe	190	11	1163	100	—	1	3
Co	1.89	94	12	185	—	4	1
Ni	0.99	1.76	157	955	—	4	1
Cu	1.42	1.76	26	98	—	4	1
Zn	129	7.42	0.52	56	—	1	3
Ga	52	8.43	0.88	1.63	—	1	4
Sr	7.48	300	1.19	1.39	—	2	4
Zr	316	175	1.48	110.	—	1	3
Nb	8.22	187	29	1.69	—	2	4
Mo	45	98	37	14	—	2	4
Cd	11	70	3.84	3.81	—	2	4
Sn	8.92	6.42	2.00	69	—	4	2

potassium and calcium intensities are exceptionally high when a scandium tube is used; this effect is due to preferential excitation of these two elements by the Sc $K\alpha$ lines. Their intensities are much lower when a rhodium window tube is used.

Instrument Settings The parameters given in Table 9.10 were optimized for a Philips PW 1400 X-ray spectrometer, equipped with a scandium side window tube, operated at 50 kV and 50 mA. For most of the elements, $K\alpha$ lines were used. The only exceptions were Pb $L\alpha$ and Bi $L\alpha$. It is inconvenient to measure $K\alpha$ intensities for tin and antimony if they are present in the sample simultaneously. In this case L lines should be used.

TABLE 9.10. Instrument Settings for Philips PW1400 X-ray Spectrometer

Element	Det	Xtal	UWL	LWL	Goniometer Angles (2θ °)	Analysis Time T_p seconds		
						Short	Long	Philips
O	F	PX-1	80	20	57.70	120	120	120
F	F	PX-1	90	26	43.86	40	90	90
Na	F	PX-1	80	20	28.22	2	6	10
Mg	F	TlAP	90	25	45.23	7	29	30
Al(1)	F	PX-1	82	20	19.65	2	3	8
Si	F	InSb	80	25	144.80	2	6	10
P	F	Ge	80	20	141.03	4	27	30
S	F	Ge	76	38	110.69	14	20	30
Cl	F	Ge	75	27	92.76	30	30	40
K	F	LiF	80	20	136.69	2	13	6
Ca	F	LiF	75	25	113.09	2	5	8
Sc(2)	F	LiF	80	20	97.70	2	2	6
Ti	F	LiF	80	06	86.14	2	8	12
V	F	LiF	80	12	76.94	2	3	8
Cr	F	LiF	75	15	69.36	2	2	6
Mn	FS	LiF	85	20	62.97	2	8	12
Fe	FS	LiF	80	15	57.52	2	4	8
Co	FS	LiF	70	15	52.80	3	5	8
Ni	FS	LiF	80	22	48.67	2	5	8
Cu	FS	LiF	80	20	45.03	2	4	8
Zn	FS	LiF	80	20	41.80	2	4	8
Ga	FS	LiF	80	16	38.92	2	3	8
Sr	S	LiF	75	20	25.15	2	2	6
Zr	S	LiF	68	20	22.55	2	2	6
Nb	S	LiF	70	25	21.40	4	4	8
Mo	S	LiF	80	26	20.33	2	4	8
Cd	S	LiF	70	25	15.31	2	4	8
Sn	S	LiF	50	10	14.04	2	2	6
Sb	S	LiF	70	25	13.46	2	2	6
Pb L α	S	LiF	68	16	33.93	2	2	6
Bi L α	S	LiF	75	20	33.01	2	2	6

Three times for peak measurement T_p (sec) are given in Table 9.10. The first time (Short) is the shortest (theoretical) time needed to obtain a 0.5% counting statistics error with a high monitor. It would also apply to a one-point drift correction when only a high monitor is measured. The second time is the (theoretical) Long, which corresponds to a Low monitor, counting statistics error better than 1.0%. In practice, the Short and Long times should be set higher to lower the counting statistical error, and permit the sample to complete at least one rotation in the spectrometer chamber. The third time is identified as Philips. It applies to the analysis of an element in all monitors used for this element in the Philips X41 software. The X41 software (version 1992) does not allow an optimum counting time to be separately declared for an element in each monitor. Consequently, a longer counting time must be selected to give the best compromise for both high and low intensity measurements. The settings outlined in Table 9.11 (ARL 8460) were used to measure intensities reported in Table 9.9.

The pulse height selection (PHS) settings, as well as the goniometer angles given in Tables 9.10 and 9.11 are only a general indication. The PHS setting and angular calibration of the spectrometer (using the monitor samples) should be carried out to determine the experimental parameters that are instrument specific.

Notes on the Data Given in Table 9.10

1. Although a PX-1 crystal was used in the analysis of Al, for a few applications, the PET or TlAP crystals would be considered as better choices. Both offer better resolution than the PX-1 crystal but the intensity is lower. Higher crystal resolution would be required for the analysis of low levels of aluminum in samples with high magnesium content.

2. Sc $K\alpha$ Rayleigh scattered intensity applies only to the scandium side window tube. It is measured to determine Sc K line overlap on the Ti $K\alpha$.

9.1.11 Frequency of Monitor Measurement

Regular measurements of the monitors, for example, on a day-to-day basis or before a sample analysis, should establish an accurate drift pattern and enable corrections to be made accurately. The monitor analytical surface should be rubbed with a clean cotton cloth shortly before the analysis. Periodically, it should be cleaned with a paper tissue saturated with alcohol. The monitors must be handled by their edges to avoid fingerprints.

The Teflon disk is the only monitor which requires any surface preparation. After a long exposure to the X-ray radiation, the irradiated surface alters, producing less fluorine $K\alpha$ intensity. Two identical Teflon disks were machined and the fluorine characteristic radiation measured. The F $K\alpha$ intensity was 32.6 and 32.5 Kcts/s, respectively. Then, the second disk was irradiated during 12 hr, and its F $K\alpha$ intensity decreased to 30.2 Kcts/s. Using the recommended analysis time of 90 s, these data translate as a 1.5% drift of the monitor signal for 100 exposures. However, once the

TABLE 9.11. Instrument Settings to Measure Intensities in Table 9.9

Element	Det	Xtal	PHS THR (mV)	WIN (mV)	Goniometer Angles (2θ°)	Analysis Time T_p(sec)	Tube Setting (kV)	(mA)
C	F	AX-16	450	1750	32.69	150	30	80
O	F	AX-06	450	1360	50.70	130	30	80
F	F	AX-06	450	1370	38.77	100	30	80
Na	F	AX-06	450	1100	24.92	10	30	80
Mg	F	ADP	450	1240	136.74	10	30	80
Al	F	PET	450	1120	144.93	10	30	80
Si	F	PET	450	1150	109.03	10	30	80
P	F	Ge	600	103	140.92	30	30	80
S	F	Ge	640	900	110.70	35	30	80
Cl	F	Ge	800	770	92.74	35	30	80
K	F	LiF	450	1230	136.54	15	30	80
Ca	F	LiF	500	1180	112.97	10	30	80
Ti	F	LiF	660	890	86.01	10	60	40
V	F	LiF	700	830	76.86	10	60	40
Cr	F	LiF	770	820	69.23	10	60	40
Mn	F	LiF	790	790	62.80	10	60	40
Fe	S	LiF	450	1260	57.43	10	60	40
Co	S	LiF	450	1180	52.71	10	60	40
Ni	S	LiF	450	1140	48.59	10	60	40
Cu	S	LiF	450	1160	44.93	10	60	40
Zn	S	LiF	450	1140	41.71	10	60	40
Ga	S	LiF	450	1170	38.83	10	60	40
Sr	S	LiF	600	940	25.15	10	60	40
Zr	S	LiF	450	1140	22.55	10	60	40
Nb	S	LiF	450	1160	21.40	10	60	40
Mo	S	LiF	450	1150	20.33	10	60	40
Cd	S	LiF	450	1150	15.31	10	60	40
Sn	S	LiF	450	1160	14.04	10	60	40
Sb Lα	F	LiF	460	1200	117.23	10	60	40
Pb Lα	S	LiF	450	1150	33.93	10	60	40
Bi Lα	S	LiF	450	1110	33.01	10	60	40

exposed surface of this disk had been manually ground with 400 grit grinding paper for over 10 s, the F Kα intensity returned to the 32.5 Kcts/s level. To minimize the irradiation time of the Teflon disk, it is recommended to use it only for fluorine.

Discussion The above monitors are practically indestructible due to their extreme durability. They have long-term stability, and therefore do not have to be replaced. As a result, accuracy of elemental analysis by XRF is not compromised, and frequent instrument recalibration is not required.

It must be realized that the monitors presented in this text are only examples of what can be accomplished in terms of element composition, depending on specific needs. If particular elements are missing in any of the described monitors, they should be supplemented accordingly.

Glass monitors are the most universal monitors, because almost all elements of interest can simultaneously be present. Compounds of elements such as phosphorus, sulfur, chlorine, and calcium, which may have low solubility in a metal matrix, are easily fused into a glass disk. In the glass monitors described, the sulfur intensity is much lower than expected (based on the raw materials used), because the initial mass of glass was reground and fused a few times in order to obtain high homogeneity. In this process volatile elements, such as sulfur, were partially lost. Chlorine may also not produce sufficient intensity. Unfortunately, for some purposes it was not added in the form of a stable compound during the glass disk manufacturing process, because at that time, another sample containing chlorine seemed a better candidate for a monitor. This sample was a KEL F disk. It is similar in appearance to Teflon, but has a different composition. The Cl $K\alpha$ intensity was 760 Kcts/s, as measured by the Philips PW spectrometer using the germanium crystal. The sample also produced 13 Kcts/s intensity for fluorine. Regrettably, the analytical surface of the KEL F disk altered much faster after prolonged X-ray exposures than a Teflon disk did, and the disk was rejected.

Each of the new monitors (except Teflon) can serve to correct the drift of more elements than a conventional monitor. Therefore, it is possible to reduce the number of necessary monitors substantially. If the spectrometer software permits more than one monitor per analytical program, a large number of unstable monitors used in many laboratories up to now should be replaced with the new generation monitors.

In the newest Philips X41 software, up to six monitors can be specified in a file called "monitor buffer" for drift correction of an element. Usually, a two-point drift correction is adequate, unless the low point does not provide enough intensity. If there is a lack of intensity, then a three-point drift correction is a better choice, especially as the third intensity is available (the X41 software measures as many points as there are monitor samples in the monitor buffer; the user declares which points are to be used in the drift correction procedure).

The drift correction procedure involves a two-point drift correction using an upper (high) and a lower (low) monitor. The choice of monitors is usually recommended in such a way that for typical applications (e.g., geological), just two monitors serve most of the elements. The least employed is monitor No. 5, which produces the highest intensity for fluorine.

It has to be said that if a monitor contains too many major elements (especially medium and heavy elements), than there is a line overlap problem. As a result, not only intensities of certain lines are affected, but their peak positions are shifted. Consequently, it would be risky to use the affected lines for the instrument angular calibration. In order to avoid or minimize the line overlap, use of more than one high setting-up sample and distribution of critical elements (for example, tin and antimony) among a few monitors would be beneficial.

9.2 RECALIBRATION

9.2.1 Introduction

The calibration of the X-ray spectrometer has been covered in detail in other sections of this text. At this point, it is assumed that calibration is complete, and that errors associated with the instrument and the specimen have been reduced to a degree that the accuracy requirements for analysis have been obtained. The errors associated with the specimen fall into two categories, specimen heterogeneity and matrix effects. Reducing specimen heterogeneity errors is the domain of specimen preparation, and the reader is referred to sections of the text that discuss in detail specimen preparation guidelines. The systematic errors associated with the instrument are dead time, pulse shift, background, line overlap, and drift. The hardware of a modern X-ray spectrometer provides corrections for dead time and pulse shift. Corrections for background, line overlap and drift are available in the software provided by the manufacturer; however, it is the task of the analyst to initiate these corrections.

Background subtraction is recommended when the concentration of the analyte element approaches the detection limit and where better calibration accuracy is improved by plotting the net, rather than the peak intensity vs. concentration. Line overlap occurs when the characteristic radiation of two or more elements occur at nearly the same two theta angle. The systematic error associated with line overlap is reduced by increasing the angular dispersion of the intensity measurement (ability to separate lines) or by performing a line overlap correction based upon the intensity or the concentration of the overlapping element. Such line overlap correction software is provided by the manufacturer.

Once the calibration has been completed, the task of the analyst switches to calibration line maintenance. The factors that effect the integrity of the calibration line are short- and long-term instrumental drift and changes in sensitivity of the X-ray spectrometer. Strategies to correct for these dynamics in the X-ray spectrometer output will be discussed in this section.

9.2.2 Quality Standard Programs

Good analytical practice dictates that each instrument be tested for accuracy. This is done by comparing the X-ray and expected value of the quality standard. The X-ray values must fall within the accepted concentration range of the quality standard in order for the instrument to be described as *in control*. The concentration range of the quality standard can be estimated as:

$$\pm \text{range}(\%) = 0.04\sqrt{C}$$

or

$$\pm \text{range}(\%)[4\sigma] = 4\frac{1}{\sqrt{ST}} \times \sqrt{(C - D)}$$

where

S = Sensitivity [c/s/%]
T = Counting time [s]
C = Concentration value of the quality standard
D = Concentration equivalent of the background.

Another definition for D is the concentration value where the calibration line crosses the concentration axis.

As a worked example, assume that the sensitivity for silicon in a stainless steel is 9,000 c/s/%, the concentration value is 1%, the counting time is 20 s and the calibration line background is -0.01%. The concentration range for Si in a quality program would be $1\% \pm 0.0024[1\sigma]$ or $1\% \pm 0.01[4\sigma]$. The instrument is in control if the Si values range from 1.01 to 0.99%.

The quality standard program tests whether the instrument is in or out of control. "Out of Range" error messages indicate that a drift correction or recalibration is needed.

9.3 INSTRUMENTAL SHORT- AND LONG-TERM DRIFT

9.3.1 Overview

Any line with a unique set of instrument operating conditions should be monitored for drift. Drift programs can be shared, as long as the measurement conditions are common. Backgrounds are not measured in drift correction. Drift corrections for the peak are also applied to the background because instrument operating conditions, except for the two theta value, are the same. Analyte line count rates in the drift correction program would be high in order to obtain small counting statistical errors in a short counting time.

Changes in the output of the X-ray spectrometer are categorized as shown in Table 9.12. Changes in the output of the X-ray spectrometer fall into categories of

TABLE 9.12. Changes in X-ray Spectrometer Output

Type of Change	Description	Corrected By
Ultra Short Term (fraction of a second)	Line "spikes" or transients	Line stabilizers and conditioners
Short Term (8 to 16 hrs)	Changes in X-ray tube flux output	Chillers to control the cooling water temperature for the X-ray tube
Long Term (weeks, months)	Changes in X-ray tube output due to tube aging and the build-up of parasite lines	Drift correction involving one or more samples

ultrashort-term, short-term, and long-term drift. Ultrashort-term drift is associated with transients in the line or improperly functioning spectrometer hardware components. The manufacturer provides site surveys in which the characteristics of the incoming line are monitored and defined. When line conditioners and stabilizers are recommended to correct inconsistencies in the incoming line, such investments should be made. Short-term drift (hours) is associated with the temperature stability of the X-ray tube water cooling system. The manufacturer will suggest a water cooling system of sufficient capacity to control the effect of changes in cooling water temperature on the intensity output of the X-ray tube.

Major contributors to changes in the X-ray spectrometer output are:

- Tube aging.
- Changes in intensity of parasite lines from the tube spectra.
- Contamination of the X-ray spectrometer by poorly prepared samples.

Short-term (typically hours) drift tends to be cyclic and can be attributed to variations in temperature of the X-ray tube, which, for the 3000 watt X-ray tubes used in wavelength dispersive systems, is often traceable to temperature variations in the cooling water. This cyclic drift is in the order of 0.1–0.15% relative in today's wavelength dispersive X-ray spectrometer. If calibration accuracy better than 0.1% relative is required, which is rare, then the frequency of drift correction is increased to compensate for this cyclic change. It is also important that counting errors be smaller than ±0.1%, which means counts in excess of 1,000,000 must be accumulated so that the error in the X-ray intensity measurement is smaller than the long-term drift.

Instruments are also subject to long-term drift (weeks). Intensity measurements for the analyte lines can increase (parasite lines from the X-ray tube or spectrometer) or decrease (tube aging). It is most important that any change in intensity have its origin in the instrument, not the specimen used in the drift correction program. All specimens used in drift correction programs must be permanent, and be inert to frequent exposure to the X-ray beam. A drift correction program is started by making an intensity measurement on the analyte line from a permanent specimen, putting that intensity value in the numerator and the denominator, to produce a drift factor of 1.000. Subsequent intensity measurements are substituted for the number in the denominator resulting in drift factor greater or less than 1.000. The most recent intensity measurement on the analyte line is multiplied by that drift correcting that measurement backwards in time to the moment that the drift correction program was started.

9.3.2 Recalibration

When the quality standard program indicates that the system is *out of control*, and drift correction does not return the system to control, then recalibration is required.

Recalibration will adjust the calibration line to reflect the current sensitivity of the X-ray spectrometer as opposed to the sensitivity of the spectrometer the day of initial calibration. Recalibration is set up by using two standards with widely differing concentration values, and assigning each an X-ray concentration value. Recalibration occurs when the intensities are remeasured and the calibration line redrawn.

9.3.3 Specimens Used for Drift Correction and Recalibration

The specimens used should have the following characteristics:

- Permanent solids inert to frequent exposure in the X-ray beam.
- Homogeneous. Capable of having the surface refreshed by cleansing with a solvent, resurfacing, etc.
- Provide high count rates for the analyte element.

The following types of standards have been successfully used: glasses, fused beads, acrylics, and epoxy doped with CP reagent compounds of the analyte elements, and aluminum metal powder in which CP reagents of the analyte element are mixed. Teflon disks have been used very successfully as zero setting standards, fixing the low end of the drift and recalibration lines.

Glasses, fused beads, and Teflon are homogeneous and their surfaces can be replenished by cleaning or resurfacing. Acrylics, epoxies, and aluminum metal powder are not homogeneous, and the surface of the sample is permanently changed by polishing, cleaning, etc. This latter category of specimens has been successfully used; however, great care must be taken to prevent surface contamination. These specimens are placed within a sample cup and not removed, thus preserving the integrity of the sample surface.

9.4 SOURCES FOR REFERENCE MATERIALS FOR X-RAY DIFFUSION AND X-RAY FLUORESCENCE

Standard samples are needed during analyses for several purposes. There are setting-up samples and alignment samples for the initial installation and calibration of equipment. There are calibration samples that may be run during analyses to obtain accuracy in the measurements. Also, there are certified samples to be run to confirm that measurements are all on the same acceptable scientific and legal scale. Many specimens in use may be prepared in the laboratory, but there are a few external sources for special materials.

9.4.1 Office of Standard Reference Data

The Office of Standard Reference Data (OSRD) at the National Institutes for Standards and Technology (NIST) is a primary source of certified samples for all types of analysis including fluorescence and diffraction. The current catalog may be obtained from OSRD by writing to the office at NIST. OSRD supplies many analyzed materials, including irons and steels, cements, and other materials for fluorescence certification and *d*-spacing and intensity samples for diffraction certification.

Perhaps the term "certified" needs some clarification in the context used here. OSRD runs multiple specimens and makes many tests to characterize the materials it supplies. The results are analyzed statistically to indicate any potential variability and to establish the limits of accuracy that can be expected from measurements on the supplied samples. This level of study makes these samples acceptable for situations where litigation may be involved. It also makes the samples more expensive, so their use is not always advisable for the more routine analyses that consume significant quantities of the standard.

There are sources for XRD standards in some other countries, but the details are not available for this book. For example, there is a standard silicon used in China.

9.4.2 Alternate Sources of Standard Materials

Many materials provided by chemical supply houses or other sources may be used as standard samples for in-house comparisons. To be really useful, such samples should be calibrated against other samples of known accuracy such as the OSRD certified samples. There are problems with some samples obtained from suppliers, however. For example, there is lots of silicon crystal available, which is used for semiconductor applications. This silicon is invariably doped, so that its cell parameter does not conform to that reported for pure silicon. The difference in cell parameter is significant (usually in the fourth decimal), so that it is easily observed with modern equipment. If such a silicon were to be calibrated against a secondary sample that had been calibrated against the OSRD silicon, the new silicon could be used as an in-house standard. Some other materials that have been used for *d*-spacing standards include NaCl, halite; silver; tungsten; SiO_2, quartz; $MgAl_2O_4$, spinel; LaB_6, and $CaCO_3$, calcite. A comment needs to be made on the spinel. One must be very careful to obtain stoichiometric spinel because there is an extensive solid solution toward Al_2O_3, corundum, which significantly changes the cell parameter. One supplier of secondary silicon and a few other standards is The Gem Dugout.

The standards for XRF are more varied and also matrix dependent. They usually are composed of one or more elements at different levels of concentrations within different matrices. Consequently, there are many potential samples that are needed for calibration of XRF analysis. Samples prepared for the analysis of cement are not suitable for analyzing alloys.

There are many sources for XRF standards. The list provided below has been developed with the assistance of several of the contributors to this book. Addresses are not available for all the sources, but are given where known (see section 9.4.4).

9.4.3 Sources for X-Ray Fluorescence Standards

AM, ANSI Committee on Chemical Analysis, USA

ASCRM, Standards, Australia

ASMW, Amt. fur Standardisierung, Messwesen, und Warenprufung, Germany

BAM, Bundesanstalt fur Materialforschung und Prufung, Germany

BAS, Bureau of Analyzed Samples, Ltd., England

BCR, Community Bureau of Reference, Belgium

BCS, Bureau of Analyzed Samples, Ltd., England

BCIRA, The British Glass Industry Research Association, England

BR, Breitlaender GMBH, Germany

CAN, Canada Centre for Mineral and Energy Technology Energy, Mines, and Resources, Canada

CXD, Praha Czech Rep.

CMSI, China Metallurgical Standardization Research Institute, China

CSB, China Standard Reference Materials, Beijing, China

CTIF, Centre Technique des Industries de la Founderie, France

DOMTAR, Domtar, Inc., Canada

FNE, Research Institute for Non-ferrous Metals, Germany

H, Aluterv-Fki, Hungalu Engineering and Development Center, Hungary

HH, HT INCO Alloys International, Canada

IMN, Institute of Non-ferrous Metals, Poland

IMZ, Institute of Ferrous Metallurgy, Poland

OSRD,USA

SCC, Sumitomo Chemical Co., Ltd., Japan

SCRM, Bureau of Analyzed Samples, Ltd., England

SGT, Society for Glass Technology, England

SS Bureau of Analyzed Samples, Ltd., England

ST, Japan

SUS, Bureau of Analyzed Samples, Ltd., England

SGRI, State Glass Research Institute, Czech Rep.

TT, 2-Theta, Germany

UNS, Institute of Mineral Raw Materials, Czech Rep.

V, Vereingte Sluminum Werke (VAW,) Germany

VS, Association of Reference Materials Producers, Russia

VASKUT, Vasipari Kutato Intezet, Hungary 11X to 215X

MBH, Analytical, Ltd., England

9.4.4 Addresses for Sources of X-ray Fluorescence and X-ray Diffraction Standards

Accustandard, Inc., 25 Science Park, New Haven, CT 06511, Phone: 1-800-442-5290

Alpha Resources, Inc., P. O. Box 199, Stevensville, MI 49127, Phone: 1-800-833-3083

Analytical Reference Materials International, Inc., P.O. Box 2246, Evergreen, CO 80439-2246, Phone: 1-303-670-1300

Brammer Standard Company, Inc., 14603 Benfer Road, Houston, TX 77069-2895, Phone: 1-713-440-9396

Bureau of Analyzed Samples, Ltd., Newhan Hall, Newby, Middlesbrough, Cleveland, UK. TS8 9EA

CANMET, EMR, Canada, 555 Booth Street, Ottawa Ontario, K1A 0G1, Phone: 1-613-995-4738

CRC Press (Geochemical Reference Materials), 3000 Corporate Boulevard, N.W., Boca Raton, FL 33431

G. Frederick Smith Company, P.O. Box 245, Powell, OH 43065, Phone: 1-900-881-5501

LECO Corporation, 3000 Lakeview Avenue, St. Joseph, MI 49085, Phone 1-616-983-5531

National Physical Laboratory, Teddington, Middlesex, UK, TW11 0LS, Phone: 01-977-3222

OSRD, Office of Scientific Reference Data, Building 202, Room 204, National Institutes of Standards Technology, Gaithersburg, MD 20899, Phone: 1-303-975-6776

The Gem Dugout, 1652 Princeton Drive, State College, PA 16803, Phone: 1-814-238-4069

REFERENCES

[1] Cali, J. P., et al. (1975), "Standard reference materials: The role of SRM's in measurement systems," *NBS Monogram 148*, NIST, Gaithersburg, MD 20899.

[2] *Annual Book of ASTM Standards* (1994), Section 14: General Methods and Instrumentation, Volume 14.02, p. iii.

[3] *NIST Standard Reference Materials Catalog, 1992–93* (1992), pp. 5–7, U.S. Government Printing Office, Washington, D.C.

[4] Brammer Standard Company, Inc. (1992, 1993), *Quality Analytical Materials from Around the World*.

[5] Sutarno, R. and Steger, H. F. (1985), "The use of certified reference materials in the verification of analytical data and methods," *Talanta*, **32**, 439–445.

[6] Taylor, J. K. (1987), *Quality Assurance of Chemical Measurements*, Lewis Publishers, pp. 33–39.

[7] Lechler, P. J. and Desilets, M. O. (1987), "A review of the use of loss on ignition as a measurement of total volatiles in whole-rock analysis," *Chem. Geol.*, **63**, 341–344.

[8] Babikian, S. K. (1993), "Raw materials analysis by XRF in the cement industry," *Adv. X-ray Anal.*, **36**, 139–144.

[9] Haswell, S. J. (1992), *Practical Guide to Chemometrics*, Marcel Dekker: New York, pp. 5–37.

[10] *Annual Book of ASTM Standards* (1994), Section 4: Construction, Volume 4.01 Cement; Lime; Gypsum, p. 88.

[11] Forte, M. (1983), "Fabrication and use of permanent monitors and standards," *X-ray Spectrom.*, **12**, 115–117.

[12] Feret, F. (1979), "Setting-up samples in the X-ray fluorescence analysis of powder materials," *Wiadomosci Hutnicze*, Katowice, No. 9, 301.

[13] Feret, F. (1981), "Metallic and glass-made setting-up standards for drift correction of X-ray spectrometer," *XXII Colloquium Spectroscopicum Internationale*, Tokyo, Japan, Sept. 4–8, 1981.

GLOSSARY

Abrasion: The reduction of size of particles by rubbing with a harder material surface.

Absorber: A material that reduces the intensity of a transmitted beam by interactions with the electrons.

Heavy: A material, composed of some or all high z elements, that significantly attenuates an X-ray beam.

Light: A material, composed of only low z elements, that does not significantly attenuate an X-ray beam.

Absorption: The reduction of photon intensity due to interaction with electrons in the sample (absorber).

Accuracy: In exact conformity with fact; without error.

Acicular: A morphological term for particle shapes that have the long dimension 15 or more times the other dimensions.

Adhesive: A material used for bonding two or more units into a single piece.

Air-sensitive samples: Samples that react with one or more components of the atmosphere at ambient temperature and pressure.

Aliquot: A representative portion of a material used for a specific analysis.

Alignment samples: Samples that are used in the alignment procedure for testing an instrument.

Allotrope: Different forms (crystalline and amorphous) of elements.

Alloy: A homogeneous metal containing two or more elements.

Amorphous: A material that lacks long-range order such that diffraction effects from periodicity are not observable experimentally.

Anisotropic: Having properties whose magnitude depends on the direction in the crystal or block.

Annealing: A thermal treatment of a sample below its melting point allowing it to dissipate strain energy and recrystallize.

Apiezon grease: A commercial laboratory petroleum-based grease useful for holding powders.

Attenuation: The reduction in the intensity of an X-ray beam due to photoelectric absorption and scattering.

Attrition: Reduction of particle size by rubbing like particles together.

Background, XRD: Scattered intensity in a diffraction experiment that is not due to diffraction from the phases in the specimen.

Background, XRF: The intensity that occurs at a 2θ angle on an analyte line when the analyte is absent.

Back-reflection region: The region in a diffraction pattern where the diffraction angle is greater than $90°$.

Beam area: see Irradiated area.

Belt sander: A type of sander with a continuous belt driven by two drums.

Binder: A material added to a specimen to hold the particles together as an integral unit.

Bladed: A morphological term for particles with one small dimension and one long dimension where the third dimension is intermediate.

Bragg equation: An equation that shows the relationship between the X-ray wavelength (λ), interplanar spacing (d), and scattering angle θ.

Calibration: Correction of a measurement scale to accurate values using measurements from a known (certified) standard.

Calibration samples: Substances that allow calibration of the experimental measurements. They may be either external or added to the specimen.

Camera: A device that uses photographic film or an image sensor to detect information.

Capillary tube: A glass (or silica glass) tube with a bore less than 2 mm diameter, which may have either very thin walls or very thick walls.

Cleavage: The breaking of crystalline fragments to yield planar surfaces due to weaker forces in the crystal structure.

Coherent: Operating together (in phase). Allows interference effects.

Collimator: Device used to limit the shape and size of a beam.

Compound: A chemical term for substances formed by the bonding of elements into a unique substance.

Confidence interval: An interval within which one estimates a given population parameter to lie.

Contaminant: An unwanted material introduced to a specimen during preparation.

Conversion factor: A factor that allows units to be converted from one system of measurement to another system.

Cow gum: see Rubber solution.

Counter (X-ray): A device used to detect the particle character of an X-ray beam.

Crystal: A continuous unit of material whose atomic arrangement is periodic. The boundaries may be planar or irregular.

Crystal substrate: A wafer of a single crystal used to support a coating or powdered specimen.

Crystallite: The coherent crystal unit that diffracts in phase. Requires no defects within the volume to disrupt diffraction.

Crystallite size: The dimensions of the coherent crystal unit that contributes to the diffracted intensity.

Culet: A small facet on the tip of a gemstone, usually a diamond.

Dead time: That portion of time in a counter where a second photon cannot be detected because the detector is still responding to the first photon.

de Jongh formula: An algorithm used in quantitative X-ray fluorescence analysis that applies theoretical matrix corrections to convert intensities to concentration.

Detector: A device for sensing radiation.

Diamond substitute: A material that substitutes for diamond in any application.

Diffraction: The process of scattering and interference of a beam by electrons or other point scatterers.

Diffraction pattern: The complete set of positions and intensities obtained in a diffraction experiment.

Diffraction trace: A graphical representation of diffraction information obtained using any of several instruments plotted as intensity versus angle.

Diffractogram: A diffraction trace obtained using a diffractometer to collect the data.

Diffractometer: A mechanical or electronically controlled device for measuring the diffraction angle of X-rays.

Disk sander: A sander that uses planar disks.

Dislocation: A linear defect in a crystal along which the structure shifts registry.

Disorder: Applied to crystals where some or all the atoms in the structure deviate from the positional periodicity characteristic of the crystalline state.

Domain: A unit of a crystal that contains no defects.

Double-coated: Tape or film that has the active layer on both surfaces.

Drift: A systematic change in a measurement that occurs with time.
> *Long-term:* Changes that occur slowly in terms of the time of the experiment.
> *Short-term:* Changes that occur rapidly in terms of the time of the experiment.

Effective thickness: The thickness of a specimen that contributes useful information to a measurement.

Elongate: A morphological term for describing particle shapes that have two short dimensions and one long dimension.

Epitaxy: Continuation of a growing crystal in perfect registry on the surface of a seed of the same material.

Equant: A morphological term for describing particle shapes that have all dimensions with similar magnitudes.

Errors: Deviations of measurements from the correct value.
> *Random:* Errors that are related to purely statistical processes.
> *Systematic:* Errors induced by experimental factors that vary with changing experimental conditions.

External standard: A standard reference material used for alignment and calibration of an X-ray diffraction instrument prior or after analysis.

Extinction: Reduction of intensity due to multiple diffraction in a specimen.

Primary extinction: Reduction of diffracted intensity due to multiple diffraction in a perfect crystal.

Secondary extinction: Reduction of diffracted intensity due to multiple diffraction in an imperfect crystal.

Extinctions (intensity): Absences of measurable intensity where the periodicity would predict occurrence.

Accidental: Extinctions that do not follow a pattern. See also Structural.

Lattice: Extinctions due to a centered lattice description.

Structural: Extinctions due to the atomic arrangement which cause the intensity to be below the observable limit.

Symmetry: Extinctions due to the translation components of the symmetry elements in a structure.

Systematic: Extinctions that follow a pattern.

Fibrous: A material that is composed of particles that are acicular.

Film: A polymeric or gelatin sheet with photosensitive emulsion on one or both sides or a very thin layer of material on a smooth substrate.

Filter (X-ray): A thin sheet of a metal used to differentially attenuate the β radiation.

Fluorescent yield: The probability that the filling of a vacancy in a specified shell will result in the emission of a characteristic X-ray photon as a result of creating the vacancy by either primary or secondary excitation.

Foil: A thin sheet of metal or plastic.

Forensic science: The application of science to criminal and legal situations.

Fracture: Breaking a particle along an irregular surface.

Front reflection: The region of the diffraction pattern where the angle is less than $90°$.

Fusion: A technique used to prepare a specimen with reduced heterogeneity effects and suitable for quantitative XRF analysis. A mixture of flux and sample are heated to a temperature high enough to melt the flux, which then reacts with or dissolves the sample to produce a one-phase glass on cooling.

Goniometer: Any device for measuring angles.

Grain: A crystal unit in a polycrystalline block.

Gum Arabic: A gum, obtained from several species of mimosaceous plants, that forms a rigid paste when mixed with water.

Habit: The external shape of particles.

Half-depth of penetration: The horizon in a specimen above which one-half the diffracted intensity originates. Magnitude of depth depends on attenuation coefficient and X-ray wavelength.

Hardness: Resistance to scratching usually described by the Moh's scale $(1-10)$ or an indentation measurement.

Heterogeneous: A substance where the properties vary as a function of position in the specimen.

Homogeneous: A substance where the properties do not vary as a function of position in the specimen.

Hygroscopic: Affinity to absorb water from the atmosphere.

Ideally imperfect crystal: A crystal composed of a mosaic of domains that may be slightly misoriented. The basic crystal concept for kinematic diffraction theory.

Ideally perfect crystal: A crystal with no defects in the periodicity. The basic crystal concept for dynamical diffraction theory. Results in multiple internal diffraction and primary extinction.

Incident: The beam path from the source to the specimen.

Incoherent: Does not operate together. Does not allow ray interference.

Inert: Chemically unreactive.

Infinite thickness: The thickness of a specimen that allows essentially 100% of the possible diffraction intensity in a reflection orientation.

Ingot: A mass of metal cast from the molten state into a convenient shape.

Internal standard: A standard reference material intermixed with the specimen during analysis.

Irradiated area (volume): That area (or volume) on a specimen which is bathed by the collimated incident beam.

Isomorphous: Substances which have the same structure and which vary continuously in chemical composition between the end compositions.

Isostructural: Materials that have the same crystal structure.

Lap: To cut or polish with a rotating flat disk.

Lattice: The geometric description of the periodicity of a crystalline arrangement (not the arrangement of the atoms in a structure).

Low atomic number (light) element: An element with $Z \leq 22$.

Line: A term in powder diffraction used to indicate the occurrence of diffracted intensity (carried over from the Debye–Scherrer films).

Line broadening: Broadening of the diffraction profiles over the instrumental effect due to crystallite size and strain.

Linear absorption coefficient: The value of the linear attenuation of the intensity of an X-ray beam in a material for a specific wavelength (units are usually cm^{-1}).

Low angle: That region of a diffraction pattern generally below $10°$.

L.O.I.: Loss on ignition.

Low-background specimen support: A specimen support made from an oriented single crystal of quartz or silicon designed to minimize the scattering or diffraction of X-rays off the support.

Macrostress: External applied stresses.

Mask: A device to control the area of irradiation by an X-ray fluorescence tube on the surface of a specimen. May be attached to the body of the X-ray tube or be part of the specimen cup assembly.

Mass absorption coefficient: Correctly termed "mass attenuation coefficient," it is a measure of the attenuation of an X-ray beam as it traverses a specimen (units usually cm^2 per g).

Matrix: That part of the specimen composed of elements or phases not being examined.

Matrix absorption: The attenuation due to that part of the specimen containing elements or phases not being examined.

Maximum: The largest value in a set of measurements.

Mean: The average value of a set of measurements.

Mesh: A measure of the opening in a screen used for sizing particles nominally implying the number of particles required to make up one inch when placed in a line.

Microabsorption: The masking of some grains in a powder specimen due to different attenuation coefficients.

Microsamples: Specimens whose weight is less than 5 mg.

Microstress: Internal stresses from defects in crystals and grain boundary interactions.

Microtome: A device for cutting very thin slices from a sample.

Mode: The most frequent value in a set of measurements.

Modeling clay: A semifirm plastic material used for supporting bulk specimens for measurements.

Monitors: see Setting-up samples.

Monolithic specimen: A specimen of material used as a solid block.

Mosaic crystal: An arrangement of small crystallites arranged with slight misorientation that comprise the whole crystal.

Muffle furnace: A type of furnace for heating materials in a flame-free environment.

Mylar (Melinex) film: A film of polyethylene terephthalate used as a specimen support (thicknesses 0.25 to 6 μm).

Packing: The mutual arrangement of particles in a composite specimen. Can also be applied to the arrangement of atoms in a crystal.

Particle: A physical fragment isolatable in space from other fragments. May contain one or more crystal units.

Particle size: The physical dimensions of an isolatable fragment in a disaggregated specimen.

Peak: A term usually employed to describe a concentration of intensity in a diffraction pattern.

Petroleum jelly: A petroleum-based grease used in specimen preparation.

Phase: A homogeneous, physically distinct portion of matter in a system which is not homogeneous.

Photon: A particle of energy equivalent to the wavelength energy of electromagnetic radiation.

Plasticine: A semifirm putty used to support specimens for analysis.

Polycrystalline: A specimen composed of many crystallites in different orientations.

Polymorph: One of several different crystalline states for the same chemical composition.

Polytype: A special form of polymorph where the crystalline differences are in the stacking of identical layers.

Precision: Measurement of experimental reproducibility.

Probability: Predictions based on mathematical models of occurrence.

Profile: The shape of a concentration of intensity in a diffraction pattern.

Recorder: A device for creating a permanent record of a series of measurements. The result may be a graph or a computer-readable file.

Reflection: Creation of a secondary beam of radiation by redirection of the incident beam by a planar surface. Commonly applied to diffraction peaks based on the model of diffraction by Bragg.

Replication: Drawing several samples from the same bulk and preparing them separately.

Reproducibility: The measure of the repeatability of a series of measurements of the same entity.

Rubber solution: A solution of latex in a solvent used for binding materials.

Sample: The material supplied by the requester to the laboratory for examination.

Sample mold: A container or die used to shape and compact specimens for analysis.

Scalar: A device for tallying a series of events (photons arriving at a detector).

Scattering:
Elastic: The deviation of direction of a particle colliding with another particle where there is no transfer of energy.
Inelastic: The deviation of direction of a particle colliding with another particle where there is a loss of energy.

Screening: Separation of powders into desired size ranges using sieves.

Setting-up samples: Materials that are used to set up and test an instrument prior to use.

Slurry: Mixture of a liquid and a powder to form a paste.

Small angle: Usually applies to angles less than 3°.

Solvent: A liquid that dissolves materials to form a homogeneous solution.

Specimen: The physical aliquot of the sample prepared for the diffraction or fluorescence instrument.

Specimen holder: The removable part of the instrument that supports the specimen during analysis.

Specimen-to-flux ratio: A ratio used in the preparation of a specimen for X-ray fluorescence analysis by the fusion method.

Spectrometer: A device that measures the energy (wavelength) distribution of electromagnetic radiation in an experiment.

Statistical significance: A mathematical measure that indicates whether the difference between two measurements may be real.

Strain: The distortions in a material in response to applied forces.

Stress: The forces applied to a material which cause deformation.

Spiking: The addition of the analyte phase to a mixture containing that phase for the purpose of quantification.

Spray drying: A process where loose particles are bonded into spherical aggregates.

Surface roughness: Degree of deviation of a surface from a plane.

Tabular: A morphological term for particle shape that has one short dimension and two long dimensions.

Tacky: A substance that remains sticky and readily adheres to particles.

Tape: A thin ribbon, usually a polymer, which has an adhesive surface on one or both surfaces.

Target: The metal block in an X-ray tube that is bombarded by an electron beam to produce the X-rays.

Texture: The description of the orientation distribution of particles or grains in a polycrystalline compact or block of material.

Topotaxy: The oriented overgrowth of a crystal on a substrate of unlike composition.

Toxic samples: A material poisonous to living organisms.

Vaseline: A commercial petroleum jelly available in pharmacies.

Vector: A mathematical entity that implies direction and magnitude.

Wavelength: The crest-to-crest dimension in a travelling wave of electromagnetic radiation. The Standard International unit to be used for the magnitudes of X-ray wavelengths is the nanometer (nm). Note 10 Å equals 1 nm.

X-ray diffraction: Analysis of crystal structure parameters by measuring the angle and intensity of coherent X-ray scattering from the periodic arrangement of the atoms.

X-ray fluorescence: Elemental analysis by X-ray excitation, dispersion and measurement of characteristic wavelengths from an element or group of elements.

Zone: In a crystal, the family of crystal faces that are all parallel to a given direction. In a diffraction pattern, it is the collection of diffraction peaks which have one Miller index zero.

BIBLIOGRAPHY

Following is a list of references which were not directly cited in the main text of this book. All of these articles contain some information on specimen preparation and are supplemental to material presented in this book.

[1] Andrew, R. W., Jackson, M. L., and Wada, K. (1961), "Intersalation as a technique for differentiation of kaolinite and chloritic minerals by X-ray diffraction," *Soil Sci. Soc. Amer. Proceedings*, **24**, 422–424.

[2] Andrews, K. W. and Hughes, H. (1958), "Microconstituents in steels—their electrolytic isolation and X-ray study," *Iron and Steel*, **31**, 43–50.

[3] Azaroff, L. V. and Buerger, M. J. (1958), *The Powder Method in X-ray Crystallography*, McGraw–Hill: New York, 342 pp.

[4] Barrett, C. S. (1969), *The Mechanisms of Phase Transitions in Solids*, London: Inst. Met., 313 pp.

[5] Barrett, C. S. and Massalski, T. (1966), *The Structure of Metals*, New York: McGraw Hill, 654 pp.

[6] Barshad, I. (1954), "The use of salted pastes of soil colloids for X-ray analysis," Proceedings of the 7th National Conference on Clays and Clay Minerals, 350–364.

[7] Bejwa, I. and Jenkins, D. (1978), "A technique for the preparation of clay samples on ceramic slides for X.R.D.A. using pressure," *Clay Minerals*, **13**, 127–131.

[8] Birks, L. S., Fatemi, M., Gilfrich, J. V., and Johnson, E. T. (1975), "Quantitative analysis of airborne asbestos by X-ray diffraction—final report on feasibility study," *Naval Research Laboratory Report 7874*, 28 February 1975, p. 11.

[9] Blount, A. M. and Vassiliou, A. H. (1979), "A new method of reducing preferred orientation in diffractometer samples," *Amer. Mineral*, **64**, 922–924.

[10] Bohor, B. F. and Randell, E. H. (1971), "Scanning electron microscopy of clays and clay minerals," *Clays and Clay Minerals*, **19**, 49–51.

[11] Bonorino, F. G. (1966), "Soil clay mineralogy of the Pampa Plains, Argentina," *J. Sed. Pet.*, **36**, 1026–1035.

[12] Booker, G. R. and Norbury, J. (1957), "An extraction replica method for large precipitates and non-metallic inclusions in steels," *Brit. J. Appl. Phys.*, **8**, 109–113.

321

[13] Booker, G. R., Norbury, J., and Sutton, A. L. (1957), "X-ray diffraction studies on precipitates and inclusions in steels using an extraction replica technique," *Brit. J. Appl. Phys.*, **8**, 155–157.

[14] Bradley, W. F. (1945), "Diagnostic criteria for clay minerals," *Amer. Mineral.*, **30**, 704–713.

[15] Bradley, W. F., Grim, R. E., and Clark, G. L. (1937), "A study of the behavior of montmorillonite upon wetting," *Z. Krist.*, **97**, 216–222.

[16] Brown, G. (1953), "The occurrence of Lepidocrocite in some British soils," *J. Soil Sci.*, **4**, 220–227.

[17] Brown, G. A. (1953), "A semi-micro method for the preparation of soil clays for X-ray diffraction studies," *J. Soil Sci.*, **4**, 229–232.

[18] Bumstead, H. E. (1973), "Determination of alpha-quartz in the respirable portion of airborne particulates by X-ray diffraction," *Amer. Ind. Hyg. Assoc. J.*, **34**, 150–158.

[19] Burtner, R. (1974), "The use of porous vicor as a substrate for X-ray diffraction analysis of oriented clay minerals," Abstract, Clay Minerals Society 11th Meeting, Cleveland, OH, p. 21.

[20] Bystrom-Asklund, A. M. (1966), "Sample cups and a technique for sideward packing of X-ray diffractometer specimens," *Amer. Mineral.*, **51**, 1233–1237.

[21] Carlton, R. W. (1975), "An inexpensive plate holder for the suction-on-ceramic method of mounting clay minerals for semi-quantitative analysis," *J. Sed. Pet.*, **45**, 543–545.

[22] Carroll, D. (1970), "Clay minerals: a guide to the X-ray identification," *Geological Society of America Special Paper* **126**, 80 pp.

[23] Clark, G. L., Grim, R. E., and Bradley, W. F. (1937), "Notes on the identification of minerals in clays by X-ray diffraction," *Z. Krist.*, **96**, 322–324.

[24] Cody, R. D. and Thompson, G. L. (1976), "Quantitative X-ray powder diffraction analyses of clays using an orienting internal standard and pressed disks of bulk shale samples," *Clays and Clay Minerals*, **24**, 224–231.

[25] Cole, W. F. (1961), "Modifications to standard Philips powder cameras for clay mineral work," *Clay Minerals Bulletin*, **4**, 312–317.

[26] Copeland, L. E. and Bragg, R. H. (1958), "Preparation of samples for the Geiger counter diffractometer," *ASTM Bulletin No. 228*.

[27] Cullity, B. D. (1974), *Elements of X-ray diffraction*, p. 166. Reading, MA: Addison-Wesley.

[28] Dana, S. W. (1943), "A pipette method of size analysis for the centrifuge," *J. Sed. Pet.*, **13**, 21–27.

[29] Devine, S. B., Farrell, R. E., Jr., and Billings, G. K. (1972), "A quantitative X-ray diffraction technique applied to fine grained sediments of the deep Gulf of New Mexico," *J. Sed. Pet.*, **42**, 468–475.

[30] Fatemi, M., Johnson, E. T., Witlock, L. L., Birks, L. S., and Gilfrich, J. V. (1976), "X-ray analysis of airborne asbestos, interim report: sample preparation," Environmental Protection Agency EPA-600/2-77-062.

[31] Fenner, P. (1966), "Clay mineral studies: results of investigations or preparation," *Proceedings of International Clay Conference, 1*, Israel, 401–405.

[32] Fenner, P. (1967), "Preliminary results of comminution effects studies on illite," *Proceedings of International Clay Conference, 2*, 241–243.

[33] Fenner, P. and Hartung, J. (1969), "Laboratory processing of halloysite," *Clays and Clay Minerals*, **17**, 42–44.

[34] Figueiredo, P. M. de F. (1965), "The effect of grinding on kaolinite and attapulgite as revealed by X-ray studies," *Esc. Geol. P. Alegre. Spec. Publ.*, **9**, 1–47.

[35] Fischer, R. M. (1953), "Electron microstructure of steel by extraction replica," *ASTM Spec. Tech. Publ. 155*.

[36] Floerke, O. W, and Saalfeld, H. (1955), "Ein Verfahren zur Herstellung texturfreier rongen–Pulverpraparate," *Z. Krist.*, **106**, 460–466.

[37] Foit, F. F., Jr. (1982), "A technique for loading glass capillaries used in X-ray powder diffraction," *J. Appl. Cryst.*, **15**, 357.

[38] Frevel, L. K. and Anderson, H. C. (1951), "Powder diffraction patterns from microsamples," *Acta Cryst.*, **4**, 186.

[39] Gibbs, R. J. (1965), "Error due to seggregation in quantitative clay mineral X-ray mounting techniques," *Amer. Mineral.*, **50**, 741–751.

[40] Gibbs, R. J. (1968), "Clay mounting techniques for X-ray diffraction analysis: a discussion," *J. Sed. Pet.*, **38**, 242–244.

[41] Gibson, M., Jr. (1966), "Preparation of oriented slides for X-ray analysis of clay minerals," *J. Sed. Pet.*, **36**, 1143–1162.

[42] Gorycki, M. A. (1976), "Rapid, reproducible mounts for X-ray powder diffraction studies," *Norelco Reporter*, **23**, No. 1.

[43] Graf, D. L. (1974), "X-ray cells for diffraction analysis of flat powder mounts in contact with liquid at elevated temperature and pressure," *Amer. Mineral.*, **59**, 851–862.

[44] Gude, A. J. and Hathaway, J. C. (1961), "A diffractometer mount for small samples," *Amer. Mineral.*, **46**, 993–998.

[45] Harward, M. E. and Theisen, A. A. (1962), "A paste method for preparation of slides for clay mineral identification by X-ray diffraction," *Soil Sci. Soc. Am. Proc.*, **26**, 90–91.

[46] Hauff, P. L., Starkey, H. C., Blackmon, P. D., and Pevera, D. R. (1982), "Sample preparation procedures for the analysis of clay minerals by X-ray diffraction," Workshop Syllabus, Denver X-ray Conference, University of Denver, CO.

[47] Hughes, R. and Bohor, B. (1970), "Random clay powders prepared by spray drying," *Amer. Mineral.*, **55**, 1780–1786.

[48] Hutchison, C. S. (1974), *Laboratory Handbook of Petrographic Techniques*, New York: Wiley.

[49] Jonas, E. C. and Kuykendall, J. R. (1966), "Preparation of montmorillonites for random powder diffraction," *Clay Mineral.*, **6**, 232–235.

[50] Keller, W. D., Pickett, E. E., and Reesman, A. L. (1966), "Elevated dehydroxylation temperature of the Keokuk geode kaolinite—a possible reference mineral," *Proceedings of the International Clay Conference*, **1**, Israel, 75–85.

[51] Kinter, E. B. and Diamond, S. (1956), "A new method for preparation and treatment of oriented-aggregate specimens of soil clays for X-ray diffraction analysis," *Soil Sci.*, **8**, 111–120.

[52] Kittrick, J. A. (1961), "A comparison of the moving liquid and glass slide methods for the preparation of oriented X-ray diffraction specimens," *Soil Sci.*, **92**, 155–160.

[53] Kuehnel, R. A. and van der Gaast, S. J. (1993), "Humidity controlled diffractometry and its applications," *Adv. X-ray Anal.*, **36**, 439–449.

[54] Leber, S. (1965), "Preferred orientation in swaged and drawn tungsten wire," *Trans. Met. Soc. AIME*, **233**, 953–958.

[55] Lerz, H. and Kramer, V. (1966), "Ein Varfahren zur Herstellung texturfrier Rontgen-Pulverpraparate vib Tonmineralen fur Zhalrohrgoniometer mit senkrechter Drehachse," *N. J. Mineral. Montash*, **50**, (no pages given).

[56] Lincoln, J. B., Miller, R. J., and Tettenhorst, R. (1970), "Random powder mounts from montrorillonite aergels," *Clay Minerals*, **8**, 347–348.

[57] McCreery, G. L. (1949), "Improved mount for powdered specimens used on the Geiger counter X-ray spectrometer," *J. Amer. Ceram. Soc.*, **32**, 141–146.

[58] Meumann, B. S. (1956), "The preparation of sealed powder specimens for X-ray analysis," *Clay Minerals Bulletin*, **3**, 22–25.

[59] Miller, W. E. and Hill, E., Jr. (1964), "A method for permanent mounting of powder samples for X-ray diffraction analysis," *J. Sed. Pet.*, **34**, 848–850.

[60] Mitchell, W. A. (1953), "Oriented aggregate specimens of clay for X-ray analysis made by pressure," *Clay Minerals Bulletin*, **2**, 76–78.

[61] Moh, G. H. and Taylor, L. A. (1971), "Laboratory techniques in experimental sulfide petrology," *N. Jb. Miner. Mh.*, **10**, 450–459.

[62] Morris, M. C., McMurdie, H. F., Evans, E. H., Paretzkin, B., deGroot, J. H., Newberry, R., Hubbard, C. R., and Carmel, S. J. (1977), "Standard X-ray diffraction patterns," National Bureau of Standards, Monograph 25, Section 14.

[63] Mossman, M. H., Freas, D. H., and Bailey, S. W. (1967), "Orienting internal standard method for clay mineral X-ray analysis," *Clays and Clay Minerals, Proceedings of the 15th Conference*, Pittsburgh, PA, 441–453.

[64] Nagelschmidt, G. (1941), "The identification of clay minerals by means of aggregate diffraction diagrams," *J. Sci. Instrum.*, **18**, 100–101.

[65] Niskanen, E. (1964), "Reduction of orientation effects in the quantitative X-ray diffraction analysis of kaolin minerals," *Amer. Mineral.*, **49**, 705–714.

[66] Oinuma, K. Kobayashi, K., and Sudo, T. (1961), "Procedure of clay mineral analysis," *Clay Sci. (Japan)*, **1**, 23–28.

[67] Owen, L. B. (1971), "A rapid sample preparation method for powder diffraction cameras," *Amer. Mineral.*, **56**, 835–836.

[68] Peiser, H. S., Rooksby, H. P., and Wilson, A. J. C. (1955), "The powder method in mineralogical research," in *X-ray Diffraction by Polycrystalline Materials*, London, UK: The Institute of Physics.

[69] Peters, Tj. (1970), "A simple device to avoid orientation effects in X-ray diffractometer samples," *Norelco Reporter*, **17**, 23.

[70] Poppe, L. J. and Hathaway, J. C. (1978), "A metal membrane mount for X-ray powder diffraction," *Clays and Clay Minerals*, **27**, 152–153.

[71] Pradzynski, A. H. and Rhodes, J. R. (1976), "Development of synthetic standard samples for trace analysis of air particulates," in *Calibration in Air Monitoring*, ASTM STP 598, 320–336.

[72] Quakernaat, J. (1970), "Direct diffractometric quantitative analysis of synthetic clay mineral mixtures with MoS3 as orienting indicator," *J. Sed. Pet.*, **40**, 506–513.

[73] Rex, R. W. and Chown, R. G. (1960), "Planchet press and accessories for mounting X-ray powder diffraction samples," *Amer. Mineral.*, **45**, 1280–1282.

[74] Rich, C. I. (1957), "Determination of (060) reflections of clay minerals by means of counter type X-ray diffraction instruments," *Amer. Mineral.*, **42**, 569–570.

[75] Rich, C. I. (1969), "Suction apparatus for mounting clay specimens on ceramic tile for X-ray diffraction," *Soil Sci. Amer. Proceedings*, **33**, 815–816.

[76] Rudman, R. (1977), *Low temperature diffraction*, New York: Plenum, pp. 161–178.

[77] Schoen, R. (1962), "Semi-quantitative analysis of chlorites by X-ray diffraction," *Amer. Mineral.*, **47**, 1384–1392.

[78] Schoen, R. (1964), "Clay minerals of the Silurian Clinton limestones, New York State," *J. Sed. Pet.*, **34**, 855–863.

[79] Schoen, R., Foord, E. E., and Wagner, D. (1972), "Quantitative analysis of clays—problems, achievements, and outlook," Proceedings of the International Clay Conference, II, 565–575.

[80] Schultz, L. G. (1955), "Mineralogical-particle size variations in oriented clay aggregates," *J. Sed. Pet.*, **25**, 124–125.

[81] Schultz, L. G. (1978), "Sample packer for randomly oriented powders in X-ray diffraction analysis," *Amer. Mineral.*, **48**, 627–629.

[82] Shaw, H. G. (1972), "The preparation of oriented clay mineral specimens for X-ray diffraction analysis by suction-onto-ceramic method," *Clay Minerals*, **9**, 349–350.

[83] Smith, E. and Nutting, J. (1956), "Direct carbon replicas from metal surfaces," *Brit. J. Appl. Phys.*, **7**, 214–217.

[84] Sorem, R. K. (1960), "X-ray diffraction technique for small samples," *Amer. Mineral.*, **45**, 1104–1108.

[85] Spoljaric, N. (1971), "Quick preparation of slides of well-oriented clay minerals for X-ray diffraction analyses," *J. Sed. Pet.*, **41**, 588–589.

[86] Stokke, P. R. and Carson, B. (1973), "Variation in clay mineral X-ray diffraction results with the quantity of sample mounted," *J. Sed. Pet.*, **4**, 957–964.

[87] Strum, E. and Lodding, W. (1972), "Statistical randomizing of preferentially oriented clay mineral particles in X-ray diffractometry samples," Proceedings of the International Clay Conference, Madrid, 807–816.

[88] Takahashi, H. (1959), "Effect of dry grinding on kaolin minerals," *Clays and Clay Minerals*, **6**, 279–291.

[89] Tien, P. L. (1974), "A simple device for smearing clay-on-glass slides for quantitative X-ray diffraction studies," *Clays and Clay Minerals*, **22**, 367–368.

[90] Towe, K. M. (1974), "Quantitative clay petrology: the trees but not the forest," *Clays and Clay Minerals*, **22**, 375–378.

[91] Towe, K. M. and Grim, R. E. (1963), "Variations in clay mineralogy across facies boundaries in the Middle Dovonian (Ludlowville) New York," *Amer. J. Sci.*, **261**, 839–861.

[92] Tucholke, B. E. (1974), "Determination of montmorillonite in small samples and implications for suspended matter studies," *J. Sed. Pet.*, **44**, 254–258.

[93] van der Gaast, S. (1991), "Mineralogical analysis of marine particles by X-ray powder diffraction," *Geophysical Monograph*, **63**, pp. 343–362, American Geophysical Union.

[94] van Langeveld, A. D., van der Gaast, S., and Eisma, D. (1978), "A comparison of the effectiveness of eight methods for the removal of organic matter from clay," *Clays and Clay Minerals*, **26**, 362–364.

[95] Vassamillet, L. F. and King, H. W. (1963), "Precision X-ray diffractometery using powder specimens," *Adv. X-ray Anal.*, **6**, 142–157.

[96] Waerstad, K. R. (1986), "Quantitative X-ray diffraction analysis of calcium sulfates and quartz in wet-process phosphoric acid filter cakes," *Adv. X-ray Anal.*, **29**, 211–216.

[97] Zwell, L., Fasiska, E. J., and von Gemmingen, F. (1962), "X-ray identification of phases in type 316 stainless steels subjected to creep rupture," *Trans. Met. Soc. A.I.M.E.*, **224**, 198–200.

INDEX